Geology

Ansel Adams

WILLIAM C. PUTNAM

GEOLOGY

Second Edition
Revised by ANN BRADLEY BASSETT

New York
OXFORD UNIVERSITY PRESS
London   1971   Toronto

I shall lift up mine eyes unto the hills
From whence cometh my help.

Psalms: cxxi, 1.

Foreword

Professor William Clement Putnam, the author of this book, was stricken and died on March 16, 1963, at the age of fifty-four. He had taught geology for thirty-two years. This book is the result of his experience as a teacher, and of his love of teaching beginning students, especially those for whom a single elementary course would be their sole contact with the science of geology.

At the time of his death Professor Putnam had completed this manuscript, prepared the drawings, and gathered the photographs. He had not made a final selection from his collection of pictures, and he had not written the commentary to accompany the illustrations. Ten of his colleagues in the Geology Department of the University of California, Los Angeles, with the devoted assistance of Mrs. Evelyn Putnam and Mrs. Genny Schumacher, have completed his work, endeavoring, to the best of their ability, to do it as he would have liked it done.

Carbon-14

An exciting scientific achievement of the post World War II period has been the identification of radioactive carbon, and the awareness not only of the role it plays in living organisms, but an appreciation of its extraordinary value in dating events in this last five minutes of earth history, as it were.

The recognition of the existence of radiocarbon goes back to research in 1939 on cosmic rays in the upper atmosphere by Serge Korff of New York University. He noted that cosmic rays produce secondary neutrons on their initial collision with nitrogen in the higher levels of the atmosphere. These neutrons, he predicted, when they collided with the abundant isotope nitrogen-14, would react to free a proton and to form carbon-14. The radioactive carbon then combines with oxygen to form $C^{14}O_2$, which is absorbed by plants and ultimately by all living things. New carbon-14 is added to the total supply on land, in the sea, and in the atmosphere about as rapidly as the old disappears.

Research on the problem of carbon-14 and its relationship to cosmic radiation was continued by Willard F. Libby, then at the University of Chicago and now at the University of California.

One of the basic premises in this method of age-dating is the assumption that when an organism dies, it no longer assimilates carbon-14, and the radiocarbon present in it decays without any new carbon-14 replacing the old. A sample of just-dead wood placed in a geiger counter yields about 15.3 disintegrations per minute per gram.

Fig. 1-4 Spiral graph representing geologic time. The numbers indicate millions of years since the beginning of each major subdivision of geologic time. From the graph it is seen that more than 85 per cent of geologic time transpired before the Paleozoic, and that the Cenozoic Era—the age of mammals—represents only about 1 per cent of geologic time.

The number of disintegrations per minute decreases with increasing age of the sample, with only one-half the original number occurring when the carbon-14 is 5568 $\pm$ 30 years old, which is the half-life (Fig. 1-5). With so short a half-life, it is obvious that here is a radioactive clock that is of tremendous value to archeologists and prehistorians, but one that is of limited significance to the geologist, because dates can be determined back with decreasing confidence to only 50,000 years or so. This leaves a gap of several hundred thousand years between the youngest rocks that can be dated by potassium-argon and the oldest material that can be dated by carbon-14. Geochronologists are working diligently to find a method which will bridge this gap.

Whether or not radiocarbon dates are true dates has stirred a controversy. There is a rather general agreement that dates well this side of the 5568-year half-life probably are true. The disputes arise over the more distant ones, for there are a number of unresolved problems still awaiting an answer. For example, in older samples the level of radioactivity is very weak; the difference between one or two counts per minute means a long difference in time. Despite elaborate shielding, with this low level of radioactivity it becomes increasingly difficult to separate the few authentic counts from the background noise of cosmic rays.

Another uncertainty arises from the assumption that the level of cosmic radiation has remained essentially constant through time. A further error requiring constant vigilance to correct is the danger of contamination of buried organic material, through the introduction of more recent carbon carried to it by ground water, by burrowing animals and roots, and by increased exposure to radioactive carbon through nuclear explosions. All these have the effect of making the measured age less than the actual age.

Despite these possibilities for error, which are being attacked as vigorously as possible, the method of radiocarbon dating holds great promise for the future in archeology, geography, geology, meteorology, and oceanography. Results have already accumulated which are little short of miraculous in establishing a chronology within various cultures or within sequences of postglacial geologic events, and relating these to one another, even across such barriers as thousands of miles of open sea.

Relative Age

We are accustomed not only to thinking of time in absolute terms such as years but also in terms of the relationship of one event to another. Thus we speak of Reconstruction, which would be the period following a particular event, in this case the Civil War.

One of the unfortunate outgrowths of the way in which we are exposed to history is our habit of thinking of it in unrelated, compartmented units. The majority of us are likely to have encountered only U.S. history, and thus to have little understanding of events in America in relation to those of the rest

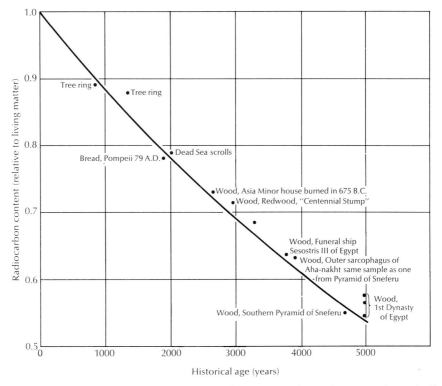

Fig. 1-5   The heavy line is the calculated decay curve for carbon-14, using a half-life value of 5568 ± 30 years, and the dots show the radiocarbon content of certain materials of known historical age.

of the world. For example, few people think of the American Revolution in its context as a skirmish in the world-wide struggle between France and Britain for supremacy.

The following passage from Palle Lauring's *Land of the Tollund Man*, which describes the prehistory of Denmark, expresses with great eloquence the contemporaneity of remote, and what at first might appear to be unrelated, events:

> But the Scandinavian Bronze Age lasted from about 1500 B.C. to 500 B.C.
>
> During this period the New Kingdom of Egypt came and went, Tutankhamen was laid to rest in his golden coffin, and in 525 B.C. Egypt fell to the Persians. The totalitarian military state of Assyria had arisen and spread death and destruction around it, crushed cultures, and perished in turn. In Greece a civilization had arisen that was to shed its lustre over centuries to come, moulding the destiny of untold generations in Europe, shaping Western thoughts and ideas, art, and architecture; the Greek states, having passed through their Archaic youth, were well on the way into the Classical period and a temple had already appeared on the Acropolis. At the Tiber ford, where an island divided the river in two, Rome had been founded and had already worsted its rival Alba Longa, and kings were reigning over the city between the hills. In Jerusalem centuries had passed since Solomon begot children with

the Queen of Sheba, and David sang psalms and seduced the wives of his officers. None of them suspected that, thousands of miles to the west, Indians were creating strange civilizations in Yucatan and Mexico and Peru. Equally far to the east the Chinese Empire was ruled by the Chou dynasty and was in the middle of the remarkable period known as the Spring and Autumn Annals. Towards the south the wise King Asoka had been buried in his Indian empire and Prince Siddhartha, called Buddha, was dead, while his thoughts were alive and spreading. The world had not stood still while the North was experiencing its Age of Bronze.

This historical analogy has a direct relevance to the second major problem we are confronted with in deciphering past geologic events, and that is how to *correlate* a succession of them in one part of the world with those in another. In human history we have the immense advantage of written records, in spite of the ravages of time, war, pillaging, fires, and floods; there are also workable calendrical systems for some parts of the world extending back for 4000 or 5000 years.

Before the advent of isotopic methods of age-dating, how were the rocks of Cornwall, England, for example, to be correlated with those of upstate New York? They happen to be of the same geologic age, and this has been known for over a century. By what method did our predecessors determine this contemporaneity? It certainly was not on the physical resemblance of the two kinds of rocks. At first they had believed this kind of evidence might be used; that all the reddish-colored sandstones of Britain were the same age, for example. It took effort and argument to establish the fact that there were several sets of red sandstone, some of which were the same geologic age, while others were separated from one another by hundreds of millions of years.

Where different kinds of rocks are in direct physical contact with one another, then by constructing a geologic map showing their mutual relationships and drawing geologic cross sections, it is possible to work out their three-dimensional geometry, and from this to determine their relative ages. Two major principles are employed in doing this. These are: (1) *superposition* of the rocks—that is, younger rocks normally rest on top of older, and (2) *cross-cutting relationships*. The latter are characteristic of rocks that were formed from a molten state and which generally intrude other rocks. The intrusive rocks which do the invading, or the cutting across of others, are the younger.

The problem becomes more acute when the effort is made to compare rocks and the sequence of events they represent in one region, with another succession of rocks or events in a region hundreds, or even thousands, of miles away with no direct contact between the two.

Fossils

The method our predecessors employed was to use the fossils contained in the rocks. In fact, working out the evolutionary succession of fossil forms and

Fig. 1-6   Geologists digging fossil bones of large animals and preparing them for shipment. Littleton, Colorado. (Photograph by J. R. Stacy, U.S. Geological Survey.)

understanding their time significance is a major creative achievement of the human mind. Fossils have excited men's interest and have been collected since before historic time. They are shown in ancient drawings; some have been employed in heraldry, and for centuries they have been cherished because of the interestingly decorative effect they give to cut and polished stone.

The word itself, a Latin one from *fossilus*, meaning something dug up, shows an awareness almost since the beginnings of language that these were things of the earth.

Although such an awareness may have existed, it was accompanied by an abysmal ignorance of the true meaning of these testimonials to the existence of life in the remote past. To some, fossils were the creation of Satan, placed in the world to confuse mortal men. To others, such as Avicenna (A.D. 980–1037), the Arab philosopher responsible for the revival of interest in the works of Aristotle, they were inorganic petrifactions that grew within the rocks as a result of the workings of a creative force, the *vis plastica*. Only by chance did they come to resemble bones or shells of living creatures. This ignorance extended to the discovery of the remains of prehistoric man, such as the Cro-Magnons and Neanderthalers, whose bones were sometimes carted to the village churchyard for a proper interment.

It remained for a singularly devoted, untutored, and eminently practical canal-builder, William Smith (1769–1839), to establish the existence of a relationship between stratified rocks and the fossils they contain. Fortunately for him, and for us, he lived and worked in the right time and place, during the period when canal building was active in England. The advent of the Industrial Revolution, coupled with the post-Napoleonic prosperity of Britain, placed a premium on the construction of such relatively cheap and modestly efficient transportation as canals. In this boom, Smith was a willing participant.

Not only is the structure of midland and eastern Britain relatively simple, but the rocks themselves are distinctive. For example, the Cretaceous strata include the white chalk of the Dover cliffs; the Carboniferous, the black coal of the midlands upon which Victorian England was to base a century of prosperity for some, grinding labor for others, and the denudation of whole countrysides as a by-product.

In the construction of canals, the recognition of various rocks and an understanding of their physical properties are matters of prime importance. If the rocks are harder than expected, or if they slump and cave readily, or if, like the chalk, they allow water to drain away, then with very little difficulty a contractor on a canal project can go broke.

Smith's approach was entirely empirical, and considering the rate at which canals were dug by hand, he was in no hurry. As a single example, he ran surveys on the Somersetshire Coal Canal for six years. All told, he tramped up and down the English countryside for twenty-four years making observations. When he was done, he had made the great fundamental discoveries that (1) the distribution of sedimentary rocks in southeastern England made a logical pattern that could be represented on a map, (2) the strata occurred in the same order—that is, the chalk beds were always found above the coal, unless disturbed structurally, and never vice versa, and (3) different layers contained assemblages of distinctive fossils. The practical result of this was

Fig. 1-7 Geologist mapping with plane table and alidade. White Canyon, Utah. (Photograph by L. C. Huff, U.S. Geological Survey.)

that, given a representative collection of fossils, he could determine from which layer of rock they came.

Smith had no awareness of the evolutionary significance of the fossils he collected so patiently over the years. To him they were little more than uniquely shaped objects. Yet, unknown to him, he had found the key by which strata could be correlated with one another, not only in a local region but with rocks whole continents apart.

This type of correlation is no simple matter, however, because we are dealing with the remains of organisms that lived in environments as varied as those of today. Furthermore, during the long span of geologic time some groups of animals and plants gradually evolved into more complex creatures, while others, having achieved perhaps an uneasy balance with their environment and their enemies, remained essentially unchanged. Still others, such as the dinosaurs, vanished completely.

This brings us to an important contribution that *paleontology* (the study of ancient life) makes to the whole realm of contemporary thought. This is the demonstration—from the fossils preserved in the thousands of feet of stratified rocks in many lands—that many creatures have evolved from relatively simple forms to complex hierarchies of plants and animals. In fact, the fossil record was one of the stronger bodies of evidence advanced by Darwin and his followers in support of the theory of evolution.

Geologic Time Scale

As knowledge accumulated of strata in different parts of the world, as well as in a single region, they could be compared with one another, or correlated, on the basis of their fossil content. It was necessary first to determine the order of deposition in one region, and when this had been done, the succession of rocks there might be compared with a different succession of rocks—perhaps a whole continent away, and not necessarily the same kinds of rocks at all, but containing fossils of the same geologic age.

The other step to be taken—and in many places the two were made simultaneously—was to determine which fossils were older than others. This is where evolutionary theory and paleontology have complemented or supported each other through the past century.

The most valuable fossils for correlating rocks in one region with those of another are ones that had comparatively short racial histories geologically speaking—and yet which in their limited span on Earth achieved a wide geographic range and underwent rapid evolutionary changes. Such chronologically ideal examples are called *index fossils*.

Geologic time is divided into units by using much the same philosophy we employ in subdividing historic time. To take a single example, in the western world one of the longer time intervals is the Christian Era. True, there is a measure of unity within this nearly 2000-year chapter of history, but were we to find ourselves transported to ancient Palestine or Rome, very likely we should be more impressed by the differences than by the resemblances.

It is a rare one-semester history course that, in attempting to cover such a broad panorama, succeeds in making it appear more lasting than the fleeting, multicolored patterns of a kaleidoscope. Fortunately, most history courses break this enormous block of time with all its crowded events into smaller units; sometimes they become very short indeed, and a semester may be devoted to a five- or ten-year period.

To carry the analogy with human history a bit further, these intensely studied parts of the record are more likely to be the later rather than the earlier part. For example, 2000 years of Egyptian history may be sketched in with broad strokes, while a large fraction of a semester may be devoted to the decades following the Treaty of Versailles.

This same distortion colors our view of geologic time. Events closer to us leave more complete and decipherable records than those grown increasingly fragmentary over a long lapse of time and through the repeated deformation rocks undergo in such processes as mountain building. The result is that more is known about events in the later parts of earth history than in the earlier, and this is reflected in the larger number of divisions of the last part of the geologic time scale.

If we consider the names of the scale, the grand divisions, comparable in significance to such human episodes as the Stone Age, are the Eras. They take their names from our concept of the dominant aspect of life during each Era; thus a free translation, beginning with the oldest, would be ancient life for the Paleozoic, medieval life for the Mesozoic, and modern life for the Cenozoic. In very broad terms, the Paleozoic was a time in which invertebrates, and relatively simple vertebrates, such as fish, amphibians, and primitive reptiles were ascendant. The Mesozoic was high noon for the reptiles, and this is the chapter in earth history when the hegemony of the dinosaurs was complete. The Cenozoic, which is the contemporary Era, is a time of mammalian dominance.

The Eras are divided into lesser units called Periods. From their names, it obviously is hard to find a common pattern. Two have a familiar aspect to Americans—the Pennsylvanian and Mississippian—and we can use them to illustrate the philosophy behind the naming of these time units. Most are named for regions where rocks containing fossils characteristic of their segment of geologic time were found. This means that strata deposited in the later Paleozoic in an inland sea covering much of the eastern United States were named for their occurrence in Pennsylvania, just as strata approximately

Fig. 1-8 Petroleum geologists in Alaska make a preliminary study, using a helicopter for transportation in a rugged mountainous area. (Courtesy of Humble Oil and Refining Company.)

20 million years older were named for their occurrence in the Mississippi valley—not the state. Incidentally, neither of these time terms is used in Europe; there the two periods are treated as a single unit, the Carboniferous, which acquired its name from the coal-bearing strata of England.

Another time term whose place of origin can be recognized readily is the Devonian, named for the rocks that crop out along the southwestern tip of Great Britain in Devonshire and Cornwall. A less familiar period is the Permian—named after the ancient Kingdom of Permia in Russia.

Only the small number of persons with a knowledge of Latin may recognize the ancient name of Wales, Cambria, as the basis for the naming of the Cambrian. Two closely related periods, the Ordovician and Silurian, perpetuate the memory of Stone Age tribes whose homes had been in Wales, the Ordovices in the north and the Silures in the south.

Some of the other periods are named, not for places, but for physical characteristics of their rocks. The Cretaceous is a good example, since it is derived from the Latin word *creta*, chalk, for the exposures of these rocks in the cliffed coast of southern England. Incidentally, this by no means is a characteristic rock type for this age. In the United States, Cretaceous rocks run the gamut from conglomerate through sandstone and shale to limestone and coal; or, as in the Sierra Nevada, may include the enormous bodies of granite intruded during this period. The Triassic takes its name from the fact that in Germany the rocks of this age are divided into three distinctive layers—a limestone in the middle, with reddish sandstones and shales above and below.

The table lists the periods in their proper order, and it is immediately apparent that most of them were named from areas in western Europe. This parochial approach—an inevitable result of the way the system was created —has led to repeated difficulties ever since. Geological events that happened in Europe very often did not happen elsewhere, and thus a system once thought to be of world-wide applicability has had to be patched up and tinkered with continuously as more and more of the world is explored geologically.

The subdivisions of the Cenozoic are called Epochs, and their rather strange names are a special case. They were established for the most part by Sir Charles Lyell on the percentage of extinction of marine molluscan fossils. Thus in Eocene (Greek *eos*, dawn + *kainos*, recent) strata about 1 to 5 per cent of the species found are still living. In the Miocene (Greek *meion*, less + *kainos*, recent) 20 to 40 per cent of the fossil species are still alive. In the Pleistocene (Greek *pleistos*, most + *kainos*, recent) 90 to 100 per cent of the fossil shells are those of species living in the seaways of the world today. Without perhaps fully realizing it, Lyell had divided this later segment of geologic time on a statistical basis—one of the earlier applications of statistics to a natural science.

The Pleistocene, the latest subdivision of the time scale, is regarded by many geologists as being coincident with the ice age, and although the length of time represented by the multiple advance and retreat of the ice sheets is unknown, it probably is between 1 and 3 million years.

The most recently accepted way of dividing the Cenozoic is into the Paleogene, comprising the Paleocene, Eocene, and Oligocene Epochs, and the Neogene which includes the Miocene, Pliocene, Pleistocene, and Recent Epochs. Many geologists, however, still use the two divisions given in the chart: the Tertiary (meaning third)—which includes all but the Pleistocene and postglacial time—and Quaternary (meaning fourth). This represents a sort of cultural lag, since we no longer speak of the Primary and Secondary rocks as our forebears did. With the first and second of a series gone, it seems strange to speak of a third and fourth, but you will see these terms employed consistently in much geologic writing today; such is the force of habit.

A Historical Science

Geologic time makes geology a historical science. How is that different from other sciences? For one thing, there are both historical and nonhistorical aspects to geology and it is important to distinguish between them. "The unchanging properties of matter and energy and the likewise unchanging processes and principles arising therefrom are . . . nonhistorical. . . . The actual state of the universe or of any part of it at a given time . . . is constantly changing" (Simpson, 1963) and is therefore historical. Physics and chemistry, then, are nonhistorical sciences, and they play an important part in geology, but added to these are the historical aspects peculiar to geology. To quote Simpson again, "The processes of weathering and erosion are unchanging and nonhistorical. The Grand Canyon or any gully is unique at any one time but is constantly changing to other unique, nonrecurrent configurations as time passes. Such changing, individual geological phenomena are historical, whereas the properties and processes producing the changes are not."

Other Differences

Geology is different from other sciences in two other ways. The first is the scale of the features being studied. To be sure, some geologists are concerned with objects of microscopic size and even with atomic structure, but for many geologists the objects of interest are far too large even to be seen from one vantage point. The second is that as the size of the objects to be studied increases, so also does the complexity and the number of variables involved in their genesis.

Consequences

Since time, scale, and complexity make geology unique, it is not surprising that some unique strategies are required in its study. We are all familiar with

Fig. 1-9 Geologist at work in the Madison River. Gallatin County, Montana. (Photograph by J. R. Stacy, U.S. Geological Survey.)

the scientific method as given in textbooks. First the scientist observes, then he formulates a problem concerning his observations. He thinks out a possible explanation, or hypothesis, which he then tests with experiments. The experiments then prove or disprove his hypothesis, and if it is proved, he can state a law.

The geologist starts with observations, too, and he, too, states his problem.

But then, the complexity of the problem and the large number of factors and processes involved lead him to formulate not just one hypothesis, but as many possible hypotheses as he can, and to consider as well any hypotheses that he can find in the literature. This is called the method of *multiple working hypotheses* which was developed by geologists in the late nineteenth century, and has since been adopted and adapted by some other sciences.

The method has a number of advantages. For one thing, it may lead to a close approximation of the truth because a complex result is likely to have a number of causes working together, rather than just one single cause. It is the geologist's job then to find out which causes operate and how important each one is to the final result. Another advantage is that it prevents the formation of a favorite theory, for with only one hypothesis the temptation is great to overlook data that would puncture it.

With his multiple working hypotheses, then, the geologist can get on with the experimental stage. Or can he? He can't build a mountain in his laboratory, and he can't wait several million years for the results of his experiments. The truth is that his experiments are "natural" experiments, rather than artificial ones; i.e. they have already been performed in nature by the time he gets to them. And so he is forced to look at the results only, anywhere he can find them, and search for any result that would negate any of his hypotheses. For while no hypothesis can, therefore, be "proved," one little piece of information is all it takes for permanent disqualification. Instead of proof we can merely say that the hypotheses that remain are more or less likely to be the real story. "Careful usage never speaks of proof . . . but only of establishment of degrees of confidence" (Simpson, 1963).

Thus the geologist attempts to deal with complexity. How can he handle the problem of scale? He cuts his problem down to size, in a manner of speaking, by making a geologic map. Such a map shows where different rock types occur on the surface and how they are arranged, i.e. their structure. This is not the mechanical, routine procedure that it may appear. In the first place he has too little data, and in the second place he has too much. This seeming paradox comes about in this way. In most areas that the geologist maps he cannot see all the rocks even on the surface. They may be covered by soil and vegetation in humid regions, or by sand and gravel in arid ones, or by man-made structures. He can usually see only occasional patches, or outcrops, of bare rock in place and from these he must infer what lies between. But in any particular outcrop he can find a wealth of detail, not all of which can fit on his map and not all of which is pertinent to that particular study. So before he can put the first line or point on his map, there is a great deal of classifying, analyzing, visualizing, and decision-making to be done. Geologic maps are highly interpretive, reflecting the knowledge, background, and experience of the map maker, as well as the climate of geologic thought that prevails throughout the profession at the time the map is made.

And time, that most important aspect of this historical science—how does the geologist deal with that? In the last quarter of the eighteenth century a remarkable man named James Hutton laid the foundation for the scientific study of the earth. At that time scientists held two beliefs: "(1) the general belief that God has intervened in history, which therefore has included both natural and supernatural (miraculous) events; and (2) the particular proposition that earth history consists in the main of a sequence of major catastrophes, usually considered as of divine origin in accordance with the first belief" (Simpson, 1963). These beliefs came to be known as catastrophism.

Hutton, a Scot, was the right man in the right place at the right time. He had the mental equipment necessary for the formation of hypotheses; he lived in Edinburgh where there was a cluster of eminent and active scientists who stimulated each other's thinking through lively discussion; the scientific atmosphere in the Age of Reason was one of doubt and skepticism, and Hutton was aware of emerging alternatives to catastrophism. At the first two meetings of the Royal Society of Edinburgh which he had helped to organize, he presented a paper, "Theory of the Earth, or an investigation of the laws observable in the composition, dissolution, and restoration of land upon the globe." In this paper he stated the *principle of uniformitarianism*, which enables the geologist to manage the time element in his investigations. In brief it stated: "(1) earth history (if not history in general) can be explained in terms of natural forces still observable as acting today; and (2) earth history has not been a series of universal or quasi-universal catastrophes but has in the main been a long, gradual development—what we would now call an evolution" (Simpson, 1963).

The catastrophists and theologians quite naturally received his ideas with something less than unbounded enthusiasm, and a vigorous controversy ensued. The outcome seems obvious to us today, for all of science depends upon the limitation that only natural explanations are permissible, but Hutton's theory was a step forward in scientific thinking of considerable importance.

Modern interpretations of the principle of uniformity do not require that all of the processes that act upon the earth must be going on at the present time. Nor is it assumed that processes have always proceeded at the same rate, or at a necessarily slow rate. What they do assume is that physical and chemical laws operate now as they did in the past; the laws that apply to matter and energy are unchanging. The conditions under which they operate are, however, constantly changing.

On the other hand, many geologic processes *have* continued for millions of years, and it is this fact that makes the principle a valuable tool in geologic investigation. It is perfectly valid, when trying to explain a completed "natural experiment" in the earth's crust, to try to find such an experiment in progress, the better to study its characteristics. In this sense the present *is*, or *can be*, the key to the past. Failure to find these processes acting today does not invalidate the principle; it merely points out that conditions today are different from what they were when that particular feature was formed.

Although many of the earth's features are the result of small forces acting through a very long period of time, the principle does not rule out those that act rapidly. A violent volcanic explosion, for example, still conforms to uniformitarianism, for none of the constant physical or chemical laws are violated.

The principle of uniformitarianism is not, of course, limited to geology, but is accepted as part of other sciences as well. It is just that geology, with its peculiar handicaps, could hardly get along without it.

Because geology is unique and requires unique methods of study, it should not be surprising that its answers also are different. Rather than "laws," they are possible explanations, hypotheses which are always working hypotheses, ever subject to change and modification to fit the growing body of information. For as Kingman Brewster, Jr., has said, "Intellectual progress is made by finding fault with the last best thought you had."

Selected References

Hubbert, M. King, 1967, Critique of the principle of uniformity, *in* Uniformity and simplicity, Geological Society of America Special Paper No. 89.

Libby, W. F., 1961, Radiocarbon dating, Science, vol. 133, pp. 621–29.

McIntyre, D. B., 1963, James Hutton and the philosophy of geology, *in* The fabric of geology, Claude C. Albritton, Jr., ed., Addison-Wesley Publishing Co., Reading, Mass.

Simpson, G. G., 1963, Historical science, *in* The fabric of geology, see above.

Fig. 2-1   Most of Africa and portions of Europe and Asia can be seen in this photograph taken from the Apollo 11 spacecraft during its translunar coast toward the moon.

2

The Planet Earth

From earliest times there have been many theories concerning the origin of the solar system and the earth, but generally they can be thought of as either catastrophic or more evolutionary. Both have been refined and elaborated as new facts about the earth have been discovered, and each has its periods of general acceptance or rejection according to its ability to fit the known facts.

At the present time the evolutionary theory of a cold, dusty origin is preferred by many scientists over a hot, gaseous, catastrophic one. Briefly, the protoplanet hypothesis, as it is called, begins with a great cloud of cold gases and cosmic dust such as we can find in the universe today. Only a slight condensation would be necessary to start many changes. The gravitational force, however slight, of this condensation would be sufficient to cause contraction of the whole mass. As it contracted and increased in density, it would start to rotate slowly, becoming more and more symmetrical and developing an equatorial plane containing about 10 per cent of the entire mass. Smaller condensations would form in this plane of material, and those about the same distance from the central protosun would probably coalesce. Eventually all the gas and dust would be attracted gravitationally to these clots of material, the protoplanets, circling the protosun. Within each protoplanet the heavier components would be attracted to the center by the ever-increasing force of gravity, and the lighter elements would form an enveloping atmosphere.

Any theory must account for the physical facts about the earth that we can observe and measure today. What are some of these attributes?

Size and Shape

Many races of men are conscious that the world about them is large, but how large is a question without relevance if their horizons are limited to a tribal territory, the confines of a mountain valley, a short stretch of the coast line, or the congested blocks of a large city. However, a number of the earlier (and nearly successful) attempts at estimating the size of the earth are of very great antiquity indeed, and one of them in worth citing here because it shows a remarkable comprehension of the size and shape of the earth as well as an understanding of part of its relationship to the sun.

Eratosthenes (*c.* 275–195 B.C.), when he was librarian at Alexandria, learned, in about 250 B.C., that at noon of the summer solstice the image of the midday sun was reflected from the water surface of a deep well at Syene—now called Aswan—yet at the same time at Alexandria, 480 miles to the north, a shadow was cast at the base of an obelisk. In fact, when the angle made by the shadow from the apex of the obelisk to the ground was actually measured, it turned out to be 7° 12′. Another quantity that was known to Eratosthenes was the distance from Syene to Alexandria, and this was thought in those days to be 500 stadia. Although there is uncertainty about the true value of this unit of measure, one interpretation is that Eratosthenes believed there was a distance very nearly equivalent to 772 kilometers (480 miles) between the two places.

By means of the geometry of that time the Greeks knew that a diagonal line cutting two parallel lines makes an equal angle with both of them. The diagram (Fig. 2-2) shows that on the basis of this simple theorem Eratosthenes knew that the angle from Syene to the center of the earth and then to Alexandria also would be 7° 12′. Since this was one-fiftieth of 360°, this means the total circumference of the earth would be fifty times the 772-kilometer (480-mile) distance from Syene to Alexandria, or 38,616 kilometers (24,000 miles)—a surprisingly close estimate to be made in a day so far removed from ours. A further, and less obvious, assumption that had to be made was that the sun was an immensely distant object.

This whole set of observations and conclusions makes another point with which scientists are familiar, and that is the distressing number of times the same discovery may have been repeated, only to be forgotten and then long years after be rediscovered anew.

Others who made estimates in the centuries after Eratosthenes reported the earth to be much smaller than it actually is, and this error, still current in Columbus's time, led him into the gross miscalculation still honored today in our using the name Indians for the pre-European inhabitants of the Americas.

Throughout Europe a tremendous surge of interest in the earth followed Magellan's ill-starred (for him) world-encompassing expedition. Not only was the earth's roundness convincingly demonstrated but an appreciation of its great size dawned on a skeptical world. Means of measuring the earth's

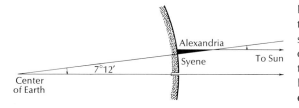

Fig. 2-2 Eratosthenes' determination of the circumference of the Earth. By measuring the inclination from the vertical of the sun at Alexandria, and by knowing the distance from Alexandria to Syene, Eratosthenes was able to calculate the earth's circumference.

dimensions are enormously more effective now than they were in their primitive state of development during the Age of Exploration. For one thing there then existed no accurate way of determining differences of longitude, or east-west distances, on the surface of the globe.

Then, much greater precision became possible through the methods of *geodesy*. This is a branch of surveying that is concerned with, among other things, precisely locating the positions of places on the surface of the earth, as well as the determination of the shape and size of the earth itself. Through patient, devoted work over the past centuries an immense amount of data has been accumulated, chiefly through the building up of elaborate triangulation networks on all continents.

Today even greater accuracy can be obtained by observing the orbits of artificial satellites. The very first, Sputnik I, amazed geodesists by showing that the earth is not nearly so flattened from pole to pole as had been calculated. Succeeding satellites have given us increasing precision, and some recent figures are: average equatorial diameter—12,753 kilometers (7926 miles); polar diameter—12,709 kilometers (7899 miles). The difference between these two figures, 49 kilometers (27 miles), shows the amount of flattening of the earth. The satellites also showed that the equatorial radius is not everywhere equal either, but has a distortion in that dimension of .14 kilometers (.09 miles).

Actually, the earth has a number of other irregularities in its shape, and for this reason its shape is known as a *geoid*, from the Greek, *geoeides*, or earthlike. An odd fact about the geoid shape is that it is exactly the shape that a drop of liquid would assume under the same conditions. We can look at the ground beneath our feet and know that it is a solid; could it be that the earth as a whole also acts as a fluid? Perhaps we will find some clues in later chapters.

Mass and Density

The land part of the earth's surface consists of rocks, or of soil derived from the breakdown of solid rock. Such materials are heavier than water; in fact, a cubic foot of water weighs about 62.5 pounds while a cubic foot of granite weighs around 168 pounds, or 2.7 times as much. We say granite has a specific gravity of 2.7, or to put it another way, a block of granite weighs 2.7 times as much as an equal volume of water.

Very early in Newton's development of the Universal Law of Gravitation (1687) he pointed out in a brilliantly intuitive generalization that the density of the earth would be five or six times as great as if the sphere consisted of water. He was not able to determine this value himself, but marked the way for others by showing that it could be found by measuring how much a mountain of known mass deflected a plumb line from the vertical. This measure would yield a value for the amount of the mountain's attraction for the plumb bob as compared to the far stronger pull of the vastly greater earth.

A simple statement of the law of gravitation is, $F = G(M_1 M_2 / d^2)$, where G is the gravitational constant (with a value of 6.673×10^{-8} in the c.g.s. system—an almost infinitesimally small number, 0.0000000667), F is the attractive force between two bodies with masses of M_1 and M_2, respectively, and d is the distance separating them. One factor making the determination more difficult then than it is for us today is that the value of the gravitational constant was not known in the seventeenth century.

Knowing the mass of a mountain from its size (volume) and the assumed density of the rocks, it then appeared to be a comparatively simple matter to solve the equation for the mass of the earth. Observations to test this hypothesis by measuring the deflection of the plumb bob were made in the Andes by Pierre Bouguer in 1783 and by Nevil Maskelyne, the Astronomer Royal, in Scotland in 1776. Maskelyne obtained a value of about 5 for the specific gravity of the earth; a notable achievement considering the primitive instruments of his day. Using the same methods of comparing the attractive force of a nearby mountain range with the attractive force of the earth, a surprising departure from the expected value was found by the so-called Trigonometrical Survey of India under the leadership of Sir George Everest. The surveyors found that the Himalayas, high as they are, failed to deflect the plumb bob from the vertical by the calculated amount that theoretically they should. The significance of this is that these immense mountains have "roots" consisting of rock that is lighter (of lower density) than the average beneath the low-lying plains of India to the south; therefore, their attractive force is not as great as if uniform density prevailed throughout (Fig. 2-3). This seemingly insignificant discrepancy actually had a profound effect in shaping the theories of mountain building that are widely held today, as we shall see in a later discussion.

Turning to the laboratory, efforts were also made at an early date to see if a value for the earth's mass might be determined. The first valid experiment was made by Henry Cavendish (1731–1810) in 1797, using two large and two small lead balls—one pair being 30 centimeters (12 inches) in diameter, the other 50 centimeters (2 inches). The small spheres were suspended from the ends of a rod, which in turn was suspended from a wire in the center. When the large spheres were brought near the small, the amount of twist (torque) in the

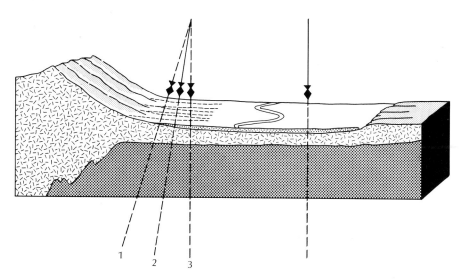

Fig. 2-3 Deflection of the plumb bob by the Himalayas. (1) Theoretical deflection that should be caused by the mountains. (2) Observed deflection (the discrepancy from the theoretical one arises from the presence of the mountain "root"). (3) Undeflected position.

wire could be measured. From this tiny displacement Cavendish determined the gravitational constant and also by means of this ingenious experiment he arrived at a value of 5.448 for the density of the earth (Fig. 2-4). In 1878, Phillip von Jolly made the first accurate determination of the gravitational constant (which can be defined as the attraction for each other of two masses of 1 gram each, spaced 1 centimeter apart). He achieved this by measuring the increase in weight of a carefully balanced 5-kilogram flask of mercury when it was brought close to a sphere of lead blocks about a meter in diameter (the sphere is still in existence and is exhibited in the Deutsches Museum, Munich). He found that the mercury flask showed an increase of 0.589 milligrams. Having determined the gravitational constant, he was then able to take the next step and compute the earth's weight *(W)*. This figure, refined since his day, turns out to be about 6.6×10^{21} tons (which is the figure 66 followed by 20 zeros). Since the size, or volume *(V)*, of the earth was already known, the simple relationship *W/V* gave Jolly a density for the earth of 5.692. More accurate determinations in modern times yield a value of 5.519.

This immediately poses a problem of fundamental importance, because typical rocks, such as granite, making up the dry land surface of the earth have a density of 2.7. Where is this additional heavy material to be found to give such a high average value for the whole earth? Does the density of the earth increase at a constant rate from the surface to the center? Or does the earth have lighter material in its surface layers and then have extremely heavy material concentrated in some kind of central core?

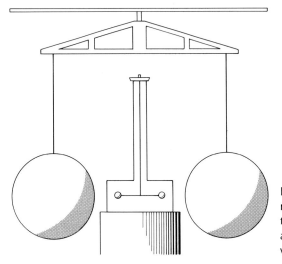

Fig. 2-4   Schematic diagram of the equipment used by Cavendish to determine the density of the earth. The small balls are free to rotate about the thin vertical wire supporting them.

The last possibility appears to be more likely, and part of the evidence for this comes from the way the earth responds to the tidal attraction of the moon. Because of the flattened shape of the earth, the axis of spin shifts slightly when the earth is subjected to such an external force, much as though the earth wobbled like a top. This shift is known as the *precession of the equinox*, and takes about 26,000 years to make a complete swing around a circle. When its nature is carefully calculated it turns out that it better fits the pattern of a planet with a dense central core. Another way of putting it is to say the earth has a smaller moment of inertia than a sphere of similar size would have with a uniformly increasing density from the surface down to the center rather than with a concentration of denser materials near the center. In fact, other lines of evidence indicate that the specific gravity near the center of the earth may be 15 and possibly as much as 18.

Gravity

In this section the term gravity is used for the property of acceleration which the earth produces in a freely falling body. This property is one that Galileo (1564–1642) established about the year 1590 despite the opposition of his contemporaries, the disbelief of onlookers who were convinced that sorcery was involved, and the prestigious authority of the immortal Aristotle (384–322 B.C.) who had taught that an object ten times as heavy as another would fall ten times as fast.

Through observations made in the many years since Galileo's day we know that in a vacuum such diverse bodies as lead and feathers accelerate at the same rate because there is no air resistance to slow down the feathers, with their large surface area and light weight. However, the rate of acceleration is not the same at all places on the earth's surface. At the equator the acceleration due to gravity is 10.03 meters (32.09 feet) per second per second, while at the pole it is 10.55 meters (32.26 feet). The reason for this difference is twofold: (1) the greater distance from the center of the earth at the equator, and (2) the greater centrifugal force which opposes the force of gravity there.

We are concerned with an enormous force here; the earth is so vastly greater than any single thing on its surface that the force with which it pulls objects toward itself is almost overpowering for large and heavy bodies. A little thought reminds us of the immense amount of energy we expend in lifting weights, in climbing mountains, or in flying airplanes, all in opposition to the force of gravity. All of us who have dropped rocks down wells or from the edges of cliffs marvel at how fast they disappear. Gravity is the force that does this; it is also the force that requires hundreds of thousands of pounds of thrust to be overcome when launching rockets into space.

How can the acceleration due to gravity actually be measured? For example, not too many of us are likely to drop iron balls from a leaning tower and time their short-lived flight with stop watches. A more practical way of determining the acceleration of gravity was discovered long ago, and also more importantly it was discovered that the acceleration varies locally over the face of the earth. Since this last factor is of the greatest geologic significance, we will discuss this in some detail. The simple device that can be used to measure the acceleration due to gravity is the pendulum. The period of oscillation (the length of time required to swing to and fro) depends upon two factors: the local acceleration of gravity and the length of the pendulum. The discovery had been made very early, by Galileo in fact, that a given pendulum makes each swing in the same length of time. By using this principle, Christiaan Huygens (1629–95) finally succeeded in perfecting a dependable pendulum-regulated clock in 1673.

During this same critical period of scientific advances, a French scientist, Jean Richer, was sent by Jacques Cassini to French Guiana to make observations of Mars at the same moment that Cassini would be making them in Paris. By making such a simultaneous observation it would be possible to use the distance from Paris to Cayenne as a base line to triangulate from earth to Mars in order to establish the distance separating the two bodies. Even considering the imperfections of such a clock in 1671, Richer was surprised to find that his pendulum clock, which was needed for keeping the time for his astronomical observations, consistently lost 2.5 minutes per day. Only when he shortened the pendulum length by 2 millimeters ($\frac{1}{12}$ inch) was the clock's accuracy restored.

Curiously enough, both Newton and Huygens, quite independently of one another, had deduced that the force of gravity would be diminished at the equator because of the earth's equatorial bulge as well as the centrifugal force set up by its rotation. In other words, the acceleration due to gravity is greater at the poles, which are 49 kilometers (27 miles) closer to the center of the earth (which is also the center of gravity) than the equator, and thus a pendulum-actuated clock gains time at the poles and loses time at the equator.

Bouguer, whom we encountered before in the Andes, knew of the relationship deduced from Richer's pendulum clock and the variations in the force of gravity. Using the same principle, he, too, found the acceleration due to gravity decreased with altitude, which is to say with the increase in distance from the earth's center which he reached on the Andean heights, the force of gravity grew less.

Since those days of the seventeenth and eighteenth centuries, the pendulum as a device for determining the force of gravity has been increasingly refined, until today gravity pendulums, torsion balances, and gravity meters are models of precision and can determine minute variations in the force of gravity over the earth's surface as well as at sea. The use of the gravity pendulum and these related instruments brings out a third reason, in addition to the first two of (1) oblateness of the earth and (2) differences in altitude, for variations in the force of gravity over the earth's surface. This third reason is the relatively slight difference in the density from place to place of the materials making up the surface layers of the earth. Here another factor entering into the equation for the Universal Law of Gravitation is important, and this is the mass of the body involved. Thus, if the density of the rocky material in a localized part of the earth's crust is less than the average for its surroundings, then the local value for the acceleration due to gravity will be diminished because the attractive force is not so great, and we say the area is characterized by a deficiency of mass, or shows a *negative anomaly*. This means it has an observed value for gravity which is less than the one that has been computed for the general region. These measurements are called Bouguer anomalies, after the French geodesist.

Geomagnetism

Imagine the confusion if the north ends of all the compass needles in the world suddenly pointed to the south. We take some of the earth's properties so for granted that such an occurrence seems incredible, and yet evidence in the rocks appears to show that such world-wide magnetic reversals have indeed taken place in the past.

The earth has a magnetic field (Fig. 2-5), just as it has a gravitational field, and behaves as if there were a bar magnet at its center. This magnetic field has an interesting effect on some rocks; they show a very weak magnetization. When lava pours out of a volcano, for example, and begins to cool and solidify, the magnetic field has no effect on it as long as its temperature remains above a critical point known as the Curie point. The Curie point is different for different volcanic rocks and can vary from 200°C. to 680°C. At this point, suddenly, some parts of the rock become magnetic. This magnetism is known

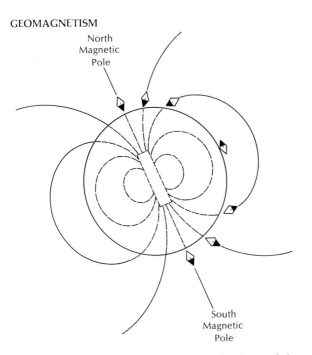

North
Magnetic
Pole

South
Magnetic
Pole

Fig. 2-5   The earth's magnetic field.

as "soft" because it is easily changed, but at a cooler point, the soft magnetism, again suddenly, becomes "hard," or permanent, reflecting the orientation of the magnetic field prevailing at that time.

In the 1940's, geologists measuring this past magnetism, or *paleomagnetism*, were startled to find that in some rocks the *polarity* of the magnetic field, i.e. the northness and the southness of the poles, was just the reverse of what it is today. After exhaustive tests to make certain that the strange magnetization could not have happened in any other way, they have had to conclude that in the past the polarity has actually reversed itself, and that at certain times, north is south and south is north.

Many of the volcanic rocks showing paleomagnetism have been dated by the potassium-argon method with several interesting results. For one thing, rocks of the same age, regardless of their geographic distribution, show the same polarity. This proves that the reversals are not local effects, but worldwide, and do indeed reflect a reversal of the entire magnetic field. For another, a definite time pattern emerges (Fig. 2-6). The present polarization is called "normal," the opposite is known as "reversed." The periods of time when the polarity is predominantly in one direction are known as *epochs*, while short reversals within the epochs are referred to as *events*. Thirdly, the reversals seem to occur quite suddenly, geologically speaking, at least within the limits of the dating method, 5000 years. Also, rocks which show a transitional polarization, i.e. somewhere between normal and reversed, are very rarely found. In addition, rocks from the bottom of the sea show this same pattern of reversals, and thus their absolute ages can be determined, a feat not easily accomplished in any other way. The study of paleomagnetism has indeed opened up new and important aspects of the earth's history.

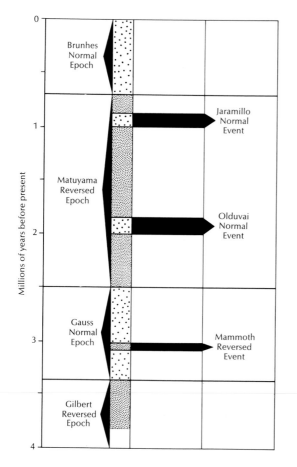

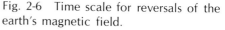

Fig. 2-6 Time scale for reversals of the earth's magnetic field.

Subdivisions of the Earth

Most of us are well aware of the commonly accepted natural division of the earth into land, sea, and air, and this tripartite separation is acknowledged scientifically by such terms as: (1) the *atmosphere*, or the gaseous envelope that surrounds the planet; (2) the *hydrosphere*, or the liquid mantle which is chiefly the sea, but that includes lakes and streams as well; and (3) the *lithosphere*, or the solid earth consisting not only of the rocky outer shell, familiar to us as the dry land on which we live, but also the deep interior. This is, in addition, the realm which provides the pattern of landscapes that make the earth the uniquely wonderful thing it is.

Geology would hardly exist as a science if we tried to limit it to the lithosphere alone. It is also a study of the interactions of the lithosphere with the atmosphere and the hydrosphere. Many of the most important processes that make our planet what it is take place at the boundaries, or interfaces, between these subdivisions. Meteorologists are finding that what happens even on a very small, or molecular, scale at the air-ocean interface determines much of what the earth's weather will be. Reactions that occur between sea water and the ocean floor are of prime importance to oceanographers. At the ocean's edge, atmosphere, hydrosphere, and lithosphere act and react together to give us a wide variety of shore features, sandy beaches as well as rocky headlands. Wind and rain and streams all bring about changes at their interface with the land. Within the earth itself interfaces are important, too. The contact of different types of rocks with each other produce observable effects which can persist through vast periods of geologic time. Reactions at boundaries deep within the earth, between the core and the mantle, between the mantle and the crust, may be the most profound of all, although they are also among the most mysterious.

All of these interfaces have one thing in common—change. The entire planet is constantly changing, from the outermost reaches of its atmosphere to its innermost core. Natural forces have conditions under which they are in balance, or in equilibrium, and anything, however small, that upsets that equilibrium sets in motion an almost endless chain of reactions acting to restore that balance. The balance of nature is usually thought of in biological terms when, in fact, it should include our entire physical environment as well. Man is probably the most active of balance-upsetters. Some of the changes he causes are beneficial to him, some are exceedingly harmful, and unfortunately, the harmful effects are not always immediately apparent. It is one of the tasks of the geologist, as it is of all natural scientists, to try to predict the long-term as well as the short-term effects of man's efforts to unbalance the huge equilibrium which constitutes the earth. So while we may turn our attention now to parts of this planet, we must not lose sight of the relationships between those parts nor our concept of the earth as a whole.

The Lithosphere

This is the realm which provides the pattern of landscapes that make the earth the uniquely wonderful thing it is. Every observant person is aware of obvious differences in the material of which the earth is made, and it is the physical nature of these differences, their meaning, and as far as we can determine, what their origin may be, that are among the major problems to which geology seeks an answer. One has only to visit a large museum and casually examine its collection of minerals, ores, fossils, and rocks to realize what a great diversity of wholly unlike substances constitutes the upper levels of the earth's crust. A monumental edifice, such as a state capitol, a city hall, a courthouse, or a large bank, usually incorporates several kinds of building stones in its façade, lobby, or halls—white marble, red granite, green serpentine, buff travertine, and variegated breccias of all kinds. All these building stones are impressive in their multitude of textures and colors, and are an indication by themselves of the heterogeneous nature of the earth's crust. Some of the explanations for this diversity of form and substance of the materials making up the solid earth are given in the next chapter, as well as in the succeeding one on rocks, and telling the story of their origin will take us on a far journey indeed.

Before becoming immersed in a sea of details concerning different rock

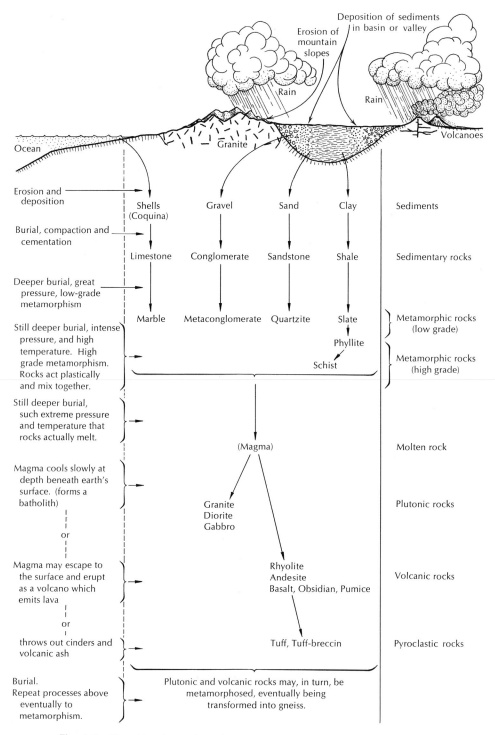

Fig. 2-7   Genetic chart of rocks.

types, it is reassuring to know that all the diversity of rocks, which are the essential substance of the lithosphere, can be grouped into three major categories. This threefold subdivision is what we call a genetic classification (Fig. 2-7) and is based on the origin, or *genesis*, of the thing described. Thus, allowing for the inevitable exceptions, borderline cases and overlapping occurrences, the rocks of the world are placed in the following broad categories—igneous, sedimentary, and metamorphic.

IGNEOUS ROCKS   These are rocks that have solidified from a silicate melt to which the name of *magma* (to knead) is given. A comparatively familiar variety of such material is *lava,* and most of us have seen photographs of this material in the craters of volcanoes or issuing as fluid streams from their flanks. Because of the high temperatures and accompanying lurid scenes associated with volcanic activity in the minds of most people, the word igneous, from the Latin *igneus,* having to do with fire, is used for rocks crystallized from the cooling lava. We use the same root in everyday language when we speak of the ignition system of a car. The varieties of igneous rocks that crystallize at or near the surface of the earth from lava when it solidifies commonly are called *volcanic rocks.*

Other igneous rocks, such as granite, crystallize at depths far below the surface of the earth, and because of the deep, inaccessible domain in which they form, such igneous rocks very often are known as *plutonic rocks,* after Pluto, Greek god of the lower world.

Granite, too, may have formed from magma, but how much or how little this may resemble lava is hard to say. After 160 years of controversy a spirited debate still flourishes as to the source of granite. But there does appear to be reasonably general agreement that the magma from which granite formed is molten, that its temperatures are high—high enough on occasion to recrystallize the enclosing wall rocks—and that although a "noteworthy part" of it is fluid, it may contain a considerable percentage of early crystallizing minerals floating in it. All are agreed that magma underground cools more slowly than magma at the surface, and for this reason the mineral crystals in the resulting plutonic rock are much larger in a granite, say, than in a typical, fine-grained, surface-cooled volcanic rock, such as basalt.

SEDIMENTARY ROCKS   Of the three rock families, these rocks are perhaps the most readily comprehended, because many of them bear a close resemblance to the materials from which they are made and because many of the processes responsible for their formation occur before our eyes or else take place in environments that are reasonably accessible.

If igneous rocks are to be construed as primary, many sedimentary rocks can be thought of as secondary, or derived rocks, in the sense that they are fragments of pre-existing rocks. Examples of sedimentary rocks of this type are (1) sandstone, which consists of sand grains cemented together; (2) conglomerate, which consists of rounded fragments the size of pebbles, cobbles, or boulders; and (3) shale, which consists of very small particles that may be comminuted down to the size of clay.

A concentration of such residues as rock salt, gypsum, Chilean nitrate, or some kinds of limestone, to name but a few of the many possibilities, may result from chemical precipitation in sea or lake water, and other sedimentary rocks of a kindred sort may result from the accumulation of a variety of organic remains.

In general, sedimentary rocks accumulate on the surface of the earth, either on land or on the floors of lakes or of the sea. Thus, they form in environments which are more susceptible to observation or to study than the depths where plutonic rocks solidify or metamorphic rocks recrystallize. Since sedimentary rocks are built up through the slow deposition of material, they typically are formed in layers. These are called *strata*; a single layer is a *stratum* (directly from the Latin—a blanket or pavement, derived from *stratus*, p.p. of *sternere*, to spread out). Individual layers may range from paper-thin sheets up to massive beds a hundred feet or more thick.

METAMORPHIC ROCKS   These are the rocks that most puzzled the first geologists, as well they might, because they do not form on the surface but appear instead to be products of the action of heat, pressure, and chemical activity operating upon rocks within the earth through long periods of time—at least long when judged by our time standards. These factors operate to produce recrystallization, either partial or complete, of the minerals of the rock. New minerals appear and they may develop a wholly new fabric or orientation with respect to each other. Instead of having a random orientation and heading every which way, as is true of many igneous rocks, in the making of some metamorphic rocks the minerals under directed stress may realign themselves parallel to one another as in a stack of silver dollars.

In some types of metamorphism, the rock may undergo little or no change in chemical composition as its minerals recrystallize. The chemical elements already present regroup themselves under these conditions of higher temperatures and pressures to form new minerals which are stable in the new subsurface environment. In other cases new minerals are formed because new material has been introduced by heated, highly charged gases and fluids circulating within the earth, very often associated with plutonic igneous activity.

The metamorphic rocks are almost certain to be complex because they have no single mode of origin but have a great diversity of origins. They can be made from all manner of rocks: igneous, sedimentary, or even from previously metamorphosed rocks. If they have any factor in common, it is crystallinity, and like the igneous rocks they consist of a fabric of interlocking crystalline minerals. Unlike many of the igneous rocks, some have a strongly banded appearance which superficially resembles the stratification of sedimentary rocks, but these bands consist of interlocking crystals segregated into layers of

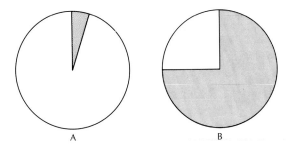

Fig. 2-8   Relative abundance of igneous and metamorphic (blank) and sedimentary (stippled) rocks. A, by volume; B, by area.

different colors—some light, some dark—rather than discrete granules deposited in laminae. Closer inspection will show that the banding, or layering, which is called *foliation* in metamorphic rocks, may be related to the parallelism of the minerals. This foliation, if only moderately well developed, is characteristic of the rock we call *gneiss*, which may look very much like streaked granite. If the foliation is better developed, then it constitutes planes of weakness running through the rock because of the parallel alignment of flat, flake-like minerals, such as mica. Because the rock splits readily along such planes as these we say it has rock cleavage, and such a rock as *slate* is a prime example.

These three families of rocks making up the lithosphere by no means have an equal distribution over the face of the earth. The two graphs (Fig. 2-8) show that only 5 per cent by volume of the earth's crust consists of sedimentary rocks, while 95 per cent are igneous and metamorphic rocks. On the other hand, about 75 per cent of the total land surface of the earth is covered with sediments. This simply means that the sedimentary rocks made a discontinuous blanket which is spread thinly over the much more abundant crystalline rocks, which are the true foundations of the continents.

Selected References

Beiser, Arthur, and the Editors of Life, 1962, The earth, Time Inc., New York.

Cailleux, Andre, 1968, Anatomy of the earth, World University Library, McGraw-Hill Book Co., New York.

Cox, Allan, Dalrymple, G. Brent, and Doell, Richard R., 1967, Reversals of the earth's magnetic field, Scientific American, vol. 216, no. 2, pp. 44–54.

King-Hele, Desmond, 1967, The shape of the earth, Scientific American, vol. 217, no. 4, pp. 67–76.

Takeuchi, H., Uyeda, S., and Kanamori, H., 1970, Debate about the earth, rev. ed., Freeman, Cooper & Co., San Francisco.

Urey, Harold C., 1952, The origin of the earth, Scientific American, vol. 187, no. 4, pp. 53–60.

Fig. 3-1   Dark hornblende crystals and white calcite, from Franklin, New Jersey. (Courtesy of the Smithsonian Institution.)

3

Rock-Forming Minerals

The rocks which make up the solid surface of the earth and that are directly accessible to us, or which can be seen to a moderate depth in deep wells and mines, show almost as many colors, patterns, and textures as there are named varieties, approximately 2,000. Most rocks—granite is an especially apt example—consist of more than one kind of material. Granite characteristically has a speckled appearance, and although most of the surface is light gray, the rock is mottled by scores of black spots sprinkled across it (Fig. 4-24). The different colored substances that make up granite are minerals; the light gray areas are mostly quartz and feldspar, while the dark specks very likely are biotite (black mica). This points up a most important distinction to keep in mind: *Rocks are made of minerals, while with only a few exceptions minerals are not rocks.* Phrased more elegantly, rocks for the most part are heterogeneous aggregates of minerals, while minerals have an essentially uniform composition.

A quartz crystal (Fig. 3-2), although as hard and inorganic as a piece of granite, obviously has a wholly different appearance. Much of this chapter and the ones to follow are concerned with the natures and differences of minerals and rocks, but the important thing to remember is that, in general, minerals are the building blocks of which rocks are made.

Fig. 3-2 Cluster of quartz crystals from Crystal Springs, Arkansas. These natural crystals are transparent and are characterized by numerous relatively smooth, flat surfaces. (Courtesy of the Smithsonian Institution.)

Before launching into a discussion of the details, we should briefly consider what are the properties of these materials of the earth's crust. Here the problem essentially is one of inorganic chemistry, and the elements involved are surprisingly few; out of the approximately ninety that have been identified in the earth's crust, eight are so much more abundant than all the others combined as to comprise practically 99 per cent of the whole. The most abundant elements are:

Element	Chemical Symbol	Percentage by Weight
oxygen	O	46.59 } 74.31
silicon	Si	27.72
aluminum	Al	8.13
iron	Fe	5.01
calcium	Ca	3.63
sodium	Na	2.85
potassium	K	2.60
magnesium	Mg	2.09

This table shows very clearly that three-fourths of the lithosphere consists of silicon and oxygen. In fact, one of the more abundant minerals is quartz, whose chemical composition is SiO_2 (one atom of silicon for every two of oxygen). When these elements are combined, the resulting compound, quartz (Fig. 3-2), is about as unlike its two components, when they are considered individually, as it could be. One, oxygen, is an invisible and highly inflammatory gas; the other, silicon, is a silvery-gray, rather metallic looking element that is never found free, or uncombined, in nature. Quartz, the compound resulting from the union of these two elements, is harder than steel; when free of impurities it is as clear as glass, and, in fact, it is the limpid material commonly called rock crystal, which, if not interfered with during its growing stage, forms beautiful, six-sided crystals.

Definitions of a Mineral

Considering the table above, it is scarcely surprising that most of the minerals that are the chief ingredients of rocks should be composed of oxygen and silicon in combination with the remaining six most abundant elements: aluminum, iron, calcium, sodium, potassium, and magnesium. Such compounds are called silicates. The feldspars and typical, and $KAlSi_3O_8$, the mineral orthoclase, is representative (Fig. 3-3). The combination of oxygen and silicon alone is called silica, and quartz (SiO_2) is an example.

Most minerals are chemical compounds; that is, they consist of two or more elements in combination. Of course there are exceptions, such as gold, copper, sulphur, and carbon (which by itself makes such dissimilar substances as diamonds and graphite), which may occur as elements by themselves as well as in chemical compounds. Minerals are naturally occurring substances. This statement rules out laboratory creations (although some, such as synthetic rubies

Fig. 3-3   Cleavage fragments of the minerals calcite $(CaCO_3)$, on the left, and orthoclase $(KAlSi_3O_8)$, on the right. Because of the characteristic regular arrangements of the atoms in these minerals the calcite breaks, or cleaves, along smooth plane surfaces in three directions. The orthoclase breaks similarly in two directions, one being the front face of the piece and the other the two sides. Quartz (Fig. 3-2), which had a different internal arrangement of atoms, has no direction of cleavage. (Courtesy of the Smithsonian Institution.)

and sapphires, are virtually impossible to tell from the natural gem stone). Minerals have a reasonably definite chemical composition. Since they are naturally occurring substances, and not laboratory products, only rarely are they chemically pure compounds. For this reason, such properties as color may vary over a range as wide as from black to white, depending on the percentage of elements present for any mineral. Then, too, some minerals belong to *isomorphous series*. That is, they may preserve about the same appearance and nearly the same crystal form even though their chemical composition may vary systematically. One variety of feldspar, known as plagioclase, is an example, and chemically every gradation exists in this particular mineral between a composition of $NaAlSi_3O_8$ and $CaAl_2Si_2O_8$. In this mineral both sodium and calcium ions exist simultaneously, but as the amount of one increases, the other decreases. Minerals also have certain physical properties, determined by their chemical composition and by the geometric arrangement of the atoms composing them. It is this atomic arrangement that determines the crystal form of a mineral. Other properties include such things as color, hardness, and specific gravity.

From these statements it might seem that water could be a mineral. There is little question about ice, the crystalline phase of H_2O, although whether or not liquid water is or is not a mineral is a topic to stir up a mild debate. Few mineralogists would subscribe to this usage of the term mineral, since such substances are commonly held to be crystalline. That is, the atoms which make up a mineral are arrayed in ordered, repetitive ranks that have fixed average positions within which they are free to vibrate. Atoms in a liquid lack the orderly arrangement characteristic of an authentic crystalline solid, and although some repetitive order exists, individual atoms are able to glide past one another in constantly changing patterns. In a gas, disorder is the ruling principle, and atoms move with nearly complete freedom on widely spaced paths. It is their frequent collisions with such a confining surface as the inner tube of a tire that produces the effect we speak of as pressure. In light of its extraordinary ability to exist in all three physical states—liquid as water, solid as ice (Fig. 3-4), gaseous as steam—water as such would lie outside the pale of any reasonable concept of what is meant by the word mineral.

In summary, then, a mineral may be defined as (1) a naturally occurring substance with (2) a fairly definite chemical composition and (3) characteristic physical properties by which it may be identified. In short, a typical mineral is a crystalline solid and is an inorganic substance. Most are chemical compounds, but a few, such as the diamond, may consist of a single element.

Crystal Chemistry

The physical properties of a mineral depend upon the relationship between the arrangement or structure of its atoms, the way in which those atoms are held

Fig. 3-4   Crystals of ice in the form of snowflakes. Note the high degree of symmetry manifested in the forms of these crystals. During the complete rotation of any individual about an axis normal to the page, there would be six positions at equal angles with respect to one another (60°) at which the appearance of the particular crystal would be essentially indistinguishable from that at any other position. This sixfold axial symmetry is a characteristic property of ice that grows in a liquid or gaseous medium. The growth forms and shapes of cleavage fragments of almost all minerals have diagnostic properties of symmetry that depend on an orderly arrangement of the atoms and groups of atoms in an array that is periodic i.e. repetitive) in three dimensions in much the same way that most varieties of wallpaper (especially the cheaper kinds!) are periodic in two dimensions. (Courtesy of Moody Institute of Science.)

together, and the chemical composition of the mineral. A study of these relationships is called *crystal chemistry*.

Structure of Atoms

The basic unit of matter is the atom, and most of us are familiar with pictures of a multicolored sphere surrounded by a host of rings, each with its own planetary sphere, and all of these circling around the central sphere like tiny planets in a minute solar system. This is a model, or a stylized concept, of how

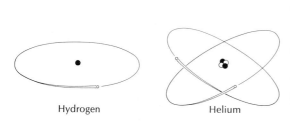

Fig. 3-5 Structures of atoms of hydrogen and helium. The large solid circles represent protons; the large open circles neutrons; and the small open circles electrons. Because a chemical element is characterized by the number of protons in the central nucleus, hydrogen is given the atomic number 1, and helium is assigned the atomic number 2.

the particles that make up an element are arranged. Very probably they do not look like this at all, but this is the so-called planetary model developed by Lord Rutherford in 1911, refined and amended to the present complex concept.

The large sphere or group shown at the center of the atomic universe (Fig. 3-5) is the nucleus, and it can be visualized as consisting of two major kinds of particles—protons which carry a positive electrical charge, and neutrons which are electrically neutral. Most of the *mass* of an atom is packed in the nucleus; in fact, 99 per cent of it or better, but most of the *volume* is in the surrounding sphere. This outer space is very thinly occupied by a cloud of negatively charged electrons circling around the nucleus in orbits somewhat analogous to those of the planets in the solar system. The electrons are infinitesimally small compared to the nucleus of the atom; an electron is only 1/1840th of the mass of the hydrogen nucleus—which is the simplest atom since it consists of only one proton as a nucleus with a lone electron whirling around it.

Most atomic structures are vastly more complex than this simple beginning. There are [104] recognized elements, and each of these is given a *name*, such as chlorine; a *symbol*, such as Cl; and an *atomic number*, such as 17. This number means that chlorine has 17 positively charged protons in its nucleus and in the surrounding electron cloud are 17 negatively charged electrons. Incidentally, the number of neutrons in the nucleus may vary; one *isotope* of chlorine has 18, the other, 20. Since the atomic weight of an element is the total number of neutrons and protons, this means there are two isotopes of chlorine: one with an atomic weight of 35 (17 + 18); the other, 37 (17 + 20). The number of neutrons present in the nucleus seems to vary with the different elements, but there usually are about as many neutrons as there are protons in the nucleus.

The electron orbits are not scattered indiscriminately throughout the space surrounding the nucleus, but are arranged in separate, unequally spaced layers, or *shells*. Since a certain amount of energy is needed to keep an electron at a prescribed distance from the nucleus, these properly are termed *energy-level shells*. The thing to remember here is that the outermost of these energy-level shells is the most significant feature of the atom from the point of view of the formation of chemical compounds, of which minerals are typical inorganic compounds. For some reason, not fully understood as yet, if there are eight electrons in the outer shell of an element it is almost completely stable and rarely combines with others.

Bonding

An atom is in equilibrium when the number of negatively charged electrons circling in their orbits is exactly the same as the number of positively charged protons in the nucleus. Should the atom lose an electron in the outer shell, then no longer are the electrons and protons in balance, but the atom has an excess of one proton and thus carries a positive charge. The element sodium (Fig. 3-6) is an excellent illustration because it is especially prone to lose an outer electron in view of the fact that its outermost shell contains but one. Its 11 electrons are grouped in three shells outward from the nucleus, as follows: 2, 8, and 1. When the outermost electron is lost and the whole structure is carrying a positive charge, we speak of it as a *sodium ion*, written Na^+. Chlorine has its electrons ranged in three shells: 2, 8, 7. If the chlorine atom can pick up an extra electron it will have achieved the goal of the maximum stability that comes from having eight electrons in the outer shell. It will then have an excess of negatively charged electrons over the positively charged protons in the nucleus, and thus it constitutes a *chlorine ion*, written Cl^-.

From this very brief discussion we learn that an electrically unbalanced atom is an *ion*, and that it may carry either a positive or negative charge. In this sub-microscopic world, perhaps unlike ours, like repels like and unlikes are strongly attracted to each other. Thus a powerful affinity can develop between a Na^+ and a Cl^- ion. The excess electron of the outer sodium shell is transferred to the chlorine shell, and added to the seven already there, it gives the eight that are needed for maximum stability. The two resulting ions are now united by an ionic bond to form a completely new compound, NaCl, which is the mineral *halite* (Fig. 3-7), or the substance we call rock salt. Such a chemical compound, produced by the transfer of electrons, is called an *ionic compound*.

Compounds formed by *ionic bonding* are very common among minerals. They are usually brittle, and their hardness can vary a good deal. While they themselves are poor conductors of electricity, their solutions are usually very good conductors.

Sometimes, in order to fill their outer shells with eight electrons, atoms will share pairs of electrons (Fig. 3-8). This is called *covalent bonding*. Covalent

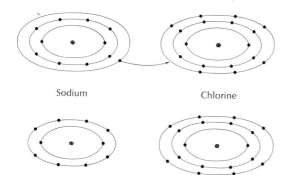

Sodium   Chlorine

Fig. 3-6 Formation of sodium ion (+) and chlorine ion (−) by transfer of an electron (solid circle) from the outermost shell of a neutral sodium to the outermost shell of the neutral chlorine atom. The process of transfer is illustrated above, and the resulting ions with the stable outer shells of eight electrons each are shown below.

Fig. 3-7 Cleavage fragment of the mineral halite (common rock salt). The three directions of cleavage are at right angles to one another. In halite the growth forms are commonly parallel to the cleavage forms. (Courtesy of the Smithsonian Institution.)

compounds may be very hard, as in diamond, because the covalent bonds are strong in all directions. On the other hand, if the bonds are strong in all but one direction, as in graphite, the mineral will break easily in that direction forming flakes or sheets. Some such compounds do not conduct electricity, but are good insulators instead. Diamond, for instance is not a conductor, but graphite is.

Metals have their own variety of bonding. In *metallic bonding* the electron clouds around the nuclei merge, resulting in a closely packed arrangement of nuclei all sharing equally a great crowd of electrons which are free to wander about and which are not "localized" around a particular nucleus, as they are in the other types of bonding. This freedom of movement is responsible for the ability of metals to conduct electricity, and the looseness of the bonding means that metals commonly bend under stress; they are malleable rather than brittle. In addition, the mobility of the electrons means that they can cause maximum interference with the transmission of light and so are opaque and have a metallic lustre.

Structure of Minerals

Because of the rapidity with which electrons revolve around their nucleus, their orbits on each of the various energy levels might be considered as forming complete spheres. This is true even for the lone hydrogen electron. It does not revolve in a path that continuously lies in very nearly the same plane, but whirls vigorously around the nucleus at the rate of 7 million billion times each second. At this incredible velocity it spins a skein of successively occupied paths, and from a practical point of view it might be thought of as forming a spherical shell enclosing the nucleus. By comparison the shells and electron orbits for the calcium and iron atoms, with 20 and 26 electrons respectively, are much more dense.

For this reason, Figure 3-9, illustrating the internal arrangement of a salt crystal, shows the individual sodium and chlorine atoms as spheres that perhaps look like tennis balls or marbles. This is unrealistic, but the device does enable us to (1) convey an idea of the spherical form of the electron shell generated by the orbiting electrons of the outermost shell, (2) show the relative size of the different atoms (in this case sodium has an ionic radius of 0.98 Angstroms, a unit that equals 0.00000001 centimeter, and the ionic radius of chlorine is 1.8 Å), and (3) most important of all, illustrate the way these atoms are arranged to make a crystal of halite. The tennis-ball diagram, as well as the skeletonized lattice diagram resembling a wire cage, show that the sodium and chlorine ions are packed together in a remarkably strong, rigid arrangement, with each sodium ion surrounded by six, equidistant chlorine ions. The wire diagram, more obviously than the tennis balls, brings out the relationship that all the lines shown connecting the atomic centers intersect at right angles.

Figure 3-10 shows diagrammatically the arrangements that atoms may form in building up crystals. These are the basic units which are repeated over and over in all directions. The tetrahedral pattern is of especial importance in the classification of the silicate minerals. Figure 3-11 shows the ways the basic tetrahedra, composed of one silicon atom and four oxygen atoms, are arranged in the different types of silicates.

Physical Properties of Minerals

Physical properties are the things we can see, or feel, or, for such minerals as halite (rock salt), taste. True enough, the chemical composition is possibly the most diagnostic property a mineral possesses, but few of us are going to pack

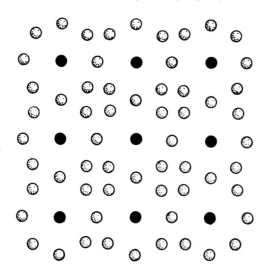

Fig. 3-8 An example of covalent bonding (in diamond).

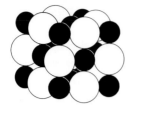

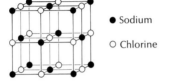

● Sodium

○ Chlorine

Fig. 3-9 Two representations of the structural arrangements of the ions of sodium (solid circles) and chlorine (open circles) in the mineral halite (Fig. 3-7). An actual visible crystal of halite would consist of a many-thousandfold repetition of the three-dimensional pattern illustrated here. The crystal structure depicted was one of the first to be determined rigorously by the use of X-rays. This development, which took place during the early part of this century, and the subsequent discovery of the structures of a myriad of other crystals, is part of one of the still unfolding heroic epochs in man's assault on the unknown, and led to the recent discovery of the structure of a large molecule in the living cell, called DNA (deoxyribonucleic acid), the carrier of heredity.

along a fully equipped chemical laboratory to be used for mineral identification on a field trip. Since one of the critical differences between minerals and rocks is that minerals are approximately homogeneous substances, and most rocks are not, this means that one piece of quartz will be about as hard as another piece, that it will have the same specific gravity, and if formed in a similar environment, it will have about the same crystal form.

The significant properties of minerals that are readily observable in the field are listed below, and a judicious use of these, together with one of the standard mineral handbooks, should enable you to identify many of the common minerals. Their collection and identification is a rewarding activity that can take you to many interesting and unfrequented places.

CRYSTAL FORM   With the obvious exception of mercury and other less familiar minerals, such as opal which is an amorphous variety of silica, nearly all minerals are crystalline substances. The crystalline state of matter is a property which has excited wonder and engendered speculation for tens of centuries. Ancient and medieval literature is filled with references to minerals, their imagined magical or curative properties, and with conjectures over their origin and nature. By many it was believed that minerals grew from seeds, or that there were mineral-generating fluids within the earth, or that sex might enliven things even within this crystalline world and that there were such entities as male and female minerals.

Interesting as these surmises of our ancestors were, we know today that they are not so. We know that the crystal form of a mineral is not a chance vagary of nature, but that its surface is the reflection of an inward orderly arrangement of the elements that constitute the chemical substance of the mineral. Nicolaus Steno (1638–86), a Dane and a true son of the Renaissance, during a life that included training as a physician, a sojourn in Paris, service in the court of Grand Duke Ferdinand II at Florence, and conversion from the Lutheran to the Roman Catholic Church in which he rose to become a prelate, found time to make remarkably perceptive observations on the geological structure and origin of the mountains of Tuscany and, most significantly for the immediate problem of crystal form, clearly demonstrated the fact that the faces of a quartz crystal always intersect at the same angle regardless of the size of the crystal (Fig. 3-2). From this beginning stems the branch of mineralogy known as crystallography.

Steno's original observation that the interfacial angles on a crystal hold constant, plus the fact that all crystals (no matter how complex their geometry may appear to be) can be placed in one of six major crystal systems is a testimonial to the orderly arrangement of their internal structure. Snowflakes (Fig. 3-4) are a familiar example of this generalization. In spite of the nearly infinite variety of forms which they may assume (it is a little too much to expect that no two are ever alike), most are some variant of a six-sided figure, and thus are placed in what is known as the hexagonal crystal system. This is the same system to which quartz belongs.

When we look at Figure 3-9, we see it is no wonder that a halite crystal, newly formed from a strong brine, should be a cube. This mineral is indeed an outstanding illustration of the principle that the crystalline form of a mineral reflects the design of the ordered ranks of the atoms within. Regardless of the size, or the imperfections of the external configuration of the crystal form that we see, the internal disciplined pattern of rigidly arranged atoms is invariant. Thus, the crystal form of a mineral is probably the most fundamental of its visible properties because for us it is the outward expression of the internal structure which is determined in part by the chemical composition.

The most convincing demonstration of the marvellously repetitive regularity of the internal geometry of crystals came about, as is so often the case, as the result of an extraordinarily fortunate and essentially intuitive experiment made

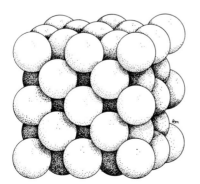

Fig. 3-10 An arrangement of atoms called cubic close packing.

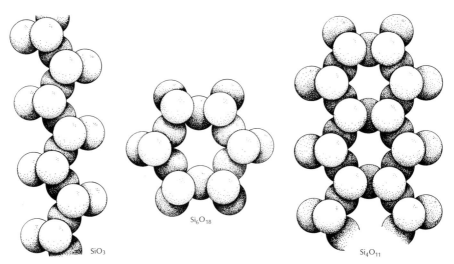

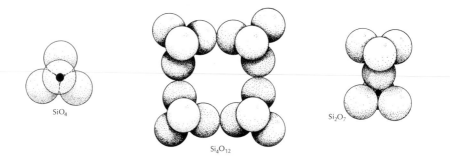

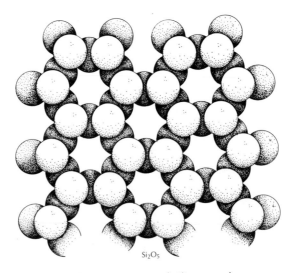

Fig. 3-11   Some examples of arrangements of silicon and oxygen atoms in silicate minerals.

in 1912 in Munich by Max von Laue and his associates. It established in one stroke the wave nature of X-rays, as well as the existence of the systematic internal arrangement of the atoms in a crystal. Laue was actually trying to find a diffraction grating (a glass plate with finely spaced parallel lines engraved on it) suitable for determining the nature of the X-ray. A diffraction grating serves to make visible light passing through the slits on its surface break up into colored spectra because the waves passing through the grating tend to reinforce or neutralize one another. Laue and his companions were convinced (correctly) that the wave length of X-rays was extremely short, and they all despaired of ever scribing a grating with slits closely enough spaced to diffract such finely structured waves. Finally, the thought occurred to them that the unknown particles in a crystal might be systematically arranged and might also be closely enough spaced that they would serve as a diffraction grating.

After the usual false starts and mishaps with their apparatus, they were successful in sending an X-ray beam first through a crystal of copper sulphate ($CuSO_4 \cdot 5H_2O$), and in a later experiment through a crystal of zinc sulphide (ZnS), discovering that an image could be obtained on a photographic plate placed behind the crystal, which in turn had been mounted between the photographic plate and the X-ray source. To their gratification the light-sensitive plate showed a pattern of dots. These were reflections from electrons that form the outer shells of the regularly arrayed atoms in the crystal. The experiment showed beyond doubt that (1) rather than being composed of rapidly moving particles, X-rays not only are wave-like in the same way that ordinary light is but have a much shorter wave length, and (2) that if the waves strike individual atoms in the crystal, the X-rays are diffracted by them in much the same fashion that light rays are diffracted by a grating. X-rays proved to be the magic key that unlocked the door to the unseen, yet ordered, world whose very existence could only be inferred as recently as half a century ago from surface measurements of interfacial angles and the geometry of crystal faces.

Deciphering the atomic structure of minerals proved to be no easy task, because, by the very nature of the evidence, mineralogists have had to work backward. If the characteristic arrangement of the atoms in a crystal were known in advance, it would be relatively easy to predict how the scattering of dots in an X-ray diffraction photograph would be patterned. However, not knowing the structure of the crystal and having to deduce it from the patterned dots involves elaborate calculations and the application of abstruse theories of wave motion. It can be done, but an immense amount of trial and error is called for in gradually constructing a model that finally fits the indirect evidence provided by the internal reflections of the atoms in a crystal. The deciphering of the labyrinthine structure of the protein molecule by this method stands as one of the great intellectual triumphs of our age. One of the proteins whose structure has recently been determined by X-ray diffraction is hemoglobin. Whereas the sodium chloride molecule contains only two atoms, the hemoglobin molecule has an incredible 10,000 atoms.

CLEAVAGE This is the most distinctive property for some minerals and is quite unlike any that is used in the identification of other solid substances. Cleavage is the ability of a mineral to break, or cleave, along rather definite planes paralleling one another, usually on a fairly close spacing (Figs. 3-3, 3-7).

The most familiar example of nearly perfect cleavage in a single direction is that of mica, especially in the light-colored variety, *muscovite*. This mineral splits in successively thinner and thinner layers until only the finest transparent sheets are left. In earlier days these were used for windows on the Franklin stove, and today this sheet-like habit of mica, together with its nonconductivity, makes this mineral ideal for use in many varieties of electrical equipment.

Orthoclase feldspar ($KAlSi_3O_8$) is an excellent example of a mineral with two directions of cleavage—one nearly perfect, the other less so—with both intersecting at an angle of $90°$ (Fig. 3-3).

Calcite ($CaCO_3$) is an example of a mineral cut by three cleavage planes. These intersect at a high angle ($74°\,55'$) to form nearly perfect rhombohedrons (a solid figure each of whose sides is an oblique-angled parallelogram with only the opposite sides being equal).

The geometrically repetitive nature of cleavage, its planar character, and the distinctive orientation of cleavage planes for many minerals are strong evidence that cleavage, like the crystal form of minerals, is a property determined by the packing, or geometric arrangement, of the atoms in a mineral.

One of the more dramatic illustrations supporting this statement is the dual expression of the carbon atom in the two wholly dissimilar guises of graphite and diamond. Both substances are composed solely of the element carbon in combination with nothing else, and for this reason they are called *atomic crystals* in contrast to *ionic crystals*, of which you may recall halite (NaCl) was used as an illustration. Graphite is a black, greasy substance that separates into flaky scales, and thus makes an ideal lubricant. Nothing could be more unlike this than the diamond, which is a most effective abrasive and, in addition, has such highly prized properties as brilliant luster, ability to take and hold a polish, and sharply defined, angular cleavages.

The reason for these profound differences in what is chemically the same substance lies in the completely unlike arrangement of the carbon atoms in the two minerals. In a diamond crystal the linkage between the carbon atoms is about as strong as it possibly can be. Each atom is in direct contact with four others, and in fact shares the four electrons in its outer shell with them. This arrangement produces a structure that is astonishingly close-knit (the centers of the carbon atoms are only 1.5 Å apart), as well as being one that stands up to stresses so well that the carbon atoms in a diamond can cut their way through any other substance. Nonetheless, there are planes of weakness in the diamond wherever centers of the carbon atoms are lined up properly. In fact, the perfect cleavage the diamond possesses gives it initially an octahedral (eight-sided) figure, which is the pattern that is employed by diamond cutters in exploiting cleavage planes.

Graphite crystallizes in thin, parallel sheets. The bonds uniting carbon atoms within each sheet are many times stronger than the feeble tie that binds one atomic layer to its neighboring sheet above or below. For this reason, graphite has a perfect cleavage in one direction—somewhat akin to that of mica—with the result that it splits readily into thin flaky scales in about the same way that mica does.

HARDNESS   This is a purely relative property, but its recognition has considerable antiquity since the scale which is still used was devised over a century ago in 1820 in the provincial Austrian city of Graz by the mineralogist Friedrich Mohs. It, too, is related to the atomic structure of the mineral, since even a small scratch requires the separation of atoms, and the ease of this separation will depend upon the kinds of atoms and the kinds of bonding. It is an easy property to determine, and is one of the first to be determined in identifying an unknown mineral in the field. A harder mineral will scratch a softer one, and minerals of equal hardness commonly will barely scratch one another.

It certainly is common knowledge that diamonds are harder than almost all other substances, and for this reason Mohs placed it at the top of his hardness scale and assigned to it an arbitrary value of 10. Other softer minerals he ranked in a descending hierarchy, perhaps without realizing that unequal degrees of hardness separate the various ranks. For example, the interval between diamond and corundum is more than all the rest of the scale combined, and if absolute values were assigned to the actual intervals on the scale, diamond would be about 42.

diamond	10	
corundum	9	
topaz	8	
quartz	7	
orthoclase feldspar	6	glass
		knife blade
apatite	5	
fluorite	4	
calcite	3	
gypsum	2	finger nail
talc	1	

LUSTER   When a mineral is viewed in ordinary light, the amount of light reflected from its surface determines its luster. The two most common lusters are *metallic* and *nonmetallic*. The first of these terms means that the surface of the mineral reflects light in about the same way that a metal such as brass, iron, or lead would. In general, this very high luster is characteristic of minerals which have covalent bonding. Minerals with a metallic luster commonly are opaque, even along thin edges held up against the light. Nonmetallic lusters range over about every other type. If a mineral reflects light to about the same degree as glass, it has a *vitreous luster*. Quartz is an excellent example. Among other terms that commonly are employed and are essentially self-explanatory are earthy, waxy, dull, resinous, pearly, and silky.

SPECIFIC GRAVITY A measure of the weight of a mineral compared to the weight of an equal volume of water taken at its maximum density at a temperature of $4°$ C. ($39.2°$ F.) is its specific gravity. At this temperature the weight of any volume of water is considered to have a value of 1. Thus, quartz with a specific gravity of 2.7 is a substance that weighs 2.7 times as much as an equal volume of such water would weigh.

The specific gravity is usually determined by weighing a mineral in the air and then weighing it fully immersed in water, then:

$$\text{Specific Gravity} = \frac{W_a}{W_a - W_w}$$

where, W_a = weight of the mineral in air, and W_w = weight of the mineral in water.

This is a property that can be estimated with surprisingly high accuracy in an entirely subjective way, after experience has been acquired, simply by hefting the mineral by hand. Pyrite (FeS_2), with a specific gravity of 5, and galena (PbS), about 7.5, are perhaps typical of metallic minerals, while the range 2.6 to 2.8 covers representative rock-forming minerals, such as quartz, feldspar, and calcite ($CaCO_3$).

COLOR This is the most obvious property that minerals possess, and for some of them it is diagnostic. An illustrative example is amethyst, which is the name given to a characteristically purple or pale violet form of quartz. The color of quartz ranges through a spectrum from absolutely colorless, glass-clear rock crystal to coal black varieties, depending for one thing on the nature and amount of impurities that are included. In short, color is an important property; for some minerals it is diagnostic, for others it is almost without significance. Unfortunately, colors cannot be used with the same confidence in identifying minerals that they can in naming birds and flowers. A considerable amount of experience has to be acquired before one learns which colors are meaningful and which are so variable as to be without significance.

Mineral Descriptions

Of the nearly 2000 minerals that have been named by now, only eleven are considered here, and these are among the more important of the rock-forming minerals. They are the building blocks of the rocks which themselves are the essential constituents of the lithosphere. Fortunately for us, all the rocks of the earth are made up of varying combinations of surprisingly few minerals.

Some Important Rock-Forming Minerals

Quartz

Orthoclase ⎱
Plagioclase ⎰ Feldspar

Muscovite ⎱
Biotite ⎰ Mica

Ferromagnesian Minerals {

Hornblende   Amphibole
Augite   Pyroxene
Olivine

Calcite

Gypsum

Halite

Quartz: SiO_2

This is the hardest of the common rock-forming minerals, with a hardness of 7 of the Mohs scale. Customarily it crystallizes in six-sided crystals (Fig. 3-2) which are terminated by a sharp-pointed pyramid at each end should the mineral have an opportunity to grow free from interference. Quartz that grew in cavities and geodes commonly has only one pyramid on the end of the crystal that extends into the opening. Crystals that grew into openings may sometimes reach dimensions of a foot or more, but in rocks such as granite quartz crystals are much smaller, seldom over one-fourth of an inch in diameter. Where they are fresh and unweathered they may sparkle like tiny fragments of glass. Quartz has a strong vitreous luster, and when pure is completely clear and colorless. In fact, the Greeks thought it was some kind of frozen water. It is also distinguished by its lack of cleavage.

Orthoclase: $KAlSi_3O_8$

Orthoclase is the potassium-bearing member of the feldspar group. The feldspars are the most abundant by far of the rock-forming minerals and are the most important constituent of the lithosphere, probably making up at least 50 per cent of its substance. They all have good cleavage in two directions, at or almost at right angles to each other.

Orthoclase has a hardness of 6, only slightly less than that of quartz. Like quartz it has a vitreous luster and may be colorless, although it is more often milky white or flesh pink. It sometimes resembles unglazed porcelain, like the dull surface exposed in a chipped dinner plate.

Plagioclase: $NaAlSi_3O_8$ • $CaAl_2Si_2O_8$

This sodium- or calcium-bearing feldspar comprises a group of minerals constituting an isomorphous series, which means that this group should be thought of as a solid solution with one end member being the sodium plagioclase, the other being the calcium-bearing form. The intermediate varieties, to which individual names have been given, comprise a continuous sequence.

Plagioclase has the same hardness as orthoclase, as well as a vitreous luster. Its color is most likely to be white or pale gray, although some varieties show a beautiful iridescence, or play of colors, much like those of a peacock's feathers. Some varieties, though, may be nearly as glass-clear as quartz. However, one means of distinguishing plagioclase from quartz or orthoclase is the presence on some of the crystal or cleavage surfaces of a multitude of very closely spaced parallel straight lines, which are almost as fine as though they had been engraved there. The lines are made by the nearly right-angle intersection of close-interval, parallel internal planes with the surface of the crystal.

Muscovite: $KAl_3Si_3O_{10}(OH)_2$

The micas, too, are a group of closely related minerals that share the property of a sheet-like arrangement of their atoms, resulting in an excellent cleavage paralleling these internal planes of weakness. The two sorts of mica that we are concerned with here, muscovite and biotite, are the most important rock-forming varieties.

The common name for muscovite is white mica, and generally it is colorless to transparent, especially when it is peeled down to thin sheets. This cleavage is perhaps its unique property, and in centuries past muscovite was used in the tiny windows of the houses of medieval Europe before the widespread use of glass brought more light to their gloomy interiors. Muscovite has a pearly or silky luster, and in a rock its tiny spangles shimmer when it catches the sunlight —the German name of *glimmer* for the mineral mica conveys an impression of this very property.

Biotite: $K(Mg,Fe)_3AlSi_3O_{10}(OH)_2$

As you can see from the chart, biotite, as well as being one of the micas, is also one of the *ferromagnesian minerals*, a group which includes a great number of the darker minerals of rocks. The formulas of hornblende, augite, olivine, and biotite show that these rock-forming minerals contain iron and magnesium. There are other ferromagnesian minerals in addition, but these are the principal rock-forming minerals within the group.

Biotite is commonly called black mica, and, as its chemical formula indicates, it includes iron and magnesium in its composition, while muscovite does not. Usually, biotite is colored dark brown or black, and thin sheets of it lack the transparency of muscovite. In rocks such as granite, it occurs as brilliant jet-black flakes that shine like satin in the sun.

Hornblende: $Ca_2Na(Mg,Fe)_4(Al,Fe,Ti)_3Si_6O_{22}(O,OH)_2$

This name is applied to the most abundant isomorphous series of a large and complex group of minerals with similar physical properties known as the amphiboles. The significant difference in composition between hornblende and its close relative, augite, is that hornblende contains some hydrogen while augite does not. Hornblende is a dark mineral, commonly dark green and black. When it is unweathered, it may be brilliant jet black, and its strong vitreous luster shines as brightly as a lacquered surface. Hornblende crystals are long and narrow as a rule—such a form is called prismatic. The cleavage pattern of hornblende is one of its more distinctive properties. There are two principal directions and their planes parallel the long axis of the crystal, but intersect each other at oblique angles of 56° and 124° (Fig. 3-12). Hornblende in a typical occurrence, such as granite, shows up as brilliant black, lath-like crystals dispersed through the rock.

Augite: $Ca(Mg,Fe,Al)(Si,Al)_2O_6$

This dark ferromagnesian mineral, like hornblende, is more common in the darker than the lighter colored rocks. Augite crystals generally are stubbier, in fact they often are nearly equidimensial. Their cleavage planes approximate a right angle in their intersection, since they are at 93° and 87°, respectively, as contrasted to the oblique cleavages of hornblende. Augite crystals seen in cross section are nearly square (Fig. 3-12). The color is about the same as that of hornblende, very dark green or black, and the luster is vitreous.

Hornblende and augite are the more abundant of the darker rock-forming minerals. The principal distinctions between the two are: (1) hornblende crystals tend to be long and narrow, while augite crystals are short and stubby; (2) hornblende has oblique cleavages parallel to the long axis of the crystal, while augite cleavage planes intersect each other at approximately right angles; (3) hornblende crystals seen in cross section approach a rhombic pattern, while augite crystals are more nearly square.

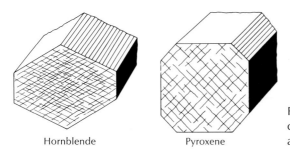

Hornblende   Pyroxene

Fig. 3-12 Illustration of the distinctive cleavage properties of hornblende and augite.

Olivine: $(Fe,Mg)_2SiO_4$

This is not an especially abundant ferromagnesian mineral, but it is a distinctive one. From its name we are likely to infer, correctly, that the color is green. Usually the mineral occurs as rounded, well-defined, granular, glassy crystals. When these are sufficiently large and free from blemishes they make the attractive, although fragile, gem stone called *peridot.* Olivine occurs most commonly in the dark, iron- and magnesium-rich igneous rocks, such as basalt. Olivine crystals in fresh basalt often look like tiny bits of dark green bottle glass, but when they weather they may alter to shades of brown and red. In some varieties of dark intrusive igneous rocks the percentage of olivine may be so great that the rock may consist almost wholly of a granular aggregation of this single mineral, as in the rock *dunite,* named for Dun Mountain in New Zealand.

Calcite: $CaCO_3$

Because of its composition calcite would be expected to have a markedly different assemblage of physical properties from the minerals described above, all of which are compounds having silicon and oxygen as essential elements (Figs. 3-1, 3-3, 3-11). Calcite is abundant enough in some circumstances to be the sole mineral in a rock. Examples of such a monomineralic rock are limestone and its recrystallized equivalent, marble. In these two rocks, both the mineral and the rock are a single compound, $CaCO_3$, and the rock is not an aggregate of minerals of diverse form and composition, as is a rock such as granite.

Normally, calcite is a light-colored (white or pale yellow) or colorless mineral, although—depending on the amount and nature of the impurities—the color may range across a spectrum including yellow, orange, brown, and black. Calcite has a vitreous luster, and is a mineral that is easily scratched since its hardness is only 3. Calcite occurs in crystals which are difficult to categorize readily since they are found in such an extraordinary variety of forms. In general, they tend to be six-sided, and sometimes, like quartz, they are terminated by a long, narrow, many-faceted pyramid. A more diagnostic property is a nearly perfect cleavage in three directions (Fig. 3-3), and the intersections of these cleavage planes almost invariably produce a rhombohedral pattern. That is, when the mineral breaks into fragments, each of the faces is a rhombus, or approximately diamond-shaped figure (like the diamond suit in playing cards).

A final test which serves to discriminate between calcite and its close relative *dolomite,* $CaMg(CO_3)_2$, is the fact that calcite effervesces, or fizzes, very strongly in cold dilute hydrochloric acid, while dolomite reacts nowhere near so readily. Dolomite, like calcite, also occurs in large enough masses to constitute whole rock layers, which have the same name as the mineral of which they are made—dolomite. Aside from its inability to effervesce as readily, a further distinction from calcite is the fact that dolomite crystals are slightly harder (3.5), have higher specific gravity, and often have crystal faces showing slight curvature instead of being sharp, well-defined planes. The origin of the rock, dolomite, is still being debated; one possibility is that magnesium-bearing solutions may alter limestone through the replacement of some of the element calcium by magnesium.

Both calcite and dolomite are typical of sedimentary rocks, or of rocks recrystallized from them. Both minerals are distinguished from quartz by their lesser hardness, both have a vitreous luster, and both are markedly crystalline.

Gypsum: $CaSO_4 \cdot 2H_2O$

This compound, like calcite, is an example of a rock-forming mineral which is not a silicate. Gypsum is the name applied to a mineral as well as to rock layers consisting of this mineral alone. Gypsum is a very soft mineral with a hardness of 2. Commonly it is white or colorless, but, like calcite, if impurities are included, it may show a wide color range. Large crystals are likely to have a nearly perfect micaceous cleavage in one direction, and less well-developed cleavages nearly at right angles to it. Thin sheets of gypsum, like muscovite, are colorless or white and are transparent to translucent. In medieval Italy, gypsum sheets were once used in windows before glass became available. A massive and essentially structureless variety of gypsum with a soft, pearly luster is known as *alabaster*, and in the Classical Period, as well as in the Renaissance, was much favored for statuary. Alabaster has the advantage of uniform texture and softness, but for this very reason it is readily scarred or mutilated.

Satin spar is a familiar type of gypsum in the arid western states, where it commonly is found as a silky, fibrous mineral filling the narrow seams between layers of shale, with the fibers standing at right angles to the stratification. *Selenite* is a variety that is white or colorless and characteristically is found in broad flat sheets that separate along a cleavage plane only slightly less well defined than that of muscovite.

Closely related to gypsum, and rather hard to distinguish from it, is *anhydrite*, $CaSO_4$. Chemically, the important difference is the absence of water of crystallization in anhydrite, which is an integral part of the formula for gypsum. The chief practical distinction between the two minerals is the slightly greater hardness of anhydrite (3-3.5).

Both gypsum and anhydrite are found in some parts of the world (west Texas is an illustration) in thick layers, or strata, fully comparable to beds of limestone, and are presumed to have crystallized from solution as the result of evaporation of what may once have been an arm of a shallow, nearly land-encircled sea. For this reason, these monomineralic rocks often are found closely associated with extensive bodies of rock salt, with which they share a common origin.

Gypsum from Imperial County, California. (Photograph by William Estavillo.)

Fossiliferous limestone. (Photograph by William Estavillo.)

Twinning in plagioclase. (Photograph by William Estavillo.)

Halite: NaCl

This mineral, which is the same substance as common salt, has been discussed earlier in this chapter. Its most diagnostic features are the most obvious—its solubility in water and its taste. Halite is a fourth example of a mineral that commonly makes a rock almost devoid of other minerals—called rock salt or simply salt. Because of the high solubility of halite, rocks containing it are seldom found at the earth's surface except in quite arid regions. Among the more spectacular occurrences of rock salt are the salt domes found in many places in the world.

Selected References

Bragg, Sir Laurence, 1968, X-ray crystallography, Scientific American, vol. 219, no. 1, pp. 58–79.

Desautels, Paul E., 1968, The mineral kingdom, Madison Square Press, Grosset & Dunlap, New York.

Holden, Alan, and Singer, Phyllis, 1960, Crystals and crystal growing, Doubleday and Co., New York.

Pough, F. H., 1953, Field guide to rocks and minerals, Houghton Mifflin, Boston.

Sinkankas, John, 1966, Mineralogy: a first course, D. Van Nostrand Co., Inc., Princeton, N.J.

Vanders, Iris, and Kerr, Paul F., 1967, Mineral recognition, John Wiley, New York.

Fig. 4-1   Eruption of Hekla, Iceland. (Photograph by Thorsteinn Josepsson.)

4
Igneous Rocks and Igneous Processes

Three quarters of a century ago captains of sailing ships beating their way through the Sunda Staits which separate the great islands of Java and Sumatra in the East Indies knew the island of Krakatoa well. Its conical, green-clad slopes rose uninterruptedly about 790 meters (2600 feet) to the summit of the central peak. The straits were important since they were on the shortest sea road for the tea clippers en route from China to England. These were dangerous, restricted waters, haunted by sea-roving Dyaks who could give the crew of a becalmed vessel a bad time. In this same seaway many years later the U.S.S. *Houston*, harried by pursuing Japanese in the early years of World War II, blew up and sank with the loss of almost all hands.

Though the island of Krakatoa had been spasmodically active since May 1883, it seemed innocuous enough to the crew of the British ship *Charles Bal*, tacking under all plain sail through the hot tropical Sunday afternoon of August 26, 1883, until they arrived on one heading at a point about 16 kilometers (10 miles) south of the island. Minutes later the mountain exploded. Seldom in the long history of seafaring has the crew of any vessel been confronted by such a satanic outburst of energy. The entire mountain disappeared in clouds of black "smoke," and the air was charged with electricity—lightning flashed continuously over the volcano, as it very often does during eruptions, and the yards and rigging of the ship glowed with St. Elmo's fire. Immense quantities of heated ash fell on the deck or hissed through the surrounding darkness into the increasingly disturbed sea. As the vessel labored across broken seas through squalls of mud-laden rain, a thundering roar of explosions continued, much like a never-ending artillery barrage, accompanied by a ceaseless crackling sound which resembled the tearing of gigantic sheets of paper. This last effect was interpreted to be the rubbing together of large rocks hurled skyward by the explosions. After an interminable night, the dawn, dim as it must have been, came as deliverance, and with the coast of Java in view and a gale rising rapidly, the *Charles Bal* was able to set all sail and leave the smoking mountain far astern.

It is well she did, for paroxysms of volcanic fury continued to shake the mountain until the final culmination of four prodigious explosions came on Monday, August 27, at 5:30, 6:44, 10:02, and 10:52 a.m. The greatest of these, the third, was one of the most titanic explosions recorded in modern times—greater in intensity than some of our nuclear efforts. The sound was heard over tremendous distances: at Alice Springs in the heart of Australia, in Manila, in Ceylon, and on the remote island of Rodriguez in the southwest Indian Ocean, where it arrived four hours after the explosion had occurred, 5000 kilometers (3000 miles) away.

The explosion seriously agitated the earth's atmosphere, and records of such a disturbance were picked up by barometers all over the world. They showed that a shock wave originating in the East Indies traveled at least seven times around the world—out to the antipodes of the volcano and back again—before it became too faint to register on the instruments of that time.

Visibly, a more impressive phenomenon was the huge cloud of pumice and volcanic debris that blew skyward. The steam-impelled cloud of volcanic ash is estimated to have risen to a height of 80 kilometers (50 miles) on August 27, and to have blanketed a surrounding area of 780 million square kilometers (300,000 square miles). The ash poured down as a pasty mud on the streets and buildings of Batavia—now Djakarta—133 kilometers (83 miles) away. Pumice in great floating rafts blanketed much of the Indian Ocean, and captains' comments recorded in logbooks of ships suddenly enmeshed in far-reaching masses of pumice far offshore make interesting reading.

Volcanic ash hurled into the upper levels of the earth's atmosphere was picked up by the jet stream—whose existence was not even suspected then—and carried with it as a dust cloud that encircled the earth in the equatorial regions in thirteen days. Incidentally, the jet stream was virtually forgotten, only to be rediscovered in our age of high-altitude flight, jet travel, and radioactive fallout. The ash continued to spread across both hemispheres of the earth and produced a succession of spectacular and greatly admired sunsets over most of the world—even in areas as remote from Java as England and the northeastern United States—for the two years that it took the finer dust particles to settle through the atmosphere.

The violent explosion of the morning of August 27 set in motion one of the more destructive sea waves ever to be recorded. It spread out in ever-widening circles from Krakatoa much as though a gigantic rock had been hurled into the sea. About half an hour after the eruption, the wave reached the shores of Java and Sumatra, and on these low-lying coasts the water surged inland with a crest whose maximum height was about 36.5 meters (120 feet). Since many of the people inhabiting such a densely populated tropical coast lived in houses built on piers extending out over the water, about 30,000 or 40,000 people lost their lives.

The sea wave, after leaving the Sunda Strait with diminishing height, raced on across the open ocean. It was registered long after it was too faint to see, as a train of pulses on recording tide gauges along the coasts of India and Africa and on the coasts of Europe and the western United States. For example, the tide gauges in San Francisco Bay showed a disturbance of about 15 centimeters (6 inches) by waves that traveled a distance of 16,642 kilometers (10,343 miles) at a speed of about 956 kilometers (594 miles) per hour, a value which seems high. In the Indian Ocean, the velocity of the wave appears to have varied between 322 and 644 kilometers (200 and 400 miles) per hour in the open ocean. This agrees with the better-timed earthquake wave that originated on the coast of Chile and destroyed the low-lying parts of Hilo, Hawaii, on May 23, 1960, 11 hours and 56 minutes later, having traveled 10,600 kilometers (6600 miles) at an average speed of 711 kilometers (442 miles) per hour.

After the explosions died down returning observers were startled to find that where the 790-meter (2600-foot) mountain had stood was now a hole whose bottom was 275 to 300 meters (900 to 1000 feet) below sea level, and that the sea now filled this large bowl-shaped depression. All that remained of the island were three tiny islets on the rim. All told, although the estimates vary, a little less than 5 cubic miles of material were hurled into the atmosphere. In popular accounts of the eruption the impression commonly is given that a volcanic mountain blew up and its fragments were strewn far and wide over the face of the earth. Were this to be the case, we should expect most of the debris covering the little islands which are the surviving remnants of Krakatoa would be pieces of the wrecked volcano, and the oversize crater, or *caldera*, now filled with sea water would be the product of a simple explosion.

Unfortunately for this seemingly plausible explanation, few pieces of the original volcanic mountain are to be found, and instead of such fragments the ground is covered with deposits of pumice up to 60 meters (200 feet) thick. You may also recall in the description of the eruption the mention of the great rafts of floating pumice in the Indian Ocean which were a source of surprise to the mariners who encountered them drifting over much of the open sea. Pumice is original magmatic material, frothed up by gases contained in the magma, and has nothing to do with the internal composition of the vanished mountain. Thus, the abundance of pumice and the absence of pieces of the mountain lead logically to the belief that the volcanic cone foundered or collapsed on itself rather than having been blown to bits.

The explanation that appears to be correct was advanced by a Dutch vol-

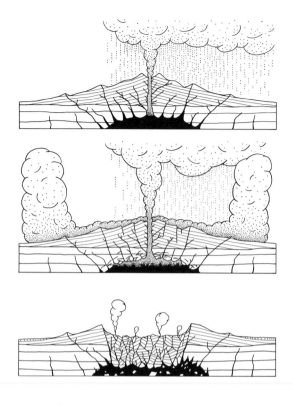

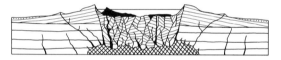

Fig. 4-2   Stages in the collapse of a volcanic mountain to form a caldera.

canologist, van Bemmelen, in 1929, and refined by Howel Williams, of the University of California, in 1941. The accompanying diagram (Fig. 4-2), adapted from Williams, shows the sequence of eruptive events which very likely were responsible for the disappearance of a 790-meter (2600-foot) mountain and the appearance of a 300-meter (1000-foot) deep caldera in its place.

Stage I. The eruptive cycle commenced with fairly mild explosions of pumice. The magma chamber was filled and the magma stood high in the conduits. With an increase in the violence of the explosions, magma was drawn off more and more rapidly and the level dropped in the magma chamber.

Stage II. The culminating explosions cleared out the volcanic conduits and rapidly lowered the magma level in the chamber. In this phase pumice was blown high above the cone, or glowing, pumice-laden clouds swept down the flanks.

Stage III. With removal of support, the volcanic cone collapsed into the magma chamber below, leaving a wide, bowl-shaped caldera.

Stage IV. After a period of quiescence new minor cones appeared on the crater floor. Some of these rise above sea level, such as Anak Krakatoa (child of Krakatoa) which appeared in 1927 and was growing as recently as 1960.

With the knowledge we have of Krakatoa's eruption, and applying the principle of uniformitarianism, we can attack the problem of the origin of Crater Lake in Oregon (Fig. 4-3). Although Crater Lake stands at an altitude of 1800 meters (6000 feet), rather than at sea level as Krakatoa does, their calderas have many similarities.

The diameter of both craters is disproportionately large compared to the dimension of the mountain of which they are a part, and in each case most of the mountain has disappeared in the making of the caldera. It was seen to have disappeared in 1883 at Krakatoa, and its disappearance is inferred on the basis of compellingly strong evidence at Crater Lake. Here the central issue is: was the mountain, Mount Mazama, blown to bits, or was it destroyed as Krakatoa was by a combination of (1) explosion, (2) violent clearing out of the magma cham-

Fig. 4-3 Crater Lake, Oregon, looking southwest. The caldera is 6 miles in diameter, the lake is nearly 2000 feet deep, and the highest points on the rim are nearly 2000 feet above the lake. Wizard Island, a small cone that erupted on the caldera floor, is seen at the right of the lake. (Photograph by Ray Atkeson.)

ber, and (3) collapse of the unsupported volcanic edifice into the void suddenly created by the emptying of the magma chamber?

The most convincing evidence at Crater Lake, as at Krakatoa, is the fact that almost all of the immense quantity of debris surrounding the site of the vanished mountain is pumice, and since it fell on land, probably not more than 5000 years ago, most of it has survived, and thus its volume can be estimated. According to Howel Williams:

> When the culminating eruptions were over, the summit of Mount Mazama had disappeared. In its place there was a caldera between 5 and 6 miles wide and 4000 feet deep. How was it formed? Certainly not by the explosive decapitation of the volcano. Of the 17 cubic miles of solid rock that vanished, only about a tenth can be found among the ejecta. The remainder of the ejecta came from the magma chamber. The volume of the pumice fall which preceded the pumice flows amounts to approximately 3.5 cubic miles. Only 4 per cent of this consists of old rock fragments. . . . Accordingly 11.75 cubic miles of ejecta were laid down during these short-lived eruptions. In part, it was the rapid evacuation of this material that withdrew support from beneath the summit of the volcano and thus led to profound engulfment. The collapse was probably as cataclysmic as that which produced the caldera of Krakatau in 1883.

Mount Pelée

In 1902, as today, Martinique in the French West Indies was one of the more picturesque links in the chain of islands reaching like green stepping stones across the Caribbean to join Cuba with the mainland of South America. The island is quite mountainous, most of the interior is garlanded with a tropical forest, and the people then, as now, lived near the coast in villages, on plantations, and in the few large towns, of which St. Pierre, with a population of 28,000, was the most important.

Mount Pelée, about 8 kilometers (5 miles) north of town, was known to be a volcano, but it had smoldered contentedly for several centuries in the mild climate of the trade winds. On April 23, 1902, it began to show signs of internal discontent, but these were little more than occasional rumblings, clouds of smoke, and spasmodic outbursts of ashes and cinders. This mildy petulant display rose to a more violent level of activity on May 4, when an outburst of hot mud, steam, and some lava broke through the crater wall, coursed down one of the radial stream canyons, buried a sugar central, and killed twenty-four persons.

By this time St. Pierre was thoroughly aroused and not even the presence of the governor and his retinue, together with the issuance of the usual tranquilizing proclamation, served to quiet the multitude. In fact, the city was kept in a continuous turmoil through the arrival of country people and villagers frightened into abandoning their homes.

The tumult and confusion were stilled forever, with appalling suddenness, early in the morning of May 8, 1902, at 7:45 a.m. According to the few eyewit-

Fig. 4-4   St. Pierre, Martinique, after the eruption of Mount Pelée in 1902. (Photograph by Brown Bros., New York.)

nesses, the top of the mountain vanished in a blinding flash, and almost immediately thereafter a rapidly moving, fire-hot cloud advancing at prodigious speed engulfed the city, whose population, swollen with refugees, probably numbered more than 30,000. All but two died in a blazing instant in a cloud whose temperature was high enough to melt glass (650°–700° C.) but not quite hot enough to melt copper (1058° C.) (Fig. 4-4). The purser of the ship *Roraima*, then approaching the harbor from the sea, left the most complete narrative of any observer. The *Roraima* was enveloped in the wall of flame that incinerated the town, was hurled over on her beam ends, the masts and stack were sheared off, her captain was blown overboard from the bridge and killed, and the ship herself burst into flames not only from the heat of the glowing cloud but, to add a bizarre touch to the holocaust, from the thousands of gallons of blazing rum that poured through the streets of St. Pierre and spread out over the waters of the harbor.

Following the explosive eruption, the top of Mount Pelée was surmounted by a great spire which started to rise from the summit in August and continued to grow until the end of the year when it towered like an immense obelisk nearly 300 meters (1000 feet) above its base. Gradually it disintegrated until by mid-1903 it had disappeared; but during its brief existence it loomed as an ephemeral memorial over the sepulcher at its feet. The lifting of this column of solidified lava by the gas pressures generated within the volcano is an eloquent testimonial to their power, since it is estimated to have weighed about three times as much as the Great Pyramid.

The escape of these entrapped high-temperature gases is the chief reason for the appearance of the catastrophically violent clouds that overwhelmed St. Pierre. Since there is no satisfactory English name for them, the French term *nuée ardente*, which might inadequately be translated as glowing cloud, seems appropriate.

These clouds move with great rapidity and have temperatures high enough to incinerate almost anything inflammable in their path. They are extremely dense clouds, heavily charged with pumice fragments and dust—so much so that in photographs of them taken following the titanic explosion of May 8, they resemble the solid, roiling sort of smoke cloud produced by burning oil tanks in refinery fires. Tremendous blocks of rock, some weighing many tons, were transported several kilometers. This was possible because of (1) the very high density of the gaseous cloud, (2) its extreme turbulence, and (3) the fact that incandescent blocks of pumice in the cloud were themselves discharging great quantities of gas. The effect of all this was to reduce surface friction and to allow these hot volcanic fragments, large and small, to be projected with tremendous velocities down the mountain slope.

The 1902 eruption of Mount Pelée was important geologically because it provided a stupendous example of a kind not too well understood before. No lava appeared in the early, violent phase of the eruption, as was possibly also true at Krakatoa, and the destruction of St. Pierre resulted from its position directly in the path of the gas-propelled cloud of incandescent volcanic fragments.

When the violently explosive phase ended, a viscous, stiff, blocky variety of lava was extruded into the summit crater, ending with the construction of a domelike protrusion of blocky lava, encrusted with lesser spires and pinnacles. By September 1903 the spire had attained a height of perhaps 300 meters (1000 feet) and a diameter about twice as great. Estimates placed its volume at 100 million cubic meters.

Such volcanic domes are more common than many people think. Lassen Peak in northern California is an excellent example: a domelike protrusion of blocky lava stands 760 meters (2500 feet) above its crater rim, with a volume of approximately three-fifths of a cubic mile. The mountain was last active in 1914–17, when steam explosions, after blasting a vent on the northern slope, melted the snow cap. The resulting mud and ash flows not only devastated the forest at the base but swept 20-ton boulders for distances of 8 to 9 kilometers (5 to 6 miles).

Among other examples of volcanic domes are the Mono Craters in east-central California. Some of the volcanoes of the Valley of Ten Thousand Smokes in Alaska, as well as the Puys of the Auvergne region of France, are domes. Of the latter, the Puy de Dôme is perhaps the best known.

Recognition of the deposits dropped by such gas-charged, highly mobile, turbulently flowing incandescent clouds as *nuées ardentes* was uncertain before the demonstration of their nature at Mount Pelée. Commonly their stratification may be chaotic. Large blocks are mixed with finer particles. Such deposits of *nuées ardentes* encircle Crater Lake, for example, and there pumice flows swept down the Rogue River canyon for 56 kilometers (35 miles) (Fig. 4-5). Their velocities may have attained 160 kilometers (100 miles) per hour, and these glowing clouds were capable of carrying pumice blocks almost 2 meters (6 feet) in diameter a distance of at least 32 kilometers (20 miles).

Types

Volcanoes show almost as much individuality as there are examples, and for this reason they are difficult to place in a rigid classification. In a general way, four major types of eruptions can be recognized, characterized as explosive, intermediate, quiet, and fissure. Two explosive eruptions have been given because each had such distinctive characteristics as to merit consideration as a subtype. In the sections to follow, a famous historic eruption representing each of the others is described.

Intermediate Eruptions

Vesuvius

Of all the world's volcanoes, none is more famous than Vesuvius, the only one active on the European mainland today (Fig. 4-6). Its renown probably results from its well-publicized eruption of A.D. 79 with the accompanying destruction of Pompeii, Herculaneum, and Stabiae. Although the mountain had been active in prehistoric times, no tradition existed among the Romans of its true nature except in a rather sketchy form. In fact, the 1200-meter (4000-foot) mountain we see today is superimposed in large part on the wreckage of the older, lower, pre-A.D. 79 crater, to which the name, Monte Somma, is given. In A.D. 63 the volcano showed some stirrings of life when a succession of earthquakes commenced and caused some of the damage still to be seen around Pompeii. This, however, was but a prelude to the historic eruption of August 24, A.D. 79.

Fortunately, one of the most complete descriptions of the event has come down to us across the intervening years through two letters from the seventeen-year-old Younger Pliny to his friend Tacitus, the Roman historian. The letters

Fig. 4-5   Tuff, or consolidated volcanic ash, with lapilli and blocks of pumice. Deposits of nuées ardentes, the eruption of which preceded the collapse of Mount Mazama to form Crater Lake. The view is of the Pinnacles along Sand Creek, Crater Lake National Park. (Courtesy of Oregon State Highway Dept.)

were written primarily to describe the death of his uncle, Pliny the Elder, a leading philosopher of the day and also, rather surprisingly, an admiral of the Roman navy.

While Pliny the Younger was suddenly impressed with the necessity of studying his books, the Elder Pliny, soon to achieve the distinction of being the world's first volcanologist, and a Roman of the old school, marched forth to his death on the mountain. Parts of Pliny the Younger's letters are cited here because they are such good examples of straightforward reporting, quite unlike the exaggeratedly impossible version in Bulwer-Lytton's novel, *The Last Days of Pompeii*, in which most of the populace dies while watching a gladiatorial combat in the arena.

Parts of Pliny the Younger's letters follow:

> Gaius Plinius sends to his friend Tacitus greeting.
>
> You ask me to write you an account of my uncle's death, that posterity may possess an accurate version of the event in your history. . . .
>
> He was at Misenum, and was in command of the fleet there. It was at one o'clock in the afternoon of the 24th of August that my mother called attention to a cloud of unusual proportion and size. . . . A cloud was rising from one of the hills which took the likeness of a stone-pine very nearly. It imitated the

Fig. 4-6 Mount Vesuvius during the eruption of 1944. The white-capped peak to the left of Vesuvius is the arcuate ridge known as Monte Somma, part of an older, prehistoric volcano. (Photograph by Brown Bros., New York.)

lofty trunk and the spreading branches. . . . It changed color, sometimes looking white, and sometimes when it carried up earth or ashes, dirty or streaked.
The thing seemed of importance, and worthy of nearer investigation to the
philosopher. He ordered a light boat to be got ready, and asked me to accompany him if I wished; but I answered that I would rather work over my
books. . . .

Ashes began to fall around his ships, thicker and hotter as they approached
land. Cinders and pumice, and also black fragments of rock cracked by heat,
fell around them. The sea suddenly shoaled, and the shores were obstructed
by masses from the mountain. . . .

My uncle, for whom the wind was most favorable, arrived, and did his best
to remove their terrors. . . . To keep up their spirits by a show of unconcern,
he had a bath; and afterwards dined with real, or what was perhaps as heroic,
with assumed cheerfulness. But meanwhile there began to break out from
Vesuvius, in many spots, high and wide-shooting flames, whose brilliancy was
heightened by the darkness of approaching night. My uncle reassured them by
asserting that these were burning farm-houses which had caught fire after
being deserted by the peasants. Then he turned in to sleep. . . .

It was dawn elsewhere; but with them it was a blacker and denser night than
they had ever seen, although torches and various lights made it less dreadful.
They decided to take to the shore and see if the sea would allow them to em-
bark; but it appeared as wild and appalling as ever. My uncle lay down on a
rug. He asked twice for water and drank it. Then as a flame with a forerunning
sulphurous vapor drove off the others, the servants roused him up. Leaning on
two slaves, he rose to his feet, but immediately fell back, as I understand
choked by the thick vapors. . . . When day came (I mean the third after the
last he ever saw), they found his body perfect and uninjured, and covered just
as he had been overtaken. . . .

Pliny's letters clearly indicate that the destruction of Pompeii and Herculaneum resulted from the fall of hot ash, and in this shroud were buried the 2000
of the 20,000 inhabitants who perished. Most of the dead were slaves, soldiers
of the guard, or people who were too avaricious to leave their worldly goods.
Most were suffocated by falling ash, by hot volcanic mud, or by volcanic
gases, and the temperature of the ash was high enough that their bodies
charred away. Centuries later when plaster of paris was poured into the cavities once occupied by their bodies, allowed to harden, and then excavated from
the ash, their shapes as well as those of dogs and cats, loaves of bread, and all
sorts of objects in similar cavities stood revealed. Hundreds of papyri were
preserved in the library, along with murals on the walls of houses, and these
give a most revealing insight into the interests and pursuits of these long-vanished Romans whose lives and preoccupations were so much like our own. The
two cities of Pompeii and Herculaneum slept undisturbed for nearly 1700 years
until the discovery of one of the outer walls in 1748 ushered in the period of
modern archaeology.

Vesuvius has continued its activity from A.D. 79 to the present; in A.D. 472
ashes drifted from its crater as far east as Constantinople. An especially violent
eruption in 1631 is estimated to have killed 18,000 people, and came after a

Fig. 4-7 The ruins of Pompeii, at the foot of Vesuvius, are now free of the volcanic ash that buried the city nearly 2000 years ago. (Photograph by Moody Institute of Science.)

period of quiescence that lasted long enough for the volcano to be once again overgrown by vegetation. A large number of minor eruptions have been recorded, but major ones occurred in 1794, 1872, 1906, and in 1944 in the midst of the Italian campaign of World War II. Lava then overwhelmed the village of San Sebastiano, but the most destructive effect, as far as the allied military effort was concerned, came from the introduction of glass-sharp volcanic ash into the moving parts of airplane engines.

Fig. 4-8 Mount St. Helens, Washington, with Mount Ranier in the distance. These two typical strato-volcanoes are members of the chain of recent volcanoes of the Cascade Mountains that extend across Washington and Oregon and into California. All are typically composed of andesite. (Photograph by Ray Atkeson, Portland, Oregon.)

The first lava is said to have appeared at Vesuvius in A.D. 1036, and its appearance has been a standard accompaniment of most eruptions ever since. Because eruptions in the current phase of the volcano's life history commonly include both the upwelling of large quantities of lava and violently explosive activity that blasts great quantities of ash, cinders, bombs, and blocks skyward, the volcanic edifice that has built up during the past 1880 years is composed in part of solidified lava from flows and internally from dikes and conduits, and in large part from pyroclastic material blown out explosively. For this reason, such a volcanic mountain as Vesuvius is called a *composite cone, or strato-volcano* (Figs. 4-8, 4-9). Its flanks stand at an angle less than those of a cone consisting entirely of cinders, and are steeper than one built up of superimposed, highly fluid lava flows, such as the Hawaiian volcanoes.

Quiet Eruptions
Mauna Loa and Kilauea

The Hawaiian Islands, surely one of the most idyllic archipelagos in the world, owe their entire existence to volcanism. They are a chain of extinct, dormant,

and active volcanoes built up from the depths of the sea and trending southeastward across the Pacific for 2500 kilometers (1600 miles) from Midway on
the north to the largest island, Hawaii, on the south in an arc bowed slightly
to the northeast. The eight larger islands are at the southeastern end, and the
relative erosional age of their landscapes generally decreases southeastward.
This means that Hawaii, the only island with active volcanoes, appeared above
the sea more recently than Oahu, the island on which Honolulu stands. One
way in which this conclusion is reached is from the more advanced state of
stream erosion, valleys, cliffs, and canyons on Oahu as well as the deeper soil
cover that has developed there.

There are five major volcanic centers on Hawaii, of which three are most
important. Two of these are the immense volcanic mountains, Mauna Kea
(4200 meters, 13,784 feet) and Mauna Loa (4170 meters, 13,679 feet) which rise
out of the Pacific depths for at least an additional 4500 meters (15,000 feet),
making them mountains as high as Everest, but of enormously greater bulk

Fig. 4-9   Occurrences of igneous rock. In the lower part of the diagram granite is
shown intruded into metamorphic rock. Above this is a thickness of layered sedimentary rocks in horizontal position into which are intruded several types of
igneous bodies shown in black. On the upper surface several volcanic features are
shown. (A) A shield volcano with a summit caldera. A fissure eruption forming a
lava flow is in progress at the left of the summit. (B) A large caldera with a minor
younger volcano within it, surrounded by a "crater" lake. (C) A lava flow eroded
so as to make a cliff in the landscape. The lava flow was fed by a dike that shows on
the upper right-hand side of the diagram. Beneath the lava flow is another similar
lava flow also fed by a dike, which has been buried by sedimentary rock and the
lava flow at the top. (D) A plug dome formed by the protrusion of a pasty mass of
lava. (E) A strato-volcano, shown in section, with lava flows alternating with pyroclastic deposits. A lava flow is to the right of the cone. (F) Three dikes radiate
from a volcanic neck, exposed because the volcano formerly formed here has been
eroded away. Below, two sills extend laterally from the feeder of the volcano.
(G) Below the point G is a sill, fed from below by a dike or tube of magma. Directly
above the feeder the sill is thickened and has lifted the surface into a dome. Such
a lens-shaped body of igneous rock is called a laccolith.

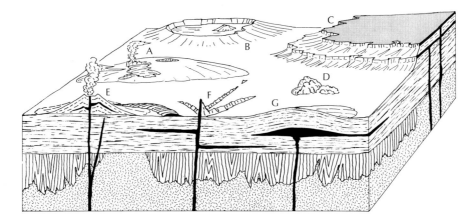

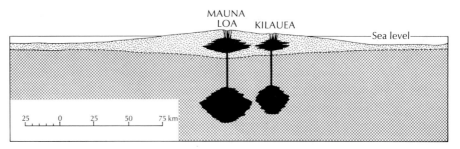

Fig. 4-10 Hypothetical cross section through the island of Hawaii. The lower black masses show where magma is thought to originate. It is thought then to move upward to form the upper masses of magma which then erupt to the surface to build the volcanoes.

since the circumference of Mauna Loa is about 320 kilometers (200 miles) at the base. Their slopes (Fig. 4-10) are extremely gentle by comparison with those of the Himalaya; seldom do they exceed 10°, and from a distance the volcanoes look like benign turtles of colossal size. With their near-circular outline and gently rounded profile they are sometimes called *shield volcanoes* because they much resemble the circular shields once mounted along the gunwales of sea-roving Viking ships.

Although some pyroclastic material is included in the mass of these huge volcanic piles, for the most part they consist of thousands of superimposed, relatively thin flows of basalt. Many of these at the time of their eruption were extremely fluid. The result is that the slopes of shield volcanoes are gentle because they are built up gradually by thousands of overlapping, tonguelike sheets of once-fluid material, rather than being loose piles of heaped-up volcanic fragments. The latter circumstance results in the building up of *cinder cones*, whose steep sides commonly stand with inclinations of 25° to 30°.

The fires of Mauna Kea are banked now, and Mauna Loa has not erupted since 1950. Most of the historic lava flows have broken out on its flanks, rather than being the result of simple overflow from the summit caldera, known as Mokuaweoweo. In fact, activity in the caldera today is at a minimum, and the caldera itself has originated as the result of foundering through removal of support from below, rather than by (1) explosion, or (2) violently explosive emptying of the magma reservoir as at Krakatoa and Crater Lake.

Mokuaweoweo is no circular crater; the very steep walls, which are about 180 meters (600 feet) high, enclose a sink approximately 5.6 kilometers (3.5 miles) long by 3 kilometers (2 miles) wide. The long dimension trends northeast-southwest and this is on the same line as the so-called Great Rift Zone, out of which so many of the historic flows have issued. The caldera itself has grown through the coalescence of several once independent pit craters on the summit of the mountain, and it is Williams's belief that this results from collapse brought about by the draining away of lava from the magma reservoir within the mountain through fissures on its flanks.

Fig. 4-11 The dissected volcanic landscape of Oahu Island, Hawaii. Gently inclined layers of volcanic rock are exposed in the fluted cliff at the left. (Photograph by Ray Atkeson.)

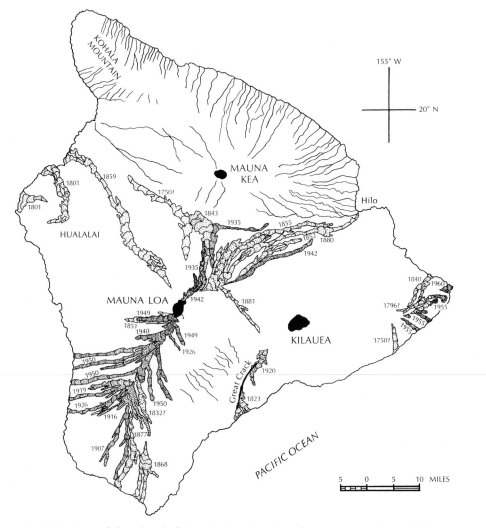

Fig. 4-12   Map of the island of Hawaii, showing lava flows erupted since 1750.

This belief is certainly supported by the pattern followed by typical eruptions. These may commence with some volcanic dust being blown from the
summit crater and a column of steam standing over it by day as well as a glow
of light that illuminates the clouds at night. Somewhat later, lava may break
out on the flanks, and almost always on the Great Rift Zone, either to the
northeast or to the southwest of the summit.

Lava flows on Mauna Loa seldom issue from a single vent, but almost always break out from great cracks, or *fissures*. The first phase of such an out-
break may be the appearance of a line of *fire fountains*, or geyserlike columns
of lava that may spurt as much as 300 meters up into the air and line up as a
nearly continuous curtain of fire along the fissure (Fig. 4-13). The basalt that
streams from these fissures is at a high temperature, and consequently may be

extremely fluid when it first pours out. It may flow down pre-existing stream courses with velocities approaching that of the rivers themselves; where there are irregularities, the lava plunges over these like a waterfall. When such a stream reaches the sea, a titanic conflict ensues between the forces of Neptune and of Vulcan, as it were. Immense clouds of steam boil upward, the sea seethes like a gigantic cauldron, and part of the lava is quenched so abruptly that it froths up as a tawny, cellular sort of volcanic glass (Fig. 4-14).

Not all the Hawaiian basalts flow in torrential streams; blocky, ponderously advancing flows are common, too. These march forward much like a tank, or caterpillar tractor, when the surface crusts over and is carried ahead by the still molten interior. The advancing crust breaks up into blocks at the leading edge of the flow, and these cascade over the front to make a carpet or track over which the flow advances. The top and bottom of such a flow will make a *volcanic breccia* when the whole flow has solidified, and the interior will be essentially uniformly textured homogeneous basalt.

Hawaii has given us two Polynesian terms to describe the surface character of lava flows, and these have now won such general acceptance that they are commonly used in the literature of geology. Basalt with a rough, blocky ap-

Fig. 4-13 Fire fountain at Kilauea, November 18, 1959. The lighter colored parts of the fountain are glowing fluid masses of basalt lava. The lava cools in the air and turns dark at the top of the fountain. (Photograph by G. A. Macdonald.)

Fig. 4-14   Basalt lava, pouring over the cliff to the right, reaches the sea. Hawaii, 1955. (Photograph by G. A. Macdonald.)

pearance, much like furnace slag, is called by the remarkably brief name of *aa* (Fig. 4-15), while the more fluid varieties with smooth, satiny, or even glassy surfaces are given the more euphonious name of *pahoehoe* (Fig. 4-16).

Kilauea is like Mauna Loa in some respects and very different in others. For one thing it is at a much lower altitude, about 1200 meters (4000 feet), and no longer is an independent mountain but is a partially buried satellite on the flank of the higher volcano. Perhaps within the next few millennia it will be inundated by flows from a flank eruption of Mauna Loa. Kilauea is the far better known of the two volcanoes since a paved road leads directly to its rim, and it has had a steady stream of visitors for more than a century.

The caldera of Kilauea is always a surprise to the first-time visitor, and for much the same reason as the Grand Canyon is. Both make such an extreme contrast with their nearly level surroundings. The almost circular caldera of Kilauea, whose dimensions are approximately 4 kilometers by 3.2 kilometers (2.5 miles by 2 miles), is countersunk with almost vertical walls into the very gently sloping surface of the old volcano. The bottom of the caldera is nearly level and is made up of only very recently solidified lava which spread like a tarry stream over the entire floor. Activity today is confined to only part of the caldera—the volcanic throat or fire pit of Halemaumau, which bears a relationship to the larger caldera much like that of a drain pipe in the bottom of a wash basin. Basaltic lava rises and falls inside the fire pit. At times it spills over Halemaumau's rim onto the caldera floor; at other times it sinks down more than a thousand feet below the surface. Then the floor of the pit is filled with long talus aprons of basalt blocks that have broken away from the vertical walls.

Commonly, lava swirls and seethes within Halemaumau without violent explosive activity, but occasionally there are impressive departures from this pattern. Such a one was the 1924 eruption, in which the sequence of events was as follows: (1) in January the lava lake was especially active and the level rose to within about 30 meters (100 feet) of the rim; (2) in February it started to subside and by May had dropped to around 183 meters (600 feet); (3) meanwhile the epicenters of a whole succession of minor earthquakes migrated steadily eastward along the line of the Puna Rift, accompanied by ground subsidence until almost certainly there was an eruption on the sea floor southeast of Hawaii; (4) immense quantities of lava blocks avalanched from the walls into the fire pit; (5) finally, these blocks and much of the debris that had accumulated on the floor of Halemaumau were hurled out in a series of violent explosions between May 11 and 27.

An interpretation of this sequence of events, which are a bit out of character for a quiet eruption, is that the lava column dropped because lava was being drained away through fissures from beneath Kilauea—whose entire level dropped, incidentally, by 4 meters (13 or 14 feet)—southeastward along the Puna Rift. This permitted ground water to move into the area vacated by the sinking column of lava. When a sufficiently high pressure was built up, and the

Fig. 4-15   Flow of a lava slowly advancing over a field. The flames and smoke are from burning vegetation. Hawaii, 1955. (Photograph by G. A. Macdonald.)

Fig. 4-16   Pahoehoe lava surface. Mauna Loa, Hawaii. (Photograph by Ansel Adams.)

lava column had subsided below sea level, the ground water was converted into steam under cover of the blocks of rock fallen into the fire pit. Then the pressure rose to a point high enough that these rocks were shattered and hurled out of the pit in what was a succession of steam explosions rather than ones produced by primary magmatic gases. To such a secondary eruption the name of *phreatic explosion* (derived from the Greek word for water well) is given, and

Fig. 4-17 Lava fountain spurts from floor of Halemaumau Crater, Kilauea Volcano, during an active eruptive phase, December 1967. (Photograph by scientists of U.S. Geological Survey.)

they are very characteristic of minor eruptions the world over; in Iceland, New Zealand, Japan, and possibly the 1914–17 eruptions of Mount Lassen.

Recent eruptions on Hawaii indicate a continuing movement of volcanic activity eastward, although the principal summit caldera of Kilauea is also active. One of the flank eruptions completely obliterated the town of Kapoho and, with others, eventually reached the sea, adding new land to the island. The 1969 eruptions were centered around one of the pit craters which extend in a chain southeast of Halemaumau.

Fissure Eruptions

Several regions on the earth's surface have been inundated by vast floods of lava that obviously could never have come from a single volcanic conduit or even from a chain of volcanoes. Prominent examples of these lava floods, which universally are basaltic rocks or closely related variants, are the Columbia lava plateau in the northwestern United States, with a surface extent of some 200,000 square miles and a volume of approximately 75,000 cubic miles, the Deccan lava sheet in western India inland from Bombay, and a broad area near the Parana River in South America (Fig. 4-18).

The Columbia lava plateau is the most thoroughly studied of the three. In places it is a mile thick, but the individual flows are much thinner, only a few being as much as 122 meters (400 feet) thick. Their composition is remarkably uniform, especially in view of the fact that such an enormous volume of basalt was not erupted all at once, but its outpouring stretches over a long span of geologic time. The surface of the lava plateau covers a very broad area—extending from the Rocky Mountains on the east to the Cascades and Pacific border to the west. The basalt had no single route to the surface, but rose through hundreds, if not thousands, of fissures that today are to be seen as dikes where they are exposed in canyon walls. The country that was buried by the lava floods was one of moderate relief. Here again, the walls of canyons cut across the basalt plateau show that individual lava flows filled valleys, overtopped ridges and ultimately coalesced to form a nearly uniform plain which buried a wholly different sort of world beneath a frozen sea of stone.

The nearest counterpart ever reported of such a fissure eruption was a minor episode by comparison, impressive as it undoubtedly must have been. This was the eruption on June 8, 1783, of the Icelandic volcano, Skaptar Jökull. There, a stream of basalt poured out from a fissure about 24 kilometers (15 miles) long.

Iceland has one of the most dramatic landscapes on earth, with over a hundred volcanic centers, of which at least twenty are active, a score of glaciers, and the all-encircling sea. In few other regions is the elemental conflict of fire, ice, and ocean more stark. In the long and remarkable cultural history of the island, extending back to A.D. 874, there have been many of these encounters between outpourings of red-hot lava and streams of ice. The usual outcome is

Fig. 4-18   Flood basalt lava flows of the Thulean Plateau, Antrim, Ireland. The Thulean Plateau formerly extended across the North Atlantic from western Scotland and northern Ireland to Iceland and western Greenland. (Photograph by Aerofilms and Aero Pictorial, Ltd., London.)

the melting of much of the ice, with the release of a sudden and devastating flood of water and mud.

This is exactly what happened in the disastrous eruption of 1783. With a fissure 24 kilometers (15 miles) long discharging basalt along its entire length, a broad tide of lava poured down the slope, filled the deep canyon of the Skapta to overflowing, and completely displaced a lake that lay in its path. The eruption continued for two years, and the two major lava torrents it produced had lengths of about 64 and 80 kilometers (40 and 50 miles), respectively. Their average depth was 30 meters (100 feet), but where canyons were filled to overflowing they were as much as 183 meters (600 feet) thick. Where the lava overtopped a stream valley and spread out across the plain it advanced along a front 19 to 24 kilometers (12 to 15 miles) wide. The flow is estimated to have covered 220 square miles and the volume discharged in the two-year period is thought to have been about 3 cubic miles, or a mass equal to that of Mont Blanc, the highest mountain in Europe.

This was one of the greatest disasters in the turbulent history of the island. The lava, blocking and diverting rivers and melting snow and ice, liberated huge floods, thus destroying much of the island's limited agricultural land. Twenty villages were overwhelmed by the lava, and many others were swept away in the floods. About 10,000 people, or 20 per cent of the population, died; 80 per cent of the sheep (190,000), 75 per cent of the horses (28,000), and over 50 per cent of the cattle (11,500) perished.

Classification of Igneous Rocks

In such dramatic fashion are the volcanic rocks of the earth formed. These are only one of the two types of igneous rocks, however; the others are called *plutonic* because they are believed to have formed deep within the crust, in Pluto's realm. This distinction between volcanic and plutonic rocks gives us a beginning in our attempts to classify and name those rocks which appear to have solidified from a fluid state.

Although classification of the objects in his environment seems to be a basic need of the human animal, in the case of rocks it is unfortunate. As you remember from the preceding chapter, minerals generally have a definite chemical composition, so that we can classify them with ease: the carbonates, the sulfates, the silicates, and so on, or the aluminum minerals, the iron minerals, the copper minerals, for example. Or we can classify them according to their crystal structure, which is also distinctive. Rocks, however, do not have such unvarying characteristics. They do show textures, which are helpful, and they do consist of assemblages of minerals, but the mineral composition is highly variable and in many cases gradational. Tom F. W. Barth, a Norwegian petrologist (a specialist in the study of rocks) has put it this way:

> Rocks are made up of definite mineral assemblages. We may use an analogy from zoological science and note that a mineral species will correspond to an

animal species, whereas a rock corresponds to a fauna. It has become customary, and probably also necessary, to name rocks according to their mineral content, and relative proportions of the constituent minerals on the one hand, and the mechanical and textural relations on the other. In zoology this would mean that one had to introduce a special name for a fauna consisting of hares, foxes, and fleas. Not only this, but new names would have to be introduced as the relative proportions of these three animals were changed, according to whether the hares are big and strong with silken fur or small and miserable from hunger and flea bites. Such analogies demonstrate how difficult it is to develop a satisfactory system of rock names and explain why just in igneous geology alone more than six hundred different rock names have been introduced. . . .

Of course, we will not be concerned here with six hundred igneous rock names, but we should be familiar with a few of the most common ones. First we should examine more closely those properties of igneous rocks which are the basis of our classification.

Texture

We must remember that there are two major occurrences of igneous rocks: (1) on the surface of the earth—the volcanic rocks, and (2) within the crust itself—the plutonic rocks. This distinction is of absolutely fundamental importance, because the mode of occurrence of an igneous rock—that is, whether the magma solidified above or below the surface of the ground—determines the rate of cooling, and the rate of cooling, in turn, is what establishes the texture of the rock.

The word texture itself is familiar to most people as having something to do with cloth; in fact, it is derived from the Latin *textura*, a weaving. Today it means the arrangement and size of the threads in a woven cloth; for example, a burlap sack has a much coarser texture than a nylon stocking. When we apply the term to such a thing as an igneous rock, we mean the size of the crystals as well as their mutual relationships.

An igneous rock in which the minerals are visible to the unaided eye is said to have a medium- or coarse-grained texture depending on the size of the crystals. An igneous rock whose texture is crystalline, as revealed by the microscope, but most of whose crystals are too small to be seen by the eye alone, has a fine-grained texture. Yet the mineral content may be nearly the same in the two rocks.

Why, then, the difference in the size of the crystals if the minerals are nearly alike? The answer lies in the way in which the magma solidified. If it cooled slowly and under relatively undisturbed conditions, then large crystals had a chance to grow around the various nuclei in the still fluid magma. They may grow to fair size, up to half an inch or more, in what is essentially a sort of crystal mush, with the last-forming minerals filling the interstices between the earlier forming minerals when the whole mass finally congeals. Should the magma cool rapidly, the same crystal growth will go on around floating nuclei as it did in the slow-cooling magma, but the whole process is halted when the magma solidifies before the minerals have a chance to grow to visible diameters. Then the rock, although crystalline, consists of a tightly knit fabric of minute crystals which are invisible to the unaided eye.

The answer to a large part of this problem of what factor controls the texture of igneous rocks is found in what is known as their *mode of occurrence*; in general terms, whether or not they solidified above the ground or below it. To phrase the statement another way, are they volcanic or plutonic rocks? Volcanic rocks cool with relative rapidity, and therefore for the most part have fine-grained textures; plutonic rocks cool more slowly, larger crystals grow as a result, and they are characterized by coarse-grained textures.

Two additional textures, important among the igneous rocks, are: (1) glassy, and (2) porphyritic.

A glassy texture is typified by the rock *obsidian*, which is also known as volcanic glass. Glass, although a solid, still possesses many of the attributes of a liquid. It is wholly noncrystalline, because it passed quickly from the liquid to the solid state without giving the ions in the original magma an opportunity to arrange themselves in ordered ranks as crystals. It is this liquid/solid state which gives obsidian many of its unusual properties. Because it is essentially textureless, it breaks or fractures in about the same way as a homogenous substance such as a black, solid chunk of roofing tar does when workmen chop open a barrel before melting the tar in the boiler. The sharp-edged, curving spall characteristic of shattered glass blocks, broken tar, and obsidian fragments makes such a distinctive pattern that it merits a name of its own, which is *conchoidal fracture*, meaning shell-like (Fig. 4-19). It is this pattern of breaking with a sharp edge that made obsidian such a deadly weapon, especially when it was flaked or chipped into the beautifully proportioned arrowheads, spear points, and knife blades used by many of the Indian peoples of the western United States and Mexico.

Fig. 4-19 Obsidian, showing the characteristic conchoidal fracture. (Courtesy of Ward's Natural Science Establishment, Inc., Rochester, New York.)

Fig. 4-20 Obsidian artifacts. (Photograph by John Haddaway.)

Porphyritic texture takes its name from the Greek word *porphyra*, which was the word used for the imperial purple—a highly prized dye extracted from an eastern Mediterranean shellfish. By extension the name was applied to a very specific kind of rock—a dark, uniformly textured igneous rock from Egypt that contains small white feldspar crystals embedded in the purplish groundmass. This is the rock that was greatly favored in the Roman world for carving busts of emperors, as well as their sycophants and the lesser dignitaries of the court. The rock makes a striking purplish bust, especially when set off against an artfully draped white marble toga. Such a colorful petrographic arrangement could scarcely fail to make even such decadent figures as Nero and Caligula look regal.

Today, by a further extension of the original word, the term is applied to any igneous rock with a duality of texture; that is, crystals of two markedly different sizes are found in the same rock. Such a texture is interpreted to mean the magma underwent two generations of cooling, perhaps an earlier slow-cooling phase during which time the large crystals grew, followed by a later more rapid phase when the smaller crystals came out of solution. In such a rock the larger crystals are called *phenocrysts* (from a Greek root, *phainein*, to show, combined with crystal) and the background finely crystalline material in which they are embedded is called the *groundmass* (Fig. 4-21).

Porphyritic texture is very characteristic of igneous rocks that solidified at shallow depths in such bodies as dikes and sills (described below). Such shallow intrusive igneous bodies are an example of a magma which may have been cooling slowly at great depths and was moved up into a shallower zone of the earth's crust by the magma stream where it cooled more rapidly and in a sense froze around the still floating larger and earlier formed crystals.

Because the larger crystals in a porphyry form while the magma is still dominantly a fluid, they grow without interference and thus may often achieve a nearly perfect crystal form. The later crystallizing minerals come out of

Fig. 4-21   Igneous rock with porphyritic texture. The dark groundmass that encloses the large white crystals is much finer grained. (Courtesy of Ward's Natural Science Establishment, Inc.)

solution more rapidly, and since the growth of each one is interfered with by its neighbors, very few have the opportunity to develop the external form of a crystal, although their internal structure may show the complete atomic pattern for the particular mineral involved.

Mineral Composition

The other property, besides texture, used to classify the igneous rocks, is the mineral composition. Obviously, the kinds of minerals to be found in any igneous rock are dependent on the chemical composition of the magma from which the rock crystallized.

Igneous rocks resemble steel in the way in which they are formed. As everyone knows, steel is usually made in an open-hearth furnace, and the properties of any batch of steel are determined largely by the addition of such elements as tungsten, molybdenum, chromium, etc., to the original charge of pig iron and scrap. To put the case very simply, if the charge contains chromium and not nickel, then the production of a chromium-bearing steel rather than a nickel-bearing steel will be the result. Or if 18 per cent chromium and 8 per cent nickel are present, stainless steel will be made. So it is with the igneous rocks. If the magma is low in potassium, it is most unlikely, then, that a potassium-bearing feldspar such as orthoclase will crystallize. If the silica content of the magma is high, then quartz should be an abundant mineral, and if it is low, then the resulting rock very likely will be quartz-free.

Not all the minerals in a rock crystallize from a magma simultaneously; rather, the evidence available to us indicates their crystallization follows an orderly sequence which was worked out many years ago by the petrologist, N. L. Bowen. If conditions are favorable, minerals may crystallize from a magma in a dual sequence in the order shown in Figure 4-22. The plagioclase feldspars crystallize in what is known as a continuous reaction series, which means that the early-formed crystals change continuously in their composition by reaction with the changing composition of the fluid magma still remaining. The other minerals, such as pyroxene, hornblende, and biotite, make a discontinuous series, so that an early-formed mineral reacts with the remaining fluid to form a completely new mineral with a quite different composition and crystal form, rather than an isomorphous series such as the plagioclases.

According to the Bowen reaction series, the first minerals to come out of solution in a magma are olivine and calcium-bearing plagioclase. Were the entire magma to crystallize at this stage, the resulting rock would be a *basalt* if it were volcanic, and a *gabbro* if it were plutonic. Should, however, these early-forming minerals settle out, then the remaining magma will have lost much of its iron, magnesium, calcium, and some silicon. With a progressive decrease in these elements the magma will be correspondingly enriched in potassium, sodium, and silicon.

The next minerals to crystallize are the plagioclases of intermediate composition along with pyroxene and hornblende. Should the magma congeal at this point, it would yield rocks of intermediate composition, such as *andesite* if fine-grained and *diorite* if coarse-grained. Should these earlier crystallizing minerals be separated from the magmatic solution, the remaining minerals to crystallize will be sodium-rich followed by potassium-rich plagioclase and the two micas. Last of all to solidify is quartz, and then only if free silica is left over in solution after all the metallic ions are used up and none are left to enter into combination. Quartz is an interstitial mineral since it fills in the voids or spaces among the earlier-forming crystals. Typical rocks with a mineral composition of quartz, mica, orthoclase, and minor amounts of sodium-bearing plagioclase and hornblende are *granite* if plutonic and *rhyolite* if volcanic.

The so-called Bowen reaction series does explain how a magma of originally basaltic composition (one that on crystallizing yields olivine and calcium-bearing plagioclase) might go through a process of differentiation to yield a granitic magma from which quartz, biotite, and orthoclase crystallize. It is

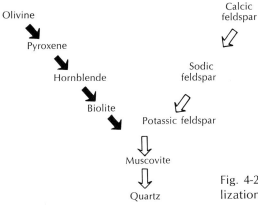

Fig. 4-22 The dual sequence of crystallization of minerals from magma.

necessary to realize, however, that an original magma may have a composition different from Bowen's idealized starting one. That is, a magma may begin with a composition about halfway through the series.

Unfortunately, the origin of the massive bodies of granite that crop out so broadly in the heartlands of the world's continents is almost certain to be vastly more complex than this simple explanation. It is difficult to conceive of granitic intrusions as extensive as those in the core of many of the earth's mountain ranges as having formed from a residual liquid representing not much more than 10 per cent of the original volume of a basaltic magma. We shall return to this problem of granite in the earth's crust later on, but it is worth emphasizing here and reiterating later what a major puzzle in the occurrence of the igneous rocks is the curious reversal of roles in that the most common of all volcanic rocks is basalt, while the most widely occurring plutonic rock is granite, yet their respective mineral compositions are at nearly opposite ends of the spectrum.

With a knowledge of the relationship existing between texture and occurrence, and the ability to recognize five or six of the more abundant rock-forming minerals, a workable field classification of the igneous rocks can be established by using the two properties of texture and mineral composition.

In the accompanying table (Fig. 4-23) which shows six of the more common igneous rocks, the fine-grained volcanic rocks, or *extrusive rocks* as they are sometimes known, are arranged along the upper row, and the plutonic, or *intrusive rocks*, are along the lower. The lighter colored and quartz-bearing rocks are to the left; the iron- and magnesium-rich, and consequently dark-colored rocks are to the right.

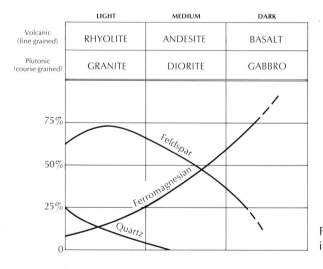

Fig. 4-23 The mineral composition of igneous rocks.

Fig. 4-24 A rough surface of granite. The white grains are feldspar, the clear glassy ones are quartz, and the black grains are mica and hornblende. (Photograph by John Haddaway.)

Description of Igneous Rocks

Of the first two rocks listed, rhyolite is a volcanic and granite is a plutonic rock. The mineral composition curves below the table show that granite (the origin of the name is lost in antiquity, but it is believed by some to be derived from the Italian adjective, *granito;* grained) is a light-colored, coarse-grained rock whose chief constituents are quartz, feldspar (commonly orthoclase and some soda-bearing plagioclase), and some ferromagnesian minerals so named from their content of iron and magnesium (biotite and minor amounts of hornblende, as a rule). Since granite is a widely used rock for such things as tombstones, monuments, government and bank buildings (in fact, it has been used for stately edifices since the beginning of Western civilization), almost all of us readily recognize the characteristically speckled appearance of this most familiar of rocks, with its white to light gray, pink, or even red background flecked with spangles of black mica or needles of hornblende, as well as its coarsely crystalline texture with some minerals (chiefly the feldspars) reaching dimensions up to perhaps one-half inch across (Fig. 4-24).

Rhyolite (named from the Greek, "lava stream" or "torrent" + stone), a volcanic rock with about the same chemical composition as granite, has a wholly different texture. Commonly the texture is porphyritic, with a groundmass so fine-grained that the individual minerals can be resolved only with a microscope. Embedded in the groundmass are small crystals of quartz, feldspar, occasional scales of biotite, and, rarely, ferromagnesian minerals. Quartz is an early-crystallizing mineral in rhyolite, and for this reason quartz crystals commonly appear completely formed and sometimes are doubly terminated. This is in contrast to its occurrence in granite where it starts crystallizing early but also finishes last, so that it appears as glassy blobs filling the spaces left between other already-crystallized minerals. Rhyolite ordinarily is light colored, and may be white, light gray, pink, and even red where it has weathered—especially in the dry climate of the southwestern United States and northern Mexico. A textural feature that is quite characteristic of the groundmass of rhyolite is a streaked pattern in the rock known as *flow banding,* which is the result of differential concentration of material in layers or bands in the still sticky, viscous lava just before it solidified (Fig. 4-25).

Fig.4-25   Flow banding caused by viscous flow in a magma that congealed as obsidian. Blocks at the base of a volcanic dome south of Mono Lake, California. (Photograph by John Haddaway.)

Diorite (whose name comes from the Greek word, to distinguish) is a coarse-grained igneous rock whose mineral composition places it about midway between granite and gabbro. For this reason it, as well as a large number of kindred rocks, is spoken of as having an intermediate composition. This is a rock with sodium-bearing plagioclase feldspar as its chief constituent and lacking quartz and orthoclase. Hornblende is its leading dark mineral, and biotite commonly is an important constituent. Pyroxene is rare, as might be anticipated from its early crystallizing position in the Bowen reaction series, and is much more abundant in *gabbro*, the next plutonic rock in the classification.

Because of the lack of quartz and of orthoclase and the approximate equality of plagioclase and the darker minerals, diorite tends to be a drab gray rock. These rocks are not used as widely for building stones as true granites, because diorites are somewhat less abundant and possibly because their somber gray color is less pleasing.

Andesite (which is named for its occurrences in the Andean summit volcanoes of South America) is a gray to grayish-black, fine-grained volcanic rock. Commonly its groundmass is too dense for any individual minerals to be resolved without the use of a strong hand lens. Phenocrysts are common, and are likely to be either transparent or light gray crystals of plagioclase or are dark minerals such as hornblende or black scales of biotite. The term andesite is now used to cover such a wide range of volcanic rocks of intermediate composition that it has lost much of its precise meaning as a rock term. However, these rocks, in general, lack visible quartz, the chief feldspar is sodium-bearing plagioclase, and the dark minerals are principally biotite and hornblende. These lavas occur more abundantly on the earth's surface than the rhyolitic rocks, but are less widespread than the basalts.

Gabbro (this is an old Italian name for many of the dark rocks—including serpentine—used in the Renaissance palaces and churches of Italy) consists typically of a coarse-grained intergrowth of crystals of pyroxene and calcium-bearing plagioclase. Many gabbros also contain olivine, and some include hornblende, too, although pyroxene dominates. Unlike the diorites, the ferromagnesian minerals are more abundant than the feldspars. There are exceptions, of course, and one variety of gabbro, *anorthosite*, consists almost entirely of a coarse-grained fabric of plagioclase crystals. Another variety of gabbro, much sought after as a decorative building stone for store fronts, banks, etc., contains large dark purplish feldspar crystals that give a wonderfully impressive iridescent play of colors, much as peacock feathers do when they catch the sunlight.

Basalt (whose name is truly one of the most ancient in geology, since it apparently dates back to Egyptian or Ethiopic usage, and one of the first references to it by name is by Pliny) is by far the most abundant of all volcanic rocks. Some regions, such as the plateau bordering the Columbia River in the northwestern United States, the vicinity of Bombay in western India, and the part of South America where Brazil, Paraguay, and Argentina are close neighbors, were inundated by vast outpourings of basalt—some covering as much as 520 million square kilometers (200,000 square miles). In addition, many of the truly oceanic islands, such as Samoa, Hawaii, and Tahiti, are basaltic volcanoes rising many thousands of feet above the sea floor. Deep-sea soundings also give us a clue to the broad expanses of basalt seemingly spread across the oceanic depths in the recent geologic past.

Basalt is a most commonplace-appearing rock for playing such a prominent role in volcanism. Ordinarily, when basalt is fresh or unweathered it is coal black or dark gray and the groundmass is too fine-grained for the minerals to be visible; but under the microscope they can be seen to be mostly pyroxene and plagioclase. If phenocrysts are present, they may be either of these minerals in addition to olivine. Some varieties of basalt crystallized in bodies thicker than thin lava flows and cooled slowly enough that the minerals in their groundmass are large enough to be visible. *Diabase* is such a rock, and its characteristic texture is a felted network of feldspar laths with the spaces between them occupied by later-crystallizing, irregularly shaped pyroxene crystals.

The distinction between andesite and basalt is largely artificial since the two types grade into one another. The separation of one from the other is based on (1) the character of the plagioclase, whether dominantly sodium-bearing (andesite) or calcium-bearing (basalt), and (2) whether or not the principal ferromagnesian mineral is hornblende (andesite) or pyroxene (basalt). These differences are determined by using a microscope and are not likely to be very meaningful in the field. A workable means of distinction in the field is to chip off a small flake of the rock. If when held up to the sun, the flake is opaque, the rock very likely is basalt; if translucent, andesite.

Basalt shows a much more interesting variability physically than it does mineralogically. Some basalt flows are as rough and jagged as so much furnace slag (Fig. 4-15). Other flows are smooth and ropy, with glazed surfaces that look as though they had once been a stream of tar that had been halted instantaneously (Fig. 4-16).

The former is the *aa*, the latter the *pahoehoe*, of the Hawaiian eruptions.

Fig. 4-26 Cellular or scoriaceous basalt. (Photograph by Hal Roth.)

Fig. 4-27   Columnar structure in basalt lava flow. The Devil's Postpile, California. (From the Cedric Wright Collection, courtesy of the Sierra Club.)

Some basalts are dense, uniformly textured rocks with no visible minerals—others are frothy and cellular and are filled with innumerable small holes called *vesicles* that were gas bubbles trapped in the still fluid basaltic lava (Fig. 4-26).

A rock structure typical of basaltic rocks—although by no means confined to them—is the development of geometrically regular columns (Fig. 4-27). Commonly these have five or six sides and are nested so closely together that when the columns are seen on end their pattern resembles that of hexagonal bathroom tiles (Fig. 4-28). This is the result of the contraction of the rock that starts when the lava has first solidified. Should crystallization take place radially outward from approximately equally spaced centers, then the most efficient pattern that can be developed between these individual cells is a hexagonal one, similar in most respects to the cellular hexagonal pattern in a beehive. Two familiar examples of columnar basalts are The Devil's Postpile in the Sierra Nevada of California and the Giant's Causeway of Northern Ireland.

Other Types of Igneous Rocks

There are several other important kinds of igneous rocks that do not fit into a classification as rigid as the one just discussed.

Obsidian (whose name comes from the Latin word, *obsidianus*, after its describer, Obsius) is also called volcanic glass (Fig. 4-19, 4-25). Actually, it should be thought of as a supercooled liquid, since it is made from magma that passed from the liquid to the solid phase so rapidly that crystals did not have time enough to form. This means that the rock is textureless, and that its appearance and fracture pattern is much like a large mass of insulator or bottle glass that solidified out of control and had to be thrown out on the dump. Actually, a natural glass, such as obsidian, is not too unlike the artificial product, the chief difference being that the careful controls used in making artificial glass eliminate the impurities, while in obsidian they are all present in the magma. The result is that obsidian commonly is black, and it may be a very striking jet black.

Pumice (an ancient name, from a Greek word meaning worm-eaten, mentioned as long ago as 325 B.C. by Theophrastus) is a rather specialized kind of obsidian which has been so dilated by volcanic gas as to become a petrified glassy froth, much as though the foam on top of a beer stein were to be instantly converted to rock. Because it has been so frothed up by gases mixed in it before it solidified, pumice is one of the lightest of rocks and the more porous varieties float on water; if blown out of coastal or oceanic volcanoes it may drift for thousands of miles before becoming waterlogged and sinking to the bottom.

Both obsidian and pumice have chemical compositions that typically are akin to rhyolite—that is, they are rocks that carry relatively high percentages of silicon, potassium, and aluminum in their composition and are correspondingly low in iron, magnesium, and calcium. Although glassy phases do exist for rocks solidified from andesitic and basaltic magmas, they are of minor importance compared to the rhyolitic varieties.

Along with lava flows, volcanoes eject great quantities of solid or semisolid material as part of their explosive activity. This material may range in size from particles as small as dust to blocks as large as houses. The products of such fragmental volcanic debris are called *pyroclastic rocks* (from the Greek; literally translated, the word means fire-broken).

Finer particles the size of dust or sand blown from volcanoes are called volcanic *ash*, and if the particles are the size of small pebbles they are some-

Fig. 4-28 Upper surface of the Devil's Postpile, California, showing the ends or cross sections of the columns seen in Figure 4-27. The grooves extending from left to right are glacial striae. The amount of weathering since the disappearance of the glacier is indicated by the loss of the striated surface. (Photograph by Hal Roth.)

Fig. 4-29   Pumice blocks and lapilli. Mammoth Lakes District, California.

times spoken of as *cinders*. These are poor terms because neither volcanic ash nor cinders have anything to do with fire in the sense that they are not the residue left from the burning of something such as coal or wood. Actually, they are small fragments of volcanic rock with the same chemical composition as the lava that is in the crater or that is flowing down the flanks of the volcano, with the difference that they have been frothed up more by the entrapped gases. Fragments an inch or so in diameter are called *lapilli* (an Italian word for little stones) and fragments that are several inches across are best called volcanic blocks if they were solid at the time of their ejection.

Blobs of lava are often blown out beyond a crater rim from the surface of the caldron within. At night they are a spectacular sight, especially if they are still brightly glowing. Their paths can be followed as they arc through the sky on about the same sort of trajectory as a mortar shell. On color photographs, their trails show as red lines in the air, and also as long streaks on the ground as they bounce and roll down the volcanic slopes. The surfaces of such lava blobs have a chance to cool in flight, and the whole blob itself may also solidify with a crust that looks much like a loaf of French bread. Such crusted-over lava blobs are called *bombs;* some are spindle shaped if they rotated in flight, while others may be nearly spherical.

Volcanic ash may blanket the countryside for many miles around a volcano like so much newly fallen snow, and, as at Paricutin in Mexico in its eruption of 1943, may accumulate to depths of more than 3 meters (10 feet). A layered rock made of such compacted volcanic ash is *tuff* (Fig. 4-5), and over the centuries it has been a widely used building stone. Much of ancient Rome was built of tuff; it is the rubble masonry one sees today in the Colosseum and in the walls of the Forum since the marble facing was stripped away centuries ago. Tuff was extensively used on the volcanic islands of Greece, where many picturesque villages were excavated in part in it, and also at Naples and Pompeii, and a whole world away at Manila, to cite but a few examples. Tuff is readily excavated, it holds its shape well, and the surface hardens somewhat on exposure to the atmosphere.

Some falls of ash, such as the one that buried Pompeii, retain their heat for relatively long periods of time, and if hot enough the ash particles may fuse together to form a *welded tuff.* Many hundreds of square kilometers of the North Island of New Zealand in the Rotorua geyser district are covered with welded tuff. It is common throughout much of the western United States and in Alaska in the Valley of Ten Thousand Smokes, which is named for the fumaroles and hot springs produced as a result of the hot ash fall in an eruption in 1912. As a welded tuff consolidates and shrinks, it may develop geometrically arranged columns much like those in a basalt flow.

If pyroclastic material consists of blocks and other large angular fragments, these may be cemented together to form a layered rock, much like tuff in its origin but consisting of visible, angular, sharp-edged blocks. Such a rock, composed of angular, indurated blocks, is a volcanic *breccia* (an Italian word meaning the "gravel or rubbish of broken walls"). Very often these breccias are formed on the slopes of volcanic cones, close to the source, or they may be made of fragmental material that accumulated in the volcanic throat and was cemented together by late-crystallizing magma to form a new rock once the volcanic fires were dead. Volcanic breccias are also characteristic of the early-congealing surfaces of lava flows whose interiors may still be fluid. Continued forward motion of this still molten mass may shatter the crust into a jumble of chaotically arranged blocks which, cemented together by lava injected into their interstices, will produce a flow breccia.

Deep-Seated Intrusive Bodies

Plutonic rocks such as granite are an enigma because no one has ever seen such rocks form. Basalt can be seen forming on the slopes and in the craters of active volcanoes, so that to achieve at least a partial understanding of the genesis of volcanic rocks is not as difficult a problem. Nevertheless, one of the most acrimonious debates in the history of geology sprang up in the late eighteenth century over a curious belief held in Germany that not only basalt, but granite as well, was a chemical precipitate formed on the floor of a universal ocean. Fantastic as this concept might now seem, it had the virtue of forcing its opponents out into the mountains and over the face of the earth to collect data to refute the belief in the aqueous origin of the igneous rocks and to demonstrate that they had indeed crystallized from molten solutions. Much of the evidence that was patiently gathered nearly two centuries ago is valid today, but even now, when we consider the lines of argument advanced then and through the succeeding years, we are still far from an answer concerning the genesis of igneous rocks. This problem continues to spur ingenious research and lively debate among geologists.

Such rocks as granite, diorite, and gabbro are all coarse-grained rocks consisting of interlocking mineral crystals without any preferred orientation. Presumably they owe this coarsely crystalline texture to slow cooling from a magma at great depth. However, no one has ever actually seen this cooling take place, and that we know of these rocks at all is only because the rocky roof which once covered them has been stripped away by erosion. The most widely exposed of these deep-seated rocks is granite and closely related rock types. Such rocks may underlie hundreds or even thousands of square kilometers in mountainous chains, or in broad areas such as Labrador or northeastern Canada.

These tremendous bodies of rocks are called *batholiths* (a word introduced about 1895 and derived from the Greek words for depth + stone). In general usage today this means a body with a surface area of at least 64 square kilometers (40 square miles). A smaller body of coarse-grained igneous rock with a surface extent of less that 64 square kilometers is called a *stock*, and it may very often be a cupola, or protuberance, sticking up above the top of a more deeply buried batholith not as yet unroofed by erosion.

Much has been learned about batholiths and stocks because erosion has cut down to different levels in them at different places and they have been penetrated in deep mining operations. In some instances, batholiths consist of granite which is uncontaminated and homogeneous right up the knife-sharp boundary of the intrusion. In such cases the granite appears to have displaced the so-called country rocks into which it has apparently advanced as a magma.

Other batholiths, or other parts of the same batholith, are surrounded by a wide marginal zone, and the transition between the invading and the invaded rocks is much less abrupt. The contact between the unquestioned granite and its enclosing shell may be blurred, and a zone of *migmatite*, or mixed rocks— partially igneous and partially metamorphic—characterizes such a contact. These are rocks in which tongues of granite may extend far out into the adjacent host rocks, and bands of granitic material may follow selected layers for long distances into the metamorphic *aureole*, or halo of country rock that has been recrystallized in place by the heat and chemical activity associated with the granitic intrusion. This is typical of the so-called *lit-par-lit*, or bed-by-bed, structure; which means that one layer of a rock may be granite, the next a metamorphic rock, the next granite, and so on. Or knots and clusters of orthoclase, or other minerals characteristic of granite, may appear in the metamorphic aureole some distance out from the main body of granitic rocks. In addition, the granite may contain ghost or relict structures, such as layering, which are characteristic of sedimentary and metamorphic rocks. These phenomena suggest that in some way the pregranitic rocks were digested, or replaced by granite—in other words, they were converted *in place* from what they were originally into a brand-new rock, in this case, granite.

Any theory that attempts to explain the origin of granite and batholiths must also explain satisfactorily these two kinds of contacts between the granite and its country rock. In addition, it must explain the so-called "room problem"—where is all the rock that used to be where the granite is now?

One explanation assumes that the granite was formed in place by a process called *granitization*. According to this theory, tremendous thicknesses of sedimentary rocks that accumulated in subsiding off-shore areas were subjected to the great pressures and high temperatures known to exist deep in the crust of the earth. These temperatures and pressures alone have no great effect on "dry" rocks, but many sedimentary rocks contain some sea water trapped during their lithification and, during subsequent metamorphism, additional water is released by the recrystallization of hydrous minerals. These are minerals that contain water as part of their molecular structure, water which they give up when they are converted to anhydrous minerals. It has been suggested that this water forms an interstitial fluid (between the rock grains) in which chemical reactions and chemical transport can take place, thus forming granite from the pre-existing rocks. In the deepest parts of the rock pile the heat may be sufficient to combine with the process of solution to produce actual melting, or at least partial melting, enough to make the whole mass capable of movement. This mobile fluid portion would rise into the shallower parts of the earth's crust where its contacts with the country rock would be sharp and definitely intrusive. At other places in the batholith, where such mobility was not attained, the end products of granitization would remain where they were formed, and here the boundaries with the country rock would be blurred and mixed and could show relict sedimentary structures. A particular bed of highly metamorphosed rock might be unusually susceptible to granitization and might be converted to granite, while "drier" or otherwise more resistant beds above and below it were not so thoroughly changed, creating the lit-par-lit structure.

In this theory then, the type of contact between granite and country rock would depend upon what part of the batholith was being observed. It neatly solves the room problem by having the rocks converted in place. Unfortunately there are other problems which it does not always solve. In the first place, some batholiths appear not to have had anything to do with the great basins of sedimentary rocks. Secondly, some batholiths are not nearly so thick as this theory would require; they do not have the tremendous "roots" that were once assumed for all batholiths.

To answer these objections, some geologists have proposed that the granitic magma forms much deeper within the earth. These hot liquid masses then rise through the rocks above them, assimilating some of the country rocks and adding their constituents to the melt, and thrusting others of these rocks outward and downward, severely metamorphosing them in the process. They believe that the magma masses may have risen like giant blobs very close to the earth's surface and that they cooled slowly enough to form their large crystals because they were covered and insulated by the volcanic rocks of about the same age with which many batholiths are associated. They further believe that batholiths are quite thin structures, on the order of 10 kilometers (6 miles) in thickness, rather than 40 to 50 kilometers (24 to 30 miles). These large masses of essentially rootless granitic magma would rise toward the surface wherever conditions were favorable, the great basins of sedimentary rocks being only one such place.

Most of the largest batholiths that we can see are of late-Mesozoic age, but large batholiths may well have formed during all of geologic time. If batholiths are originally as thin as these geologists believe, it is quite possible that older batholiths did exist, but have been removed by erosion operating through a greater length of time. The few large Paleozoic and Precambrian batholiths which still remain may have been thicker than 10 kilometers or have been less eroded. We must have more evidence than we have at present before we can determine which of these theories, or which parts of them, are correct.

Intrusions of Intermediate Depth

These are much smaller intrusive bodies than their immense relatives, the mountain-inhabiting batholiths, and although along their length some are measurable in scores of miles, their breadth is more commonly measured in feet. These are called *hypabyssal rocks*, a word that means "of intermediate depth."

Hypabyssal rocks generally have textures intermediate between those of volcanic rocks and those of plutonic rocks. If they are part of bodies of moderate to large size they will have cooled sufficiently slowly that their minerals, although smaller than in batholithic rocks, will be visible to the unaided eye. If they are part of a small intrusion, whose walls were close together, the heat loss is comparatively rapid and crystal sizes are diminished accordingly, to the point where the texture of such rocks may be indistinguishable from their volcanic counterparts. In other words, the texture of hypabyssal rocks approaches those of volcanic rocks on the one hand and at the other extreme it may be about the same as those of plutonic rocks—depending very largely on the size of the intrusion and the composition of the magma. Incidentally, porphyritic textures are very typical of the group because there may have been a considerable amount of moving of magma from one environment to another in these shallow intrusive channels, with correspondingly different rates of crystallization before final solidification occurs (Fig. 4-21).

Dikes are a common variety of hypabyssal intrusive. They are tabular bodies, which means that of the three dimensions possessed by a solid body, two are very large compared to the third—in the case of dikes they have about the same geometry as that of a thin pad of notepaper. Another attribute of dikes is that they are *discordant*—that is, they cut across the layering, or stratification, or the grain of the rocks they invade (Figs. 4-9, 4-30).

Dikes are seldom more than a few tens of feet thick, but some are hundreds of miles long. An exceptionally long one is the so-called Great Dike of Rhodesia in southeast Africa. It is more than 480 kilometers (300 miles) long but has an average width of only about 8 kilometers (5 miles).

Dikes commonly occur in groups known as dike swarms. Sometimes they may be aligned on roughly parallel courses, or they may radiate from centers, such as the host of basaltic dikes that lace the northern part of Great Britain, with some individual ribbon-like intrusions reaching lengths of 160 kilometers (100 miles) or so (Fig. 4-31). The focal point for such a radial set of dikes very likely will be the throat, or conduit, of a now extinct volcano, and subsequent erosion may reveal its connection with the dikes.

Depending on their resistance to erosion, relative to their host rocks, dikes may be distinctive features of the landscape. If they are more resistant they stand up somewhat as continuous walls, much like the two conspicuous ones radiating outward from Ship Rock, an eroded volcanic conduit in New Mexico. Should they be weaker, they may be etched out, especially if they crop out in a sea cliff exposed to the full fury of the sea.

Columns also characterize dikes, but instead of standing vertically, they may lie horizontally if the dike itself is vertical, and then may look much like an immense stack of cordwood. Such columns grow at right angles away from a cooling surface—in a lava flow this will be upward from the ground and downward from the top of the flow, and commonly these inward-growing columns meet with a rather ragged matching up of the columnar sets along a line close to the center of the flow. In a vertical dike the columns grow inward horizontally from the vertical side walls which here are the cooling surfaces.

Hypabyssal rocks which have been injected into the earth's crust as tabular, sheet-like bodies essentially parallel to the stratification of the enclosing rocks are called *sills* (Fig. 4-9). These are *concordant* intrusions, while dikes, with their cross-cutting relationships, are discordant. This does not mean that each sill follows only a single stratum, because it is not at all uncommon for them to cross up or down from one stratum to another and to follow it for perhaps hundreds or thousands of meters before making another cross-over (Fig. 4-32).

Fig. 4-30 Dike in Precambrian gneiss. Jefferson County, Colorado. (Photograph by J. R. Stacy, U.S. Geological Survey.)

Some sills run for great distances across the country and have lengths comparable to those attained by dikes. An excellent example is the Great Whin Sill in Northumberland in noreastern England. It looms as a dark, north-facing ledge dominating the surrounding country for much of its course, with the result that the Romans, with their experienced eye for the military potentialties of the terrain, seized upon this natural barrier as a foundation for Hadrian's Wall, which was built to keep the Picts from ravaging northern Britain.

Certainly the most familiar of all sills in the United States is the abrupt cliff of the Palisades that follows the Jersey shore of the Hudson and is clearly visible to the millions of inhabitants of Manhattan, let alone the multitudes of commuters who see it every working day. The Palisades sill is a large one, as these things go, and is over 304 meters (1000 feet) thick. It cooled slowly and without much internal disturbance, with the result that the magma within it had a chance to differentiate. That is, the heavier early-crystallizing minerals, such as olivine and pyroxene, sank to the lower levels of the large pancake-shaped magma chamber while the lighter plagioclase crystals were concentrated near the top.

A more renowned group of hypabyssal rocks, although perhaps not often considered in this light, are the dike and sill that played a decisive role in Pennsylvania during three hot summer days a century ago in July of 1863. One is the thick basaltic sill that underlies Cemetery Ridge and the other is the narrow but persistent dike that supports Seminary Ridge, both of which were the dominant elements of terrain at the Battle of Gettysburg. The Union Army held the former and the Confederate forces the latter. Both sides made the most of the so-called ironstone boulders which had weathered from basalt outcrops along the top of each ridge. Piled into fences these made excellent defensive positions against the round shot and Minié balls of that day. Cemetery Ridge, along which the Union brigades were deployed in full strength on the fateful third morning, makes such a continuous rampart, with a nearly unbroken forward slope up which Pickett's command had to charge, that the assault was virtually foredoomed to failure. This was doubly so when they were called upon to dislodge men sheltered behind a practically shot-proof basaltic barricade and supported, in addition, by one of the greatest concentrations of artillery assembled in the war.

Because contraction occurs in the solidifying magma of sills as well as in dikes and lava flows, columnar joints are characteristic of them, too; since

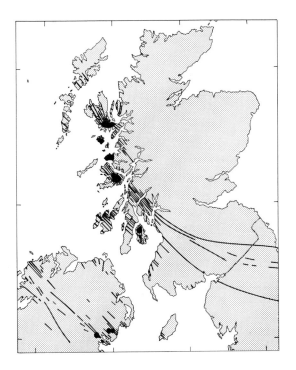

Fig. 4-31 The dike swarms of Scotland and northern Ireland. The black areas are centers of igneous intrusion to which the dikes are related.

most sills tend toward a horizontal position at the time of their origin, the columns in such an intrusion are likely to be vertical, having developed at right angles to the cooling surface.

Discriminating between buried lava flows and sills intruded into sedimentary rocks is not one of the simpler field problems in geology. Igneous rocks in both categories may have about the same texture, and both may have very well developed columnar joints. The most likely place to look for the evidence needed to decide between the two possibilities is at the contact between the sill, or buried lava flow, and the enclosing rocks.

In many sills there very commonly will be a chilled zone in the igneous rock at the margins where crystallization was the most rapid, and here the texture of the rock will be fine-grained, in some cases almost glassy. Additionally, there may be a baked zone in the invaded rocks immediately adjacent to their boundary with the sill. They will have been fired or hardened in much the same way that clay is fired in a kiln to make bricks or pottery. In a sill, fragments of the overlying host rock may be included in the igneous material. However, if the igneous body is an interbedded lava flow, then the baked contact will be lacking on the upper surface of the flow, and the chilled zone can be present only on the bottom. In addition to this, the strata overlying these volcanic rocks will have been deposited after the lava solidified, and more likely than not fragments of rock eroded from the flow will be found incorporated in the covering strata. Sills are more commonly of denser volcanic-type rocks, while lava flows often are vesicular—with many small holes formed by gas bubbles.

A third type of hypabyssal intrusive body is the channel or conduit that once served to connect a volcanic vent on the earth's surface with the magma reservoir at depth. In an erosional sense, volcanoes are vulnerable features; much of their interior consists of loosely consolidated ash and pyroclastic material, their steep slopes lend themselves to gullying, and if they are high enough they may intercept more snow and rain than the surrounding countryside. The result of this augmented runoff working on such an exposed structure is that, geologically speaking, the interiors of many volcanoes are bared to view shortly after they lapse into dormancy.

Their internal skeleton of radial dikes and central conduit often proves to be made of sterner stuff, with the result that dikes stand up as partitions and the solidified magma of the conduit forms a central tower, known as a *volcanic neck*. A well-known example is the Ship Rock in New Mexico.

More often than not, volcanic necks are found in clusters, rather than standing isolated. Prime examples in the United States are the volcanic buttes of the Navajo-Hopi country near the so-called Four Corners area. Here more than a hundred volcanic centers interrupt the surface of the plateau.

Another group of volcanic necks that have achieved a measure of notoriety are the so-called pipes of *kimberlite*, a hypabyssal rock composed dominantly of ferromagnesian minerals, which are a source of the diamonds of the Republic of South Africa. The ancient volcanic conduits containing the diamonds are deeply weathered near the surface into what is called the "blue ground," a sticky clay from which the diamonds are separated by washing. The pipe-like columns of hypabyssal rock were mined in the Kimberley pit to a depth of 1060 meters (3500 feet) before the workings were abandoned—now they are idle and stand water-filled to within 183 meters (600 feet) of the surface. Since the pressures and temperatures necessary for diamonds to crystallize from solution are not fulfilled short of a depth several miles below the ground level, diamonds are one of the better geologic indicators which demonstrate that (1) volcanic magma can originate at considerable depths within the earth's crust, and (2) because Kimberley diamonds are found at levels far above the depths where pressures necessary for their formation prevail, the magma of which they once were a part streamed upward in a pipe-like conduit to reach the higher levels of the earth's crust, even discounting the removal of some of the overburden by erosion.

Volcanism

Of all the geologic phenomena visible on the earth's surface, none is more alien to the everyday American world than volcanism. Glaciers are logical constructions of such familiar substances as ice and snow, just as the erosional work of streams, the wind, and the sea are part of our daily experience. Volcanoes play a more exotic role, and also have the enchantment that distance lends, since none is active today within the contiguous United States. To see them within our country it is necessary to go to Alaska or Hawaii, and many of the world's more striking examples are in far-off lands, such as Kamchatka, Indonesia, or on the truly oceanic islands. Fortunately, from a scientific point of view, but not from that of the people whose lives are occasionally disrupted, some of the outstanding examples are in densely populated areas, such as the central Mediterranean and the main islands of Japan.

Distribution

This brings up the interesting and significant point that volcanoes are not scattered randomly over the earth, but many are concentrated within fairly well-defined zones or bands (Fig. 4-33). Perhaps the most renowned of these chains is the so-called ring of fire that girdles much of the Pacific Ocean. The map showing the distribution of active and only very recently dormant volcanoes demonstrates that they are spotted along most of the length of the eastern and western margins of the Pacific Ocean. They are also the reason for the existence of the mountains on the higher islands; Mauna Loa and Mauna Kea on the island of Hawaii are familiar examples.

Fig. 4-32   Diabase sills and dikes, showing the relationship between the two. The sill in the center of the face sent tongues into its floor; it becomes a dike toward the left side. Victoria Land, Antarctica. (Photograph by W. B. Hamilton, U.S. Geological Survey.)

The Mediterranean world has its share of volcanoes, too. Best known, perhaps, are Vesuvius near Naples and Etna on Sicily. Because of the long and literate span of history encompassed in this ancient world, these volcanoes have played a role in mythology and theology. G. W. Tyrrell wrote, "In the Middle Ages the Mediterranean volcanoes were appropriated by the theologians who regarded them as places of the eternal punishment of certain great sinners. Thus the Adrian emperor Theodosius was assigned to Vulcano itself; while Etna was regarded as the place of torment of unhappy Anne Boleyn, the innocent cause of Henry the Eighth's secession from the Faith."

The map (Fig. 4-33) shows, when compared with a similar one for earthquakes (Fig. 16-14), the very close relationship in the distribution of these two disturbing elements of the earth's crust. They do not coincide, but the resemblances are greater than the differences, and this leads most geologists to believe that some kind of relationship exists between the two. This is *not* to say that volcanoes cause earthquakes, or vice versa, but the forces operating within the crust to cause the one are also very likely producing the other as well.

The map showing volcano distribution also points up the fact that most of them are within sight of the sea (Fig. 4-34). In earlier days this led to the

Fig. 4-33   Distribution of the active volcanoes of the world. Each black dot represents one or more presently or recently active volcanoes, of which there are more than 470.

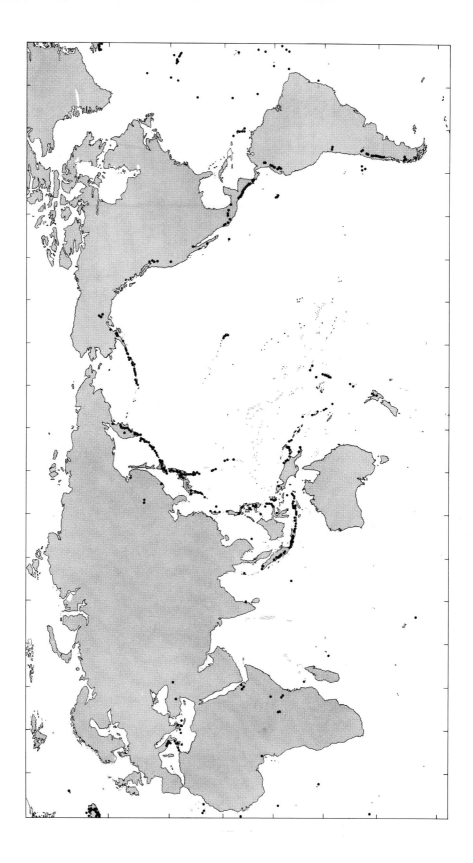

plausible explanation that eruptions were the result of downward-percolating sea water coming in contact with molten material in the earth's crust, with the resulting steam explosion being responsible for the volcanic eruption. This seemed all the more reasonable in view of the fact that men recognized from early days that steam was the most important of all the volcanic gases. Unfortunately, as in the case of many other splendid hypotheses that have a brief vogue only to vanish into oblivion, this belief is not true. A number of volcanoes are far inland, many miles distant from the sea. Kilimanjaro, the peak of the celebrated snows, is a good example since it is close to 1200 kilometers (800 miles) from the coast. Insofar as the truly oceanic islands are concerned, more convincing evidence is that the required physical and chemical conditions under which their magma originated prevail at depths far greater than could ever be reached by sea water percolating downward through interstices in the rocks.

A more significant control over the distribution of volcanoes is that the greatest number of them are in places where fracturing appears to be active in the earth's crust, and where there is strong evidence of a good deal of crustal unrest. Such fractures may provide the channels along which magma rises from great depths, ultimately to reach the surface. This explanation seems to work well for those volcanoes ranged in long lines or chains, such as those of the Aleutian Islands or Java. It is not as satisfactory for those standing alone and well apart from the more typical volcanic lineaments. Volcanic centers on the Colorado Plateau in the southwestern United States are perhaps good examples of this latter type as they are characteristic of eruptive conduits which broke through to the surface by penetrating essentially flatlying, undisturbed strata.

The total number of active volcanoes shown on the map (Fig. 4-33) should be 476. This number is based on information compiled in 1934. A few minor volcanoes have appeared since then, chiefly off the coasts of Japan, Mexico, in the Azores and Tristan da Cunha, and, most recently, Iceland, but the general picture remains unchanged. According to Tyrrell there have been about 2500 eruptions since the beginning of recorded history, and of these nearly 2000 were in the Pacific Basin—an impressive demonstration of the dominant role this area plays in the volcanology of the earth.

Prediction and Control

Will it ever be possible to protect man and his property from the destructive forces of volcanoes? Certainly it would help if we could know when and where volcanic eruptions were going to occur. The location of possible eruptive

Fig. 4-34   A volcano erupts from the sea. Myojin Reef, about 170 miles south of Tokyo, Japan. (Official U.S. Navy photograph.)

Fig. 4-35 The tiny settlement on Tristan da Cunha was temporarily abandoned because of the threats posed by the eruptions of 1961. (British Admiralty Official photograph. British Crown Copyright reserved.)

activity can be predicted in only the most general way, the most likely being areas of unrest in the earth's crust and where volcanic activity is now taking place or has recently done so.

About the timing of eruptions, geologists can be a bit more precise. Earthquakes give us one clue because many eruptions are preceded by earth tremors, although these may continue for twenty days, as at Paricutin in Mexico, or sixteen years as in the A.D. 79 eruption of Vesuvius. These quakes, some of which may be easily felt and some of which show up only on sensitive instruments, are caused by the swelling of the volcanic area as the magma beneath rises to the surface. This swelling can be measured by tiltmeters placed on the surface of such an area, and is one of the most accurate methods of prediction. The rising of the hot nonmagnetic magma also causes local changes in the earth's magnetic field and in its electrical currents, both of which are detectable.

Once an eruption has begun, measures can be taken to warn persons who might be endangered and to deflect the lava flow in order to decrease the amount of damage done. In 1669, the town of Catania, Italy, was directly in the path of a flow from Mount Aetna. The Catanians quickly started to dig a channel to divert the lava, but they were stopped by the inhabitants of a neighboring town who were understandably perturbed when the flow was turned in their direction. This illustrates one of the problems that still plague attempts to divert lava flows. Unless they can be allowed to flood government-owned waste land, the legal problems are virtually unsolvable.

In Hawaii attempts have been made to control eruptions of Mauna Loa by aerial bombing. The theory is that the banks of a lava stream can be destroyed so that the lava spreads out and the pressure of the main stream is lessened. Bombing should also stir up the lava so that the gases within it are dissipated, making the lava much more viscous and slow-flowing. Finally the cinder cone itself could be shattered so that many small flows would replace a single large one.

In Java both populated and agricultural areas are endangered from time to time by volcanic mud flows. Diversion dams channel the mudflows into places where they will cause least destruction. Artificial hillocks are sometimes built as islands of refuge for villagers where volcanic mudflows are a constant menace.

The Challenge of Igneous Processes

Volcanism is one of the more dramatic of all geological phenomena, one that impinges directly on human affairs, that has been the source of comment and discussion from the days of Strabo and Pliny to ours, yet little is known of its true nature. We see but part of the picture, only the surface aspect. Nobody can do more than speculate on the connection between volcanoes and the origin of plutonic rocks within the earth. Reasoning tells us there probably is a relationship between the two, but of this we have no direct proof. Nor, for that matter, do we know the source of heat, the depth of burial of the volcanic hearths, or even more fundamentally, what the actual cause of volcanism may be. Some of these questions we shall take up again. Through what we do know of volcanoes, active and extinct, and of the nature of their gaseous, liquid, and solid products, we have some basis for making a few controlled guesses about the origin of the atmosphere, the ocean, and what may very likely have been the composition of the original dry land surface of the earth.

What, then, are some of the things actually known about volcanoes and the products of volcanism that can be summed up here and kept in mind as they bear on other fundamental problems of the earth?

As we have seen, volcanoes are not randomly distributed over the surface of the earth, but are found in the following environments: (1) along the margins of continents, especially the somewhat mountainous coasts bordering the Pacific; (2) within the ocean basins where the truly oceanic islands—such as Iceland and Hawaii—are dominantly volcanic; (3) in the regions bordering, but not actually within, mountain ranges in the continental interior (although this is a less familiar occurrence)—in this category are the volcanic provinces of central France and southern Italy; and (4) along large fracture zones or *rifts*, such as the remarkable system that extends through much of eastern Africa.

Geologically speaking, there are two kinds of areas where active volcanoes are not found, and these are (1) the heart of mountain ranges whose rocks have been intensely compressed—such as the Alps and Himalayas—and (2) the broad expanses of deeply eroded, very ancient, and intensely deformed rocks which are spoken of as *shields*—most of Labrador and northeastern Canada are typical of such a region. This is not to say, however, that these regions do not have volcanic rocks; they do, but they are very ancient and extremely metamorphosed.

Trying to find a common factor to explain both the occurrence and non-occurrence of volcanism is perhaps beyond our capabilities today, but we have found that volcanoes appear to shun areas of strongly compressed rocks and to favor regions where some kind of fracture system extends from the surface down to the depths where magma can form. The importance of this appears to be twofold: (1) fractures provide a channel by which magma can reach the surface, and (2) their presence indicates that under favorable circumstances there may be a release of pressure at depth. This is important because even at fairly moderate depths within the earth, the temperatures presumably are high enough that rocks ordinarily would be expected to melt, but the high pressures prevailing there prevent this (for most substances, the higher the pressure, the higher the temperature required for melting to occur). With a release of pressure, melting takes place, and fracturing of the rocks in the earth's crust may be one of the ways by which this release is achieved.

This raises the next question; what is the depth from which the magma rises in a volcanic eruption? Here, the evidence is mixed. Some lavas appear to have had shallow sources. At Tahiti, where the central conduit is bared through deep erosion, the depth is perhaps a mile; at Vesuvius the nature of fragments brought up from subvolcanic basement led the German volcanologist Rittman to the belief that magma probably rose about 6 kilometers (4 miles). Diamonds in the pipes of Kimberley probably crystallized at a depth of several kilometers below the surface, judging from what we know of the temperatures and pressures required to make diamonds in the laboratory.

The tremendous outpourings of basalt in fissure eruptions very likely come from subcrustal depths, in Hawaii perhaps 56 kilometers (35 miles). The reasoning here is that higher temperature is required to liquefy basalt compared to other lavas, that enormous volumes are involved, which would appear to require more than a local source, and that these plateau basalts are of remarkably uniform composition both in space and time.

Many additional problems await solution. In an earlier day, when it was believed that the earth had solidified from a fiery, molten sphere that on cooling developed a rocky shell, the magma reservoirs within the crust could be interpreted as hot spots left behind when all else had solidified. Today, through other lines of evidence, we know that this is not so, but we are still at a loss for an answer. Radioactivity has been appealed to as a source of heat, but serious objections intervene here, too. The lack of radioactive constituents in the lavas with the highest temperatures, the plateau basalts, militates against this, together with the lack of helium — a product of radioactive decay — among the associated volcanic gases.

So the list grows. Even the mechanism by which a column of lava and its entrapped gases tunnels its way upward through the crust is unknown. In part it must involve fluxing, in part wedging or shouldering aside, and in part stoping or piecemeal engulfment; but what the precedence of these relative roles is remains speculative.

The even more fundamental problem of what processes are responsible for the wide variety of compositions of volcanic rocks still awaits solution. Magmas of many compositions may be erupted, even from the same vent or very closely related conduits; an outstanding example is a recently extinct volcano in Oregon which has simultaneously erupted lavas as unlike as basalt and obsidian.

Differentiation, the process described in the discussion of the Bowen reaction series, almost certainly plays a role, but this process moves only one way. It is possible to have a basaltic magma go through a series of reactions so that a rhyolitic magma results, but it cannot go in the opposite direction.

Another problem is the relationship between volcanic rocks and plutonic rocks. Why are most plutonic rocks granite while most volcanic rocks are basalt? Why is much continental igneous rock granite, while the ocean floor is almost entirely basalt? Why are volcanic rocks sometimes, but not always associated with batholiths?

In fact, the origin of magma itself is unknown, and finding an answer to that problem would put us well on our way toward finding a solution to the fundamental riddle of the origin of all igneous rocks.

Selected References

Hamilton, W. B., and Myers, W. B., 1967, The nature of batholiths, U.S. Geological Survey Professional Paper 554-C.

Jackson, Kern C., 1970, Textbook of lithology, McGraw-Hill Book Co., New York.

Read, H. H., 1957, The granite controversy, Interscience Publishers, Inc., New York.

Walton, M., 1960, Granite problems, Science, vol. 131, pp. 635–45.

Williams, H., 1951, Volcanoes, Scientific American, vol. 185, no. 5, pp. 45–53.

Fig. 5-1 Ripplemarks on the Dakota Sandstone. Jefferson County, Colorado. (Photograph by J. R. Stacy, U.S. Geological Survey.)

5

Sedimentary Rocks

Widely spread over the surface of the earth is a relatively thin blanket of sediment which has been consolidated into rock through slow-acting processes that are relatively simple to understand when compared with those responsible for the origin of igneous and metamorphic rocks. These sedimentary processes operate in environments on land or in the sea at temperatures and pressures much more like those familiar to us than the 1370° C. (2500° F.) needed to keep basaltic magma molten. True, pressures on the floor of the ocean, the final repository of much of the waste of the land, rise to six tons to the square inch, but these pressures are still slight when compared with the crushing burdens prevailing in the crustal realm where processes operate to produce the metamorphic rocks.

Sedimentary rocks, for the most part, are secondary or derived rocks. One important category of them consists of layers made up of clay, sand, or gravel particles which are derived from the disintegration or decomposition of pre-existing rocks. Layered rocks made of such fragmental material are called clastic sedimentary rocks.

Another large and economically important group of sedimentary rocks is chemically precipitated in water such as evaporating lakes or shallow embayments of the sea. Perhaps the best known example of this category is rock salt. Closely akin to it in origin are such well-known substances as gypsum and borax—both of which are chemically derived.

Organic sediments are a third category, and an enormously important one. Coal, a vitally significant fossil fuel, is in this group, as are the so-called oil shales, a possible reserve for the future. Another familiar kind of organic sedimentary rock is limestone, and of its many forms several represent the slow accumulation over many centuries of the deposits made by lime-secreting plants and animals.

Environments of Deposition

Sedimentary rocks can accumulate in a wide variety of environments on the earth's surface—about as many as there are kinds of landscapes or different sorts of climates. Two major realms of sedimentation commonly are recognized, and these are (1) in the sea, or marine, and (2) on land, or continental. Like most classifications, there is much that is arbitrary about this one, and several occurrences might as well be placed in one category as in the other; for example, the silts and muds in the deltas of large rivers could be assigned readily to either province.

Marine

In the sea at least two of the factors controlling the distribution of sediment are (1) the distance from land and (2) the depth of water. To simplify the story, there are four leading zones where sediments accumulate that have sufficiently unlike characteristics to merit setting them up as separate units.

Seaward from the land, the first of these zones is the *shore zone,* and for all practical purposes this is where the surf breaks against the shore. On many coasts where the tidal range is large a very broad expanse of sea floor adjacent to the land may be laid bare at low water.

The *continental shelf* is a much broader zone and normally extends seaward to a depth in the general neighborhood of 183 meters (600 feet). On some coasts this depth may be only a few miles offshore; on others, such as the coast of Siberia, it may be 322 kilometers (200 miles) or more. As a rule the continental shelf is the region where land-derived sediments are deposited after being winnowed and shifted about by waves and currents of the sea. This is the zone where most of the sediments were accumulated that we see exposed as marine sedimentary rocks on the earth's surface.

The *continental slope* and the deep floor of the sea, or the *abyss,* are inaccessible by ordinary means of observation. It is now possible to take photographs of the ocean bottom with underwater cameras, and, by means of the bathyscaph, to visit this dark, silent realm, almost as remote in its way as the world of space. Here, on the floor of the open sea, for the most part accumulate the finest sediments, the ooze composed of the remains of free-floating and swimming microscopic plants and animals and the extremely divided clays that carpet the abyssal plains.

Continental

On the land many of us are aware of the large number and variety of possibilities available for the trapping of a multitude of different sorts of sediment
and of their ultimate conversion into rocks. Among the many examples, the
following are typical:

LAKES   Some of these natural settling basins, such as the Great Lakes, the
Caspian and the Aral seas, have such great size that in a sense they may be
thought of as small oceans. All lakes, however, large or small, serve as local
traps in which sediment transported to them by streams, moving ice, or the
wind may accumulate.

FLOOD PLAINS AND DELTAS   Flood plains are flat surfaces adjacent to streams,
especially in lowlands, but in mountainous regions as well, over which streams
spread in times of flood. During each flood a new layer of sediment is deposited. These depositional sites range all the way from plains bordering the
Nile, the Yangtze, and the Mississippi down to narrow strips bordering small
streams.

Deltas form where sediment-laden streams enter bodies of relatively still
water, such as lakes or the sea. Although they form mostly below sea level,
their top surfaces rise above sea level and so become flood plains.

ALLUVIAL FANS   When a stream comes rushing from mountains or hills
carrying a great deal of rock debris and suddenly reaches a nearly flat inland
basin, its sediments are dumped in a spreading fan-like form. Excellent examples of these fans are found bordering the mountains of our southwest
deserts.

SAND DUNES   These deposits testify to the effectiveness of the wind in those
parts of the world where such factors as an abundant supply of sand, little
vegetation with which to stabilize it, and strong winds to move the sand about
occur together. Such combinations are likely to be encountered in deserts,
along many of the world's coasts, and along the flood plains of large rivers,
of which the Volga is an excellent example. Wind also sweeps lighter material
than sand before it and this may pile up in vast windrows of dust, or silt size
particles. Such a thick blanket of feebly consolidated dust is a dominating
element of the tawny landscape of northern China near Peking.

GLACIAL DEPOSITS   Deposits left by glaciers are a final category, and these
will be discussed in much greater detail in Chapter 12. Glaciers, which today
are confined to higher mountains or to far distant Arctic and Antarctic shores,
were once more widespread than they are today, and their deposits—usually
more disordered than those laid down by streams or in the sea—blanket

much of North America and northern Europe. A good example of a typical glacial deposit is boulder-clay, which is literally that—rocks the size of boulders set in a clayey matrix with very little sorting of particles according to size.

Features of Sedimentary Rocks
Stratification

Most of these rocks are made of particles, ranging from very large down to submicroscopic, that settled out through such a medium as air or water. In addition to this, the majority are layered (Fig. 5-2), and these layers, too, show a great range in their dimensions from laminae whose thickness is measurable in millimeters up to ones that are measured in hundreds of meters.

Such depositional layers in sedimentary rocks are called *strata*; an individual layer is a stratum. In everyday language such layers are commonly called beds if their dimensions are fairly large. If the layers are very thin, they are better called *laminae* (from the Latin, *lamina*, for thin plate, leaf, or layer), and the term is used here in much the same sense that we speak of the laminations in plywood.

Among the many reasons for the rhythmic layering in sedimentary rocks is discontinuous deposition, with slight differences in coloration or grain size to mark the new laminae when deposition starts up after an interruption. An especially striking kind of rhythmic deposition is the annual layering characteristic of the very fine-grained laminae deposited on the bottom of cold-climate lakes that freeze over in the winter. These uniform layers are called *varves* (Fig. 5-5), and the thicker, light-colored layers are generally interpreted as having been deposited during the summer when the lake is open and streams are free to sweep comparatively coarse sediment into the lake. The finer dark band is thought to represent fine-grained material, in large part organic, that settled out through the still water of the lake under the ice during winter.

Some sedimentary rocks deposited in the sea, especially some kinds of marine clay, show similarly repetitive laminae, too, and these are interpreted as annual layers, as are also the remarkably regular, paper-thin layers deposited on the floors of large nonfreezing lakes; perhaps the best known examples being the oil shales of the western United States, especially in the vicinity of Green River, Wyoming (Fig. 5-6).

Some sediments, ranging from coarse- to fine-grained, show a very different sort of stratification. In this category an individual layer, instead of having particles of the same, or even of different sizes distributed uniformly throughout, will have the larger particles concentrated at the bottom, the smaller at the top. Such a layer is said to have *graded bedding*. This feature occurs when a mass of sediment is discharged suddenly into a relatively quiet body of water. The larger particles drop out quickly, followed by the medium-sized ones, and finally the finest particles are deposited.

Fig. 5-2   A monastery at Meteora, Greece, surmounts bluffs of stratified conglomerate of Tertiary age. (Photograph by Jean B. Thorpe.)

An excellent example is the excessively muddy Colorado River where it flows into Lake Mead, which is backed up behind Hoover Dam on the boundary between Nevada and Arizona. The muddy river water seems to disappear as if by magic, and anyone who has seen the dark, blue-green water of Lake Mead cannot fail to be impressed by the contrast it makes with the turbid

Fig. 5-3   These colossal images of Rameses II (1301–1235 B.C.), next to the Nile are carved from horizontal sandstone layers that extend through the statues. (Photograph by Jean B. Thorpe.)

river. An explanation for the disappearance of the muddy water is that with its higher density it sinks below the surface of the lake and moves as an under-flow along the bottom. This is known as a density current.

Cores taken from the abyssal plains of the ocean floor show great thick-nesses of graded beds, in which the cycle from coarse to fine sediments has been repeated many times with the result that many gradational sequences are stacked one upon another. The currents which formed them so far from land are still somewhat of a puzzle to geologists, and this problem will be discussed in Chapter 11 where it properly belongs with its related problems of submarine canyons and slumping of marine sediments.

Color

Igneous rocks, unaltered by exposure to the atmosphere, typically are shades of gray or black, since these are the prevailing colors of their most abundant constituents, feldspar and the ferromagnesian minerals. Sedimentary rocks may be much more colorful. Some kinds are made up of large fragments of

Fig. 5-4   Effects of erosion of horizontal sedimentary strata in an arid region. The outcropping edges of more resistant strata simulate contour lines. (Photograph by William Garnett.)

Fig. 5-5 Varves from an ancient and extinct lake. The darker bands contain the most organic matter. Garfield County, Colorado. (Photograph by W. H. Bradley, U.S. Geological Survey.)

other pre-existing rocks, and if a wide variety of these is present, the resulting sedimentary rock will be correspondingly variegated.

In addition to the possibility of a variety of colors in a sedimentary rock resulting from the great range of colors in the rocks that comprise it, an important source of coloring matter may be the very fine interstitial material that fills the space between the individual grains. If this should contain hematite (iron oxide, Fe_2O_3), the resulting rock is likely to be colored red. This is the source of most of the red color in the walls of the Grand Canyon. Other forms of iron may stain a rock brown, or even shades of pink and yellow. Iron possibly may be responsible for much of the purple, green, or black colors of some sedimentary rocks, but what the true nature of some of the coloring matter may be is not known.

Many of the darker sedimentary rocks owe their color to the organic material they contain. Coal is an excellent illustration of this. Its composition is entirely organic and the very name is a synonym for black. With varying amounts of organic material, sedimentary rocks may have a color range from shades of light gray to black. In some cases, however, black muds owe their color to finely-divided iron sulphide dispersed through them rather than to carbonaceous matter.

The range in colors that sedimentary rocks may display is one of their more intriguing properties, and in dry countries where vegetation is lacking and the soil cover is sparse, the true color of these rocks stands revealed in striking fashion, as in Grand, Zion, and Bryce Canyons, Monument Valley, Canyon de Chelly, and the Painted Desert. It is the brilliant coloring of their sedimentary rocks as much as any other attribute that makes these places so renowned.

Special Features

MUD CRACKS   When wet, clayey mud that is exposed to the air dries, it shrinks, and on shrinking, cracks, generally with the formation of a nearly uniform pattern of hexagons and pentagons—much resembling the tops of lava columns. In lava, the reason is contraction upon cooling; in wet muds, it is contraction

Fig. 5-6   Very thinly laminated shale from near Green River, Wyoming, containing a fossil fish.

Fig. 5-7 Color banding. Bryce Canyon National Park, Utah. (Photograph by W. H. Bradley, U.S. Geological Survey.)

resulting from dehydration. On continued drying, the mud layers on the tops of the polygons may curl up at the edges, so much so at times as to make complete rolls, much like a cardboard tube.

Mud cracks indicate that the sediment of which they were once a part was alternately wet and dry, and thus these cracks are very typical of mud-bottomed, shallow lakes that on occasion dry up. They are not so characteristic of muddy tidal flats because the time of exposure at low tide is too brief for much drying out.

FOSSILS No other property is so distinctively a characteristic of sedimentary rocks as fossils. These are the remains of once living things that on their death were buried in sand, silt, lime, or mud. Much of the organic matter that some of them originally contained gradually was replaced over the centuries by inorganic matter, until, to use petrified wood as an example, many of the woody fibers and cellulose have been replaced by silica. Representatives of just about everything that crawls, walks, swims, or flies among the animals, or that simply stands in place, such as the plants, have been preserved as fossils. This includes such improbable creatures as jellyfish, whose composition must be more than 95 per cent water, or such fragile things as the compound eyes of flies, as well as the delicate tracery of dragonfly wings. Such things are the exceptions, however, because the organisms most commonly preserved as fossils are those that already have durable elements in their make-up, such as shells, bones, and teeth. In fact, most fossils are the remains of shells or skeletons. In some instances the entire rock may consist of organic matter. A layer of coal is made up of plant fragments—chiefly spores—and some limestones may be composed of the remains of coral or of calcareous algae, or may be a felted mass of sea shells, in which case the rock is called a *coquina* (Fig. 5-8). In addition to the remains of organisms, footprints, tracks, trails, and burrows may be considered as fossils, too.

RIPPLE MARKS   Nearly everyone has seen the characteristic corrugated surface made by a stream or tidal current flowing across a sandy bottom, or has seen

Fig. 5-8 Coquina, from Saint Augustine, Florida. (Courtesy of Ward's National Science Establishment, Inc.)

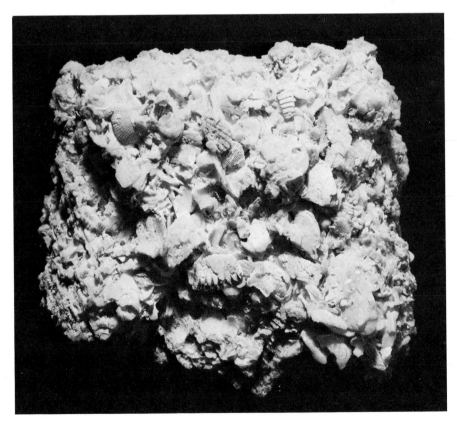

photographs of virtually the same pattern produced by the wind blowing across a desert sand dune. Such ripples are likely to be asymmetrical with the gentle slope on the upstream, the steep slope on the downstream side. This pattern results from sand grains being rolled by the current of water or air, up the upcurrent side, and then sliding down the downcurrent slope which stands at an inclination known as the *angle of repose*. This term means the maximum slope at which sand grains will stand without sliding down this so-called *slip face* by gravity. Almost all such ripples form at right angles to the current that made them, and thus when they are converted into solid rock they can be used to establish the direction once taken by long-vanished currents in the atmosphere or under water. It was once thought the *current ripples* in water-laid sediments were an indication of shallow depth, but underwater photographs recently taken of the tops of submarine ridges show ripple patterns on the sea floor at a depth of 1800 meters (6000 feet).

Another type of ripple has symmetrical sides, sharper crests, and more gently rounded troughs than current ripples do. These symmetrical corrugations are called *oscillation ripples*, and presumably they are the result of surface waves of a type known as waves of oscillation stirring up the sandy bottom of a shallow water body.

CROSS-BEDDING   Earlier in this chapter the point was made that sedimentary rocks customarily are deposited in essentially parallel layers known as strata. But as with most generalizations there are almost always bound to be exceptions, for there are several varieties of stratification in which the laminae are inclined at steep angles to the horizontal (Fig. 5-9).

One kind of cross-stratification forms in sand dunes. Each layer inside the dune at some time past was part of the surface, and since the dune's configuration was established largely through a balancing of wind transport upslope and gravity gliding downslope, most of these layers are sweeping curves which more typically than not are concave upward. Since sand dunes are ephemeral land forms whose position and orientation change with the inconstant wind, it is not surprising that these sweeping, shingled layers may intersect one another in complex patterns such as are to be seen in the sandstone in the walls of Canyon de Chelly, Arizona, or in Zion National Park (Fig. 5-10).

Another kind of cross-bedding is made in deltas by steams carrying a fairly large load of moderately coarse debris, and then forced to deposit this sediment rapidly when their current is checked upon reaching a water body, such as a lake. Here, the sediment dropped by the stream constructs a leading edge out into the lake, much as a highway fill is built out into a canyon by end-dumping from gravel trucks. The outer slope of such a delta, like the slip-face of a sand dune, also stands at the angle of repose. When these sediments are consolidated into rock, three distinctive layers may result. At the top and bottom of a deltaic deposit will be horizontal strata, which are known as *top-*

Fig. 5-9 The body of the Sphinx is formed from nearly horizontal layers of rock, but the head and neck are carved from cross-stratified layers. The cross-strata are inclined toward the pyramid. (Photograph by A. E. L. Morris.)

set and *bottomset* beds respectively, while the steeply inclined layers that once were the delta front as it advanced out into the lake are *foreset* beds.

CONCRETIONS   Round or almost round solid bodies are sometimes found in sedimentary rocks. These concretions, as they are called, are formed during the process of lithification of the surrounding sediments. Any small particle, a sand grain, a piece of shell, even a small insect, can act as a nucleus for the concretion. The cement, which will eventually bind all the particles together, collects around the nucleus, and gradually enlarges the concretion until it may reach several feet in diameter, although most are much smaller. The nucleus is usually a fragment unlike the rest of the sediment—an alien particle, so to speak.

GEODES   These also are nearly spherical in shape but are much more spectacular as beautifully terminated quartz crystals commonly grow inward from the walls projecting into the hollow interior. The walls themselves are also of quartz of the variety called chalcedony, which is derived from a silica gel. This gel probably forms around a pocket of water in the sediment while the rock-forming processes are in progress. Eventually the walls become hard, and

cracks develop through which later mineral-laden water can penetrate and precipitate the inner crystals.

Conversion to Rock (Lithification)

Most of the discussion thus far in this chapter has had to do with sediments and the process of sedimentation, and very little has been said about the way in which these are converted into solid rock. What process is it, for example, that converts loose sand, which at the beach can be idly sifted through the fingers, into a rock such as sandstone which may be almost as unyielding as granite?

Is it pressure? The answer to this is an emphatic, No. To apply enough pressure to force sand grains to adhere to one another would be to crush them into smaller and smaller particles. Pressure does play a role, however, in the process of *compaction*, which is the squeezing together of the particles in a sediment, with the result that the *porosity*, which is the fraction of the total volume that is pore space, is reduced. If, for instance, enough pressure is applied to fine-grained muds, such as clay or silt, most of the interstitial water is squeezed out, the sediment shrinks markedly, and if clay is a dominant consituent the particles tend to adhere to one another.

The closing up of the space between the particles through compaction is an important precursor for the most significant process involved in the conversion of sediments into sedimentary rock. This is *cementation*. Fundamentally it involves the deposition from solution of such a soluble substance as $CaCO_3$ and its building up as a layer of film on the surface of sand grains, silt particles, or clay flakes, as the case may be, until all the pore space separating them is filled. Such a limy cement is precipitated in much the same way, although at a lower temperature, as the scale that forms inside a kettle or a hot-water bottle.

Calcium carbonate ($CaCO_3$) is one of the more abundant of natural cements. It is among the more soluble of the common substances that may be dissolved in water in the ground and then be precipitated out of solution to fill the voids separating the mineral grains and ultimately to bind these grains together to make a solid rock. Obviously it will be most effective in regions where a large amount of lime is available, most typically from the solution of limestone. Another important natural cement is silica (SiO_2), which is also soluble, although less readily than $CaCO_3$. Iron oxide (Fe_2O_3), too, is a cementing agent, and, as mentioned earlier, wherever it is present the whole rock is correspondingly iron-stained or rust-colored.

Types of Sediments and Their Related Rocks

In the opening section of this chapter, the point was made that there are three major categories of sedimentary rocks: *clastic*, or fragmental; *chemical* precipi-

Fig. 5-10 Giant cross-strata at Checkerboard Mesa, Zion National Park, Utah. (Photograph by Ray Atkeson.)

Fig. 5-11   Two large limestone concretions removed from shale by stream action. A larger one in the center is still in place and has a pit weathered out of its top. Delaware County, Ohio. (Photograph by A. M. Bassett.)

tates; and *organic* deposits. Like many classifications of natural phenomena, these categories are more rigid than the actual state of affairs. There not only are gradational types from one category to the other, but there are also varieties that might just as logically be placed in one pigeonhole as the other, as well as a few that fit into none.

Clastic Sedimentary Rocks

The clastic rocks truly are secondary rocks since they consist of particles that are fragments of pre-existing rocks and these may range in size from blocks the size of boxcars down to colloids so fine as to remain in suspension almost indefinitely. Since these clastic rocks consist of fragments of other rocks, they are very likely to show a wide range of composition. So much so, in fact, that in setting up the classification of the clastic rocks the first property to be considered is the *size* of the particles that are cemented together to make a sedimentary rock, rather than the *material* of which they are made.

Take the word sand, for example. To most people sand has a double connotation: (1) it is a size term—all of us are conscious of the grittiness of sand in a bathing suit or between our teeth; and (2) for most of us it has a compositional meaning—the beach sand most of us think of ranges from white to a tawny yellow, and is likely to be thought of as consisting of quartz grains. In actuality many beach sands contain mostly feldspar grains as well as a liberal sprinkling of other sand-size rock particles and mineral grains. Sand can consist of almost any substance of sufficient durability. Along some of the rivers of the Atlantic states, sand bars are made of coal fragments. On some of the beaches of Hawaii the sands are coal black, too, but are composed of ground-up basalt. In the islands of the South Seas, the straw-colored sands of their fabled shores are made of fragmented coral heads, pieces of shells, and other organic debris.

The size terms that follow are in fairly common usage, though here they are arranged in a sequence and are defined in a more rigorous sense than is ordinarily employed. A major difficulty in trying to establish a hierarchy of sediment sizes is that there are no sharply defined, arbitrary boundaries between such things as sand and silt, for example, for these actually are part of an unbroken series.

A classification that has won wide acceptance is one that was originally proposed in 1922 by C. K. Wentworth and that has undergone some modification since. It has the advantages that almost all the terms used are everyday words, and that the size ranges are close to the ones in common usage, yet the actual dimensions are so arranged that they are in a geometric progression.

Classification of Clastic Sedimentary Rocks

Sediment		Grain Size (in mm.)	Rock
Gravel	Boulder	256	
	Cobble	64	
	Pebble	4	Conglomerate
	Granule	2	
Sand	Very coarse sand	1	
	Coarse sand	1/2	
	Medium sand	1/4	Sandstone
	Fine sand	1/8	
	Very fine sand	1/16	
Mud	Silt particle	1/256	Shale or
	Clay		Mudstone

Almost all of the clastic sedimentary rocks are commonplace over much of Europe and the United States, and in centuries past they were widely used as building stones. The White House and the Capitol are both built of sandstone

Fig. 5-12   These water-rounded rocks may someday become a conglomerate. (Photograph by William Estavillo.)

quarried a short distance down the Potomac from Washington, D.C. In the Victorian Era—especially the General Grant period—one of the favorite construction materials was the so-called brownstone—a drab red sandstone that regrettably will long outlast most of us. Many of Europe's celebrated landmarks are made of clastic sedimentary rocks—the castles at Heidelberg and Salzburg and most of the great ducal palaces of Florence are a few from among scores of famous examples. Sedimentary rocks were greatly preferred over granite by builders in those distant days because such stratified rocks split more readily along their bedding planes and also could be worked far more easily with the primitive hand tools of the time.

CONGLOMERATE   These are cemented gravels, and the larger fragments may range in size from boulders with diameters of several feet down to particles the size of small peas (2 millimeters). More commonly than not, the interstices or pore spaces between the larger boulders, cobbles, or gravel are filled with sand or mud and then the whole mass of sediment is cemented together to form a single rock (Figs. 5-12, 5-13).

Breccia is a variety of conglomerate with angular rather than rounded fragments. The same word was used for pyroclastic volcanic rocks in Chapter 4. Here the same principle applies; if most of the large fragments in the rock are angular rather than rounded, the rock is a breccia—the adjective sedimentary or volcanic is usually added to indicate its origin.

SANDSTONE These sedimentary rocks consist of cemented sand grains and, as the table of sediment sizes shows, these are particles whose diameter ranges between 2 and $\frac{1}{16}$ millimeters. Because this size occupies a middle ground of the classification, it is not surprising that gradations exist between sandstones and conglomerate om the one hand and shale on the other.

Sandstones very commonly include shale layers, or beds of sandstone may alternate quite regularly with beds of shale (Fig. 5-14) or with lenses of conglomerate. Pure, well-sorted sandstone, as mentioned before, was often used as a building material before the advent of prestressed concrete or of lightweight aggregate. Quite a number of college campuses are adorned with pseudo-venerable examples of academic gothic—more often than not inhabited by the geology department—which were hewn out of sandstone blocks, and one in particular, at Stanford University, is a reincarnation in tawny sandstone of the Mission Era.

The cement is what determines the degree of induration, or hardness, of sandstones. In some the cement is weak, and individual grains separate readily from their neighbors; in others the cement may actually be tougher than the grains and when the rock breaks it breaks across them. When the cement is

Fig. 5-13 Conglomerate. (Photograph by William Estavillo.)

Fig. 5-14   Beds of hard sandstone alternate with beds of softer mudstone. Tyee formation, of Eocene age, Oregon. (Photograph by Parke Snavely.)

soluble it may dissolve readily and then the rock may seem to melt away, leaving a residue of sand grains behind.

Compositional differences affect the appearance of sandstone, too. Among the innumerable kinds of sandstone, two leading varieties are *arkose* and *graywacke*.

Arkoses are sandstones that are made up dominantly of quartz and feldspar grains, and therefore commonly are red or pink. As a rule, their grains are moderately angular, and their porosity may be high. Arkoses typically result from the erosion of granitic rocks, and for their formation they also require relatively rapid transportation and deposition without too much abrasion and rounding of the individual sand grains.

Graywackes were originally named for distinctive sandstones in the Harz Mountains of Germany. They are darker than arkoses, and although they commonly contain quartz and feldspar minerals, they have a much higher content of rock fragments—chiefly of the darker varieties of igneous and metamorphic rocks. These, too, are quite angular and unweathered, but unlike arkoses these sand-size particles are set in a clayey or silty matrix which at the time of deposition was essentially a muddy or clayey paste. Characteristically, these are dense, tough, well-indurated rocks whose colors are dark green, or gray, or black. Some graywackes appear to have been deposited in the sea, close in to a steep mountain range, and to have been in an environment where muddy water was carrying a large volume of sediment, including sand, which was moved but a short distance from its source and deposited so rapidly that little weathering and rock decay occurred.

This statement of the general characteristics of graywacke and its origin covers many of the points where a measure of agreement exists. The term is an unsatisfactory one, however, because a score of definitions now exist and almost as many interpretations of the origin of graywackes have been made as there are geologists engaged in studying these perplexing rocks.

SHALE   This is a fine-grained rock whose original constituents were clay flakes and silt particles, and typically is now a laminated rock that splits readily into thin layers. Shale is an ancient term in our language; it comes from the Old English word *scealu*, meaning scale or shell. In geological terminology when we use a word as ancient as this it usually means we are dealing with a property so distinctive that it was recognized early enough to make its way into the rootstock of our native tongue.

Since these are rocks made of clay flakes and of individual mineral grains or rock particles less than $\frac{1}{16}$ of a millimeter in diameter, few of the constituents can be distinguished by the unaided eye. Under the microscope they can be resolved, and most shales are made of minute fragments of quartz, feldspar, and mica, and of rock fragments along with the ubiquitous clay flakes. Despite the small size of their individual grains, these are most important rocks since shales constitute very nearly half of the total of all sedimentary rocks.

Many shales are shades of dark gray or even black, especially if they contain organic matter. Other shales are dark red or green or parti-colored, depending on their iron content or upon the presence of other kinds of coloring matter.

Although *fissility*, or the ability to split along well-developed and closely spaced planes, is a leading property of shales, it is by no means characteristic of all of them. Some varieties whose composition and grain size appear to be comparable are not fissile at all, but break in massive chunks or small compact blocks. These are best given the descriptive name of *mudstone*.

Precipitated Sedimentary Rocks

In addition to the clastic rocks consisting of fragments and of mineral grains derived from pre-existing rocks, there is a second large clan of sedimentary rocks made of chemically precipitated materials. In the following pages these chemically formed rocks are discussed according to their composition as well as according to their mode of origin. This, unfortunately, makes for confusion since some varieties of rocks—specifically, the carbonates—may have similar compositions but unlike origins and thus of necessity the same term appears more than once in the classification.

EVAPORITES These are rocks that result primarily from the evaporation of water which contained dissolved solids. As the water becomes concentrated these ions separate out from solution until a crystalline residue is left.

Most familiar of all such rocks is *salt* (NaCl). Commonly it is formed when evaporation in an arm of the sea dominates over the inflow of water from outside. Judging from some of the renowned salt deposits of the world this must have been a process repeated many times over in order to have built up the great thicknesses that are found. The evaporation of an inland waterbody, such as Great Salt Lake, can also produce the same result, as anyone knows who has seen the nearby Bonneville Salt Flats—widely known for the ideal surface they provide for speed trials.

Layers of salt deposited in the geologic past are sometimes interbedded with other sedimentary rocks, and where these are near the surface, salt springs or "licks" may be found. From earliest times salt has been a highly prized commodity. Today we take it for granted, but in ancient times men gave their lives in battle to win control over salt deposits or to seize the trade routes over which it moved. Famous among these historic deposits were those of northern India—the locus of a flourishing trade before the time of Alexander—as well as those of Palmyra in Syria from whence salt moved by caravan to the Persian Gulf. The salt mines of Austria are deservedly famous, and in the Salzkammergut region around Salzburg they were in operation at least as early as 2000 B.C.

Gypsum ($CaSO_4 \cdot 2H_2O$) is closely related to salt in its origin. Like a great deal of the rock salt of the world, it, too, is a product of the evaporation of sea water. Gypsum is less soluble than salt and thus is precipitated earlier when sea water is evaporated. Along with it is also found an anhydrous (water-lacking) calcium sulphate ($CaSO_4$), *anhydrite*. Both gypsum and anhydrite come out of solution when about 80 per cent of the sea water has evaporated, and salt appears when 90 per cent has gone. Following the precipitation of salt, the very soluble halogens appear in such forms as NaBr (sodium bromide) and KCl (potash).

According to Pettijohn the evaporation of a 304.8-meter (1000-foot) column of sea water would leave a residue of 2.8 meters (9.4 feet) of gypsum and anhydrite, 3.6 meters (11.6 feet) of salt, and .9 meters (3 feet) of potassium and magnesium-bearing salts. Considering the fact that gypsum and anhydrite make up strata many hundreds of feet thick in West Texas and New Mexico, an immense quantity of sea water must have been evaporated there in the geologic past. This is not meant to imply that an ocean thousands of feet deep was dried up leaving a thin layer of gypsum behind. This is an unreasonably difficult answer to the problem, nor does it take care of the question that immediately arises were simple evaporation to be the answer, and that is the absence of the extensive bodies of salt that should be associated with the gypsum beds.

An explanation advanced by P. B. King for the gypsum of West Texas is that water in a shallow, sun-warmed lagoon might reach the temperature and concentration where calcium sulphate would be precipitated, and this would be settled out, allowing the NaCl-rich residual water to flow back to sea before the stage would be reached where salt would come out of solution. Then more gypsum-carrying water could come in again and the process would be repeated. Were such a basin to be a subsiding one, an immense thickness of evaporites could accumulate without the water necessarily being deep. Studies by L. I. Briggs show that the saline deposits of Michigan may have been formed by a continuous inflow without the necessity of emptying and refilling if the proper balance were maintained between evaporation and influx of sea water.

There are many other kinds of evaporites, of minor significance in volume but of major consequence economically. Among these are *borax* ($Na_2B_4O_7 \cdot 10H_2O$), a compound of sodium-boron-oxygen and water, and *potash* (KCl), both of which are found in lakes or lake deposits of desert regions, such as the Mojave Desert in California.

CARBONATE ROCKS These are rocks that are chiefly compounds of calcium or magnesium with carbonate, generally in the form of calcite ($CaCO_3$) or dolomite ($CaMg(CO_3)_2$). These two rocks also have an organic origin as well, but the particular varieties described here appear to be primarily chemical deposits.

Travertine is a good example of a limy deposit that appears to have been deposited from spring waters saturated with calcium carbonate. It is of no great geologic significance, but plays a disproportionately large role in human affairs since it is so greatly favored as an architectural material. It is soft and readily worked, has an interesting array of colors—generally pale yellow or cream colored if pure, brown and darker yellow if it contains impurities—and often shows pronounced banding in wonderfully complex, curving patterns. *Tufa*, or *calcareous tufa* as it is sometimes called to distinguish it from volcanic tuff, also forms in springs and lime-saturated lakes, although to some degree its deposition seems to be fostered by the work of lime-secreting algae. Tufa and travertine when cut and polished make a building stone much favored for the lobbies of banks, building and loan associations, and the large railway terminals of a past era. Great quantities of tufa are imported from Italy and, as might be anticipated, much of monumental Rome is built of tufa, including Bernini's columns that nearly encircle the piazza in front of St. Peter's.

In dry countries, such as West Texas, the ground surface may be mantled with a crust-like cap of lime rock known as *caliche*. This was precipitated through the evaporation of ground water carrying $CaCO_3$ in solution which was drawn to the surface by capillarity.

No unequivocal evidence exists for the direct chemical precipitation of lime-

Fig. 5-15 Bedded chert in the Franciscan Formation, Marin County, California. The layers have been crumpled into nearly recumbent folds. (Photograph by Mary Hill.)

stone from sea water, although a strong case can be made for the snow-like blanket of white, limy ooze on the sea floor of the Great Bahamas Bank. This appears to be a direct precipitate from the shallow, sun-heated, saturated sea water covering this shoal, and is a finely divided, mud-like deposit of microscopic crystals of aragonite, a chemically unstable form of calcium carbonate.

Another curious type of direct limy precipitate is the variety of limestone known as *oölite*. This is a limestone made of minute spherical grains of $CaCO_3$ the size of fish roe, from which it derives its name from the Greek word *oö* for egg + *lithos* for stone. Although the origin of this curious variety of limestone is debated, little doubt seems to remain that it results from the chemical precipitation in water of layers of $CaCO_3$ around a nucleus—perhaps in much the same way that layers of pearl shell are built up.

SILICEOUS ROCKS These are rocks made largely of chemically precipitated silica. A representative, although minor, type is *sinter*. This is a spongy or porous deposit of silica (SiO_2) that accumulates around hot springs or that builds up pedestals at the base of active geysers, such as those at Yellowstone.

A far more widely occurring siliceous rock is *chert*, a name serving as a blanket to cover a host of varieties of very dense, hard, nonclastic rocks made of microcrystalline silica. One familiar form is *flint*, which occurs in dark-colored siliceous nodules. These very often are found embedded in limestone. Since flint is uniformly textured, has a conchoidal fracture much like obsidian, and is easy to chip and at the same time retains a sharp edge, it proved to be the ideal strategic material for arrow- and spear-points in the Stone Ages of Europe and the east and central United States. In what was perhaps a braver day than ours, flints were essential for survival on the frontier, not only to strike sparks from steel for fire but also to fire the flintlock gun of the eighteenth and nineteenth centuries. Red varieties of the same rock commonly are called *jasper*.

Sometimes chert is found by itself in bedded deposits, thin-bedded as a rule and generally dark-colored. These chert beds are composed of very dense, closely fractured rocks that break up readily into small angular blocks (Fig. 5-15).

The origin of chert remains a vexatious problem. Much of the difficulty may be explained on the same ground as the unlike interpretation of the elephant which was touched by the blind men in the fable, each one of whom held a different part. Unquestionably chert formed in more than one way and this may make for spirited arguments between people holding different views.

Among the preferred hypotheses are (1) that chert forms from direct chemical precipitation of SiO_2 on the sea floor, (2) that the silica is introduced after the rocks in which it is found were deposited and that this silica brought in by solutions has *replaced* parts of the original host rock. This is a process much like the one involving the replacement of woody fibers by silica in the making of petrified wood.

The source of the free silica from which the cherts are made is not clear. In part it may be supplied by springs on the sea floor, in part from magmatic sources such as submarine lava flows, or possibly from silica leached out of beds of volcanic ash, or from layers of organically formed silica, such as strata containing shells of microscopic marine plants as the *diatoms* or animals as the *radiolaria*, or finally from the weathering of silica-rich rocks.

Organic Sedimentary Rocks

These are rocks that are made of the remains of organisms, both animals and plants. *Coal* is an excellent illustration since it consists of partially decomposed remains of land plants. Much coal contains finer plant remains, such as spores, in spite of the popular view that it is a chaotic jumble of fallen trunks and

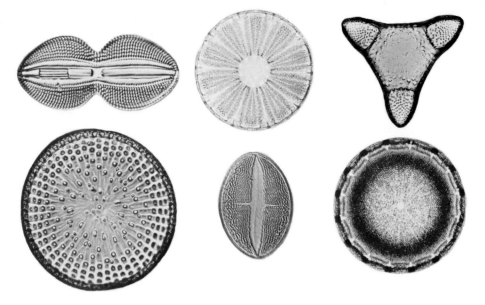

Fig. 5-16 Diatoms, greatly magnified. The average size of these siliceous plant remains is about 50 microns—that is, about 2/1000 of an inch. (Photograph by G. Dallas Hanna, California Academy of Sciences.)

tangled roots which were once set in a miasmic marsh peopled with monsters winging their way through a canopy of bizarre trees or slithering over the floor of the swamp.

With loss of hydrogen, coal moves along a progression from lignite (brown coal) to bituminous, to anthracite, and finally to graphite or pure carbon (these last two forms are regarded by many as metamorphic rocks).

The most abundant of the organic sedimentary rocks is limestone, and probably most examples of this particular rock are truly organic rather than being chemically precipitated. Some limestones have been built up by organisms as lime-secreting algae or the patient coral—builder of the great calcareous edifices of the tropic sea and whose greatest monument, the Great Barrier Reef, stretches for 1930 kilometers (1200 miles) along the coast of eastern Australia.

Beyond any reasonable doubt, some limestones are made of the tiny skeletons of such animals as coral, still preserved substantially in the positions of growth, or of $CaCO_3$ deposited directly by other lime-secreting organisms. Other limestones consist of fragmental calcareous debris and are comparable in many respects to sandstone, only the grains here are small pieces of fossil shells or fragments of coral rather than quartz or feldspar. Such clastic limestones, by their very nature, are likely to grade into limy shale on the one hand with increasing muddiness of the original sediment and into calcareous sandstone on the other if significant amounts of sand were present.

An interesting and perplexing accompaniment of many limestones is the very closely related rock, *dolomite*, which is an example of a monomineralic rock since it consists of the mineral dolomite, $CaMg(CO_3)_2$. Both limestone and dolomite look very much alike; the most practical field distinction between the two is that limestone will effervesce, or fizz, if cold hydrochloric acid (HCl) is dropped on it, while dolomite remains inert.

Limestone grades imperceptibly into dolomite when increasing amounts of magnesium enter into its composition. In some places dolomite occurs as widely spread layers or beds interbedded with ordinary limestone strata. In other occurrences, dolomitic masses cut across limestone layers or follow fracture patterns cutting the limestone in very much the same fashion that some hypabyssal igneous rocks do. For such dolomite masses, the belief is rather widely held that they are the result of partial replacement of calcite in the limestone by magnesia-bearing solutions. For the interbedded, rhythmically alternating layers of limestone and dolomite the evidence is less clear-cut. Some geologists believe that the dolomite layers were precipitated directly on the sea floor. Others take the view that the dolomite layers represent selectively replaced layers of limestone, and here there is an opportunity for further debate: (1) was the original limy material replaced by magnesia very shortly after deposition, or (2) did this chemical alteration occur long afterward when the limestone was completely lithified? To none of the queries can an absolute yes-no answer be given, but then geology would not be much of a challenge if there were no problems left to solve.

An interesting, although relatively minor, type of organically derived sedimentary rock is *diatomite*. Typically, this is a finely laminated, light-colored, sometimes brittle shale which includes myriad remains of diatoms. These are microscopically ornate, single-celled plants that proliferate by the uncounted millions in the surface waters of the colder seas of the world. This floating pasture of nearly invisible protoplasm is the chief food supply for the far-ranging Antarctic whales.

Unlike the plants with which most of us are familiar, these minute, free-floating, single-celled organisms are encased in tiny shells shaped much like an old-fashioned round pillbox and made of glass-like silica extracted from sea water (Fig. 5-16). When these plants die, their microscopic remains sift down through the water to accumulate on the sea floor. There they ultimately harden into shale with an above-average silica content. Such organically derived shales are typical of the central part of the California Coast Ranges and of some of the lands bordering the Mediterranean, where this type of deposit, to which we give the name of diatomite, is called tripoli.

Chalk is roughly comparable in its origin, in that it consists of organically formed calcite, and is a relatively pure deposit containing the remains of the minute, free-floating, single-celled animals, the *foraminifera*, whose tiny shells are made of calcium carbonate. Some chalk deposits, such as those near Dover, England, are 100 million years old, and the truly remarkable thing about them is how little alteration or recrystallization they have undergone in all this time.

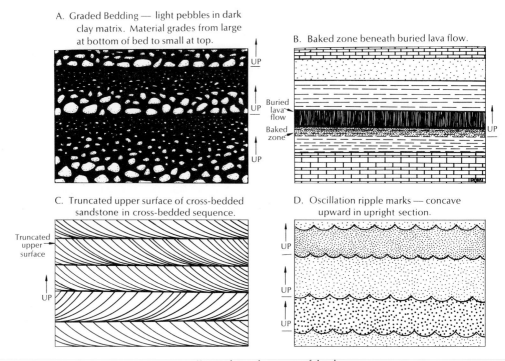

A. Graded Bedding — light pebbles in dark clay matrix. Material grades from large at bottom of bed to small at top.

B. Baked zone beneath buried lava flow.

C. Truncated upper surface of cross-bedded sandstone in cross-bedded sequence.

D. Oscillation ripple marks — concave upward in upright section.

Fig. 5-17   Some ways to tell top from bottom of beds.

Which Way Is Up?

Basic Assumptions

In order to reconstruct the history of an area which contains sedimentary rocks, the geologist assumes two statements to be true. One has already been mentioned in connection with relative geologic time, the assumption of superposition. This states that in a sequence of layered rocks, the oldest are on the bottom and the youngest are on top. This assumption is, in turn, based upon another which states that sedimentary strata were originally deposited in a horizontal or nearly horizontal attitude.

Top and Bottom of Beds

Sedimentary beds do not always remain horizontal, however. They may be tilted, turned up vertically, or even overturned. It becomes clear, then, that there must be some way to tell the top from the bottom of a bed. Actually there are quite a few, some of which are diagrammed in Figure 5-17.

Sedimentary Facies

After a geologist has examined a rock in the field, it is part of his challenge to determine as much as he can about its environment of deposition. For example, he might break off from a cliff face a piece of conglomerate. This specimen consists of pebble-sized, well-rounded pieces of other types of rock, in a matrix of fine, clean quartz sand, and the whole is cemented with calcium carbonate and contains a few shell fragments. All this suggests strongly a beach environment of deposition. The size of the pebbles would indicate an origin fairly near land, probably land with considerable relief. The roundness of the pebbles combined with the cleanness of the sand suggests a beach where constant wave action wears off the sharp corners and washes away mud-sized particles, leaving the quartz sand behind. The calcium carbonate cement indicates a marine origin as it was probably derived from dissolved shell fragments, some of which still remain.

Suppose that we turn from the hand specimen now and observe the whole cliff. We can trace this conglomerate bed a considerable distance, and as long as it maintains this same appearance and composition, we can refer to the conglomerate as a *sedimentary facies*. But as we get farther away from our starting point, the rock begins gradually to change. There are fewer and fewer pebbles, and a greater proportion of the rock is sand until we have a nearly pure quartz sandstone. This would be another facies, and the transformation of the one rock type to another is called a *facies change* (Fig. 5-18). Although the conglomerate and the sandstone are the same age, their appearance and composition are quite different. It seems, logical, then, to infer that their environments of deposition were also different, and such is the case. For seaward from the beach conditions under which the conglomerate was deposited, we might find that wave action on the bottom is much less, and the pebbles cannot be carried that far from their original position. At a certain point only sand can be shifted back and forth, and this would be the changed environment that gave rise to our sandstone. Still farther along we might expect to find a shale with its finer and lighter particles which could be carried to that greater distance from the land.

Fig. 5-18   Cross section showing facies change in sedimentary rocks.

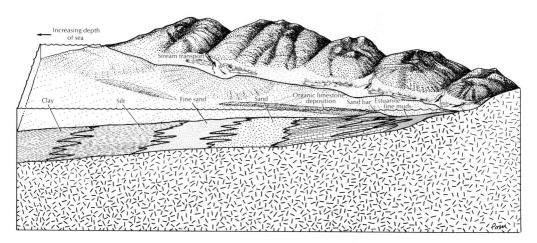

Fig. 5-19 "But beauty is in the eye of the beholder, George—and they can't see Hell's Canyon now. Give 'em a beer and a motorboat and they'll forget anything."

The important thing to remember is that all these rocks were deposited at the same time. A vertical change in rock type indicates a change in environment of deposition with time, a lateral facies change indicates a change in environment of deposition with location.

Selected References

Dunbar, C. O., and Rodgers, John, 1957, Principles of stratigraphy, John Wiley and Sons, New York.

Laporte, Leo F., 1968, Ancient environments, Prentice-Hall, Inc., Englewood Cliffs, N.J.

Pettijohn, F. J., 1959, Sedimentary rocks, Harper and Bros., New York.

Shrock, Robert, 1948, Sequence in layered rocks, McGraw-Hill Book Co., New York.

Fig. 6-1 Differential weathering of interbedded mudstone and sandstone. The more resistant sandstone forms the ledges and the cap. Egyptian Temple, Capitol Reef National Monument, Utah. (Photograph by J. R. Stacy, U.S. Geological Survey.)

6
Weathering and Soils

In the days of the Pharaohs a cherished status symbol was the obelisk. These hieroglyph-bedecked stone columns early became collector's items for a procession of conquerors of the Nile, beginning with the Caesars and ending with Napoleon. Or more properly speaking, ending with us, because we collected an obelisk in 1879. After prodigies of effort, involving among other things the cutting of a loading port in the bow of one of the primitive steamships of the time (whereupon it nearly foundered), our obelisk was finally set up in Central Park (Fig. 6-2) to take its place among similar far-wandering artifacts in cities such as Paris, London, and Rome—which alone holds twelve of them.

The climate of New York is considerably more stimulating than that of Egypt. A mixture of cold winters, sleet- and snow-laden winds, and hot, steamy summers, in addition to an atmosphere laden with coal smoke and gasoline fumes, is bound to add a certain amount of zest. No wonder that in about seventy years many of the hieroglyphs spalled off and the whole surface of the obelisk started to come apart, while its counterparts still standing in Egypt have survived nearly unscathed beneath the desert sun for almost 4000 years (Fig. 6-3).

This brief story makes the point that climate is one of the leading factors in determining the rate and manner in which rocks disintegrate or decompose, or as we say *weather*—using the word in about the same way we do when we speak of a weather-beaten face. Another critical factor in determining the ef-

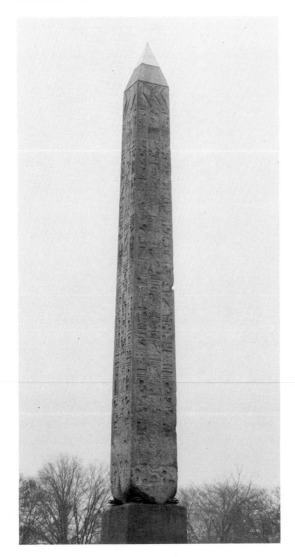

Fig. 6-2 The obelisk of Thothmes III, from the Temple of Heliopolis, Egypt, now in Central Park, New York. The lower part of the column shows a loss of detail due to weathering. The monument was brought to New York in 1879. (Courtesy of the Metropolitan Museum of Art.)

fectiveness of weathering is the kind of rock which is exposed to atmospheric attack. Evidence for this can be found in New England graveyards, where slate headstones carrying the salty epitaphs beloved by some of our forebears survive from the 1700's, while the words carved on limestone or marble markers of much more recent vintage may be wholly obliterated.

In general, limestone and marble are more susceptible than slate to weathering, because they consist of the soluble mineral calcite ($CaCO_3$). Many slates are made from recrystallized clay flakes, which are about as durable as any material can be. However, as with most general statements, there is the inevitable exception. Limestone and marble are extremely resistant rocks in an arid climate, where there is a lack of moisture or of plants to provide a source for carbonic acid (H_2CO_3). Limestone also may hold up surprisingly well even in a moderately semi-arid climate, as is demonstrated by Mayan temples and carvings, overgrown these many years by the jungles of Yucatan.

Often different types of rocks are interbedded, and their differing resistances to weathering can be conspicuous. The resistant rocks will form steep-fronted ledges, while the less resistant will form gently sloping surfaces between the ledges. This is known as *differential weathering*. Resistance to weathering is determined in part by the rock's chemical composition and in part by the ease with which it disintegrates. As mentioned above, the resistance of any given rock will depend also upon the climate in which it occurs, so that in Figure 6-4 the ledge-forming beds might be limestone in an arid region, and sandstone in a humid area. Basalt layers are very resistant and often form the capping of

Fig. 6-3 The surface of this obelisk, still standing in Karnak, Egypt, is scarcely marred by four millennia of weathering. (Photograph by Jean B. Thorpe.)

Fig. 6-4a   Differential weathering in an arid climate. Limestone and sandstone are the cliff-formers; shale is a slope-former, often covered by talus.

Fig. 6-4b   Differential weathering in a humid climate. Here, too, sandstone is a cliff-former, but limestone weathers by solution to form irregular slopes. Again, shale is a slope-former, often covered by a thick soil horizon.

hills because everything above them has been removed by erosion while the basalt protects the underlying layers.

Every observant traveler has probably noticed how different the surface of the ground appears in various parts of the world. In some, the soil may be dark colored and deep, in others there may be only the thinnest veneer of sterile, stony soil and original colors of the rocks dominate, as they do throughout the deserts of the southwestern United States.

Climate is perhaps the leading, although far from being the only, reason for many of these differences. In humid, warm regions where vegetation flourishes and organic acids are abundant, chemical processes are dominant and rocks are prone to *decompose*, or to decay. In harsher climates where frost action may dominate, rocks break up mechanically, or *disintegrate*, without undergoing chemical alteration. In other words, when rocks decompose, they are changed into substances with quite different chemical compositions and physical properties than they started with. If they disintegrate, they are simply broken up into smaller fragments, much as if they had been struck a hammer blow. There are few areas where only chemical weathering or only mechanical weathering operates to the exclusion of the other process, but there are many where one or the other rules.

Decomposition or disintegration of rocks produces a mantle of rock particles on the surface of the earth thick in some places, thin in others, or even totally lacking in still others. Such a mantle is called the *regolith*, and the formation of a true soil is an end-product of its development. In fact, soil is by far the most valuable of all the mineral resources of the earth. Without it, life—such as we know it—would be impossible. While we could survive without a number of other substances such as gold or diamonds, which are admittedly more attractive than plain, ordinary dirt, we would never make it without the latter.

Mechanical Weathering

Some aspects of mechanical weathering are irritatingly familiar to all of us, such as the wedging apart of sidewalks, foundations, and walls by the roots of grass, trees, and shrubs (Fig. 6-5). The same thing goes on in mountains, and a common sight high on their slopes is an isolated pine clinging to a sheer granite ledge. With no soil in which to take hold, its roots succeed in forcing their way into crevices, springing the rocks apart. This is a process much like the one used millennia ago by Egyptian slaves, when they utilized water-soaked wooden wedges to pry out granite blocks to make obelisks.

Almost all rocks are cut by cracks, large and small; sometimes as closely spaced as a fraction of an inch, at other times they may be scores of feet apart, as they are in the stupendous cliff of El Capitan in Yosemite Valley. These cracks are called joints, and they provide an ideal path for roots, organic acid-bearing solutions, and water to penetrate far into the rocks (Fig. 6-6).

Freezing and Thawing

Water is a highly unusual substance. The property of the greatest significance to us here is the expansion water undergoes when its temperature drops from 4° C. (39.2° F.) to 0° C. (32.0° F.), its freezing point. Water expands by about 9 per cent when it is chilled in this range.

Should water freeze in a confined space, then it is capable of delivering an enormous outward thrust against its containing walls. Everyone who has glumly contemplated a cracked engine block or ruptured radiator knows this fact too well. Few realize, though, how great this force actually is. At the minimum, it probably is as least 907 kilograms (2000 pounds) per square inch. No wonder the need for repairs in the wake of water frozen in the plumbing can devastate a household budget.

This outward pressure continues to build up at temperatures below 0° C. (32° F.), since a certain amount of expansion continues to take place at subzero

Fig. 6-5   The results of root wedging at Angkor Wat, Cambodia, where a temple built of laterite blocks is gradually being destroyed. (Photograph by Leonard Palmer.)

temperatures, at least down to $-22°$ C. $(-7.6°$ F.). Here a pressure with a theoretical maximum of 13,605 kilograms (30,000 pounds) per square inch is possible. Probably this figure is never reached, because few rocks could stand up to such a pressure without rupturing. In addition, for such a tremendous stress to build up it is necessary to have a completely enclosed system with air excluded, and this is seldom realized in an environment such as a crack in a rock.

Actually the pressure made available by the freezing of water is enough to sunder most rocks exposed in high mountains. Water freezing in an open crevice freezes from the top down and thus is sealed in with a cover of ice. Then, if confined in the lower part of the crevice, it can act as a wedge to spring rocks apart along planes of weakness such as joints.

The process of *frost wedging* is most effective when it is repeated the greatest number of times. In other words, if the temperature swings from 3.9° C. to 0° C. (39° F. to 32° F.) each day, the amount of frost riving will be many times greater than in the Arctic, where everything is in a deep freeze through the winter.

For this reason, freezing and thawing reaches its peak effectiveness above timber line on high mountains outside of the Arctic. Temperatures on them rise above the melting point by day and drop below at night. As a result summit uplands may be carpeted with a pavement of frost-shattered angular joint blocks (Fig. 6-8). Sometimes these are packed so tightly together they resemble an artificial pavement, and in fact will support a light plane for landings and takeoffs.

The accumulation of joint blocks found in a long apron at the base of a steep

Fig. 6-6   Vertical and horizontal joints in granite in the Sierra Nevada, California, provide avenues for moisture, which may aid in mechanical disintegration when it turns to ice, and also enable plants to send down their roots, which wedge apart blocks. (From the Cedric Wright Collection, courtesy of the Sierra Club.)

Fig. 6-7   Weathering along joints. Point Loma, California. (Photograph by William Estavillo.)

slope is called *talus* (Fig. 6-9), a term borrowed from medieval military engineering for the slope at the base of a fortification wall. In Scotland such an accumulation is called a *scree*, and this incisive word, derived from Old Norse, occasionally is used in this country (Fig. 6-10).

Should water freeze in the pore spaces, or interstices of a soil, a surprising amount of damage may result. The principal effect of the growth of soil ice is a phenomenon known as *frost heave*. This is a familiar problem to people in northern lands, even though it may manifest itself in forms of no greater severity than garage doors that stick in winter. Frost heave may cause serious damage to concrete which contains water. Should this water freeze, it may lead to the breaking up of such things as runways, roads, and foundations. Agriculturally, frost heave is an unmitigated nuisance in areas with stony soil, such as New England, where each year it seems almost as though a new crop of boulders had been heaved out of the ground.

Frost heave is more effective in finely textured soils, such as clay and silt, than it is in coarse ones such as sand and gravel. For frost heave to operate, more is required than having the water initially contained in the ground freeze.

To be more effective the ice should continue to grow in the ground; this is best achieved when water is added continuously and then frozen. This can be accomplished best where the pore spaces or interstices are thread-like, because then the force known as *capillarity* can operate. All of us are familiar with its workings—it is, for example, the force that draws water up into a sponge, or ink into a blotter. As an illustration of the effectiveness of capillarity, water will rise about .3 meters (1 foot) in a glass tube 1 millimeter in diameter, but if the diameter is reduced to 0.1 millimeter, it can rise about 3 meters (10 feet).

Ice wedging can be a most destructive process. Silt layers are wedged apart, the surface of the ground is heaved up differentially, and buildings, roads, etc., are cracked or thrown out of alignment.

Bad as this may be, things rapidly become worse in the spring when such ice wedges melt. Then the ground literally falls apart because the binding or cementing action of the ice is destroyed. A thawed silt becomes spongy, and if it is water-saturated it quickly becomes a boggy morass in which even four-wheel-drive vehicles flounder helplessly.

Permafrost

An important kind of ice widely distributed throughout the Arctic is in ground which remains frozen from one year to the next. The coined name of *perma-*

Fig. 6-8 The summit of Mount Whitney (alt. 14,495 ft.) is composed of frost- and ice-shattered granite which is closely jointed. Most mountain crests above timberline are blanketed with mechanically shattered blocks resulting from frost wedging. (Photograph by Tom Ross.)

Fig. 6-9 Accumulations of ice-shattered joint blocks at the base of steep slopes are termed talus or scree, shown here at the base of Kearsarge Pinnacles, in the Sierra Nevada, California. (Photograph by Tom Ross.)

frost was applied to this ground ice during World War II, and all efforts to substitute terms that sound a little less like a trade name for a refrigerator have been successfully resisted.

Permafrost is more widely distributed than many people realize, since it underlies almost 20 per cent of the *land* surface of the earth, including about 80 per cent of Alaska and perhaps half of Canada. The actual distribution of permafrost is shown on the accompanying map (Fig. 6-11). Interestingly enough, areas underlain by permafrost today are not the same as those glaciated during the Ice Age which presumably ended about 12,000 years ago.

Permafrost reaches its maximum thickness around the margins of the Arctic Ocean in Alaska, Canada, and the Soviet Arctic. It may extend to depths of as much as 305 meters (1000 feet) below the surface of the ground along the northern margins of North America and Eurasia. In a general way the thickness decreases southward until finally it thins to zero about along the southern boundary shown on the map (Fig. 6-11).

Overlying the permafrost is a soil layer in which ground ice thaws in the spring and freezes in the fall to remain frozen throughout the winter. This is the so-called active layer, and in areas where permafrost is widespread it may be only .3 to .9 meters (1 to 3 feet) deep. The upper surface of the permafrost in the ground is called the *permafrost table*.

Below the permafrost table the available pore spaces are filled with ice, and the surface water in the active layer cannot sink underground. This is one reason why much of the tundra in the Arctic is so boggy and water-soaked. Although precipitation over large segments of the Arctic is very slight, the available water stands in lakes and muskegs on the surface because there is a relatively low evaporation rate and no chance for it to sink underground.

Before 1942 little attention was given in this country to permafrost and its manifold problems, although Russians had studied this phenomenon intensively for more than half a century. Visitors to the interior of Alaska, especially in the vicinity of Fairbanks, are invariably surprised at the tilted houses and

Fig. 6-10 A scree slope below a rocky summit in the Sierra Nevada. (From the Cedric Wright Collection, courtesy of the Sierra Club.)

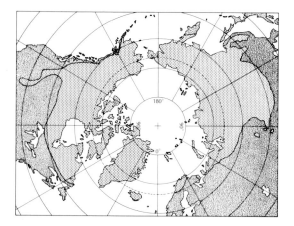

Fig. 6-11 The area of permafrost, or frozen ground (coarse dots), covers about 20 per cent of the land surface of the earth at higher northern and southern latitudes.

the "drunken forests" with their tipsy looking trees, thrown out of line by melting of the permafrost or by upward growth of permafrost into the active zone.

Out of the scores of problems that permafrost can produce, only a few need be mentioned. Heated buildings in the Arctic are likely to thaw out the permafrost beneath them and melt their way down into the soggy, unstable ground under the foundation. Usually such houses sink unevenly, so that floors sag, walls tip, and doors stick. The most practical answer to the problem seems to be to put houses and barracks up on stilts. This allows cold air to circulate under them and makes a minimum of disturbance of the permafrost.

Unheated buildings, especially large ones such as hangars or warehouses, insulate the ground surface below them so that the active layer does not thaw in the summer. The result is that the permafrost table rises, and may constitute a dam which blocks the flow of ground water through the active layer. If the ground water is forced out onto the surface where it freezes, this is the phenomenon known as *icing*. It may be spectacular indeed. For example, the interior of an unheated building into which ground water forces its way may be converted into a block of ice, with ice cascading out of the doors and windows.

Roads and airfields have about the same effect on the permafrost table as an unheated building. If ground water is forced to the surface it may form an ice field, or an "icing" to translate the Russian word, which can block the highway for thousands of meters. Two methods of alleviating this are (1) to extend broad aprons, called *berms*, on either side of the highway or airfield to prevent the permafrost table from rising abruptly under the highway, and (2) to excavate a broad, shallow ditch, perhaps 9 meters (30 feet) wide, on the upslope side so that ground water moving within the active layer will be intercepted. Ice sheets will form harmlessly in the ditch whose embankment keeps them from spreading across the pavement.

The list of problems permafrost can create seems endless; the degree to which

Arctic pioneers have solved them is a testimonial to their ingenuity and perseverance. Even such a simple thing as developing a water supply in a permafrost area can become a major frustration. Ground water in the active layer is available only during the summer, and usually is at so shallow a depth that it is readily contaminated by surface wastes. Although there may be ground water below the permafrost, it is deep, and the part of the well that is within the frozen ground is almost certain to freeze. If water pipes are buried underground, they freeze; if placed above ground, they freeze, too, and are likely to be thrown out of line as the ground under them either heaves when it freezes or sinks when it thaws. Fire fighting is an exceedingly difficult enterprise when all surface water freezes, and if buildings are destroyed by fire their occupants may freeze, too.

The Russians ran into permafrost difficulties when a far northern dam started to leak shortly after the reservoir behind it was filled. The dam was built on volcanic rock whose tiny cracks were permanently filled with veinlets of ice. Ordinarily, ice-filled rock below the permafrost level can be treated as solid rock. In this case, however, the filled reservoir with its insulating layer of ice on the surface acted as a heat trap. The ice veinlets in the rock melted, and the bottom of the dam acted like a sieve. Newer dams built in such cold areas have to be refrigerated by pumping cold air into them to prevent such melting. As the air leaves the dam it may be $10°$ to $12°$ warmer than the atmosphere.

Sewage disposal is perhaps the ultimate problem. Septic tanks freeze, and in the absence of bacteria, decay does not dispose of waste as it does in warmer climates. At Point Barrow the unsightly, but practical, solution is to heap everything up in a pile on an ice floe in the winter. In the summer this cake of ice floats out into the Arctic Ocean, melts, and the waste sinks. Waste disposal will become even more of a problem now that oil has been discovered in northern Alaska. The rapid development of the area will make this method and its attendant pollution intolerable.

Permafrost and the Alaskan oil discovery have also precipitated the controversy over the Alaskan pipeline construction. The Trans Alaska Pipeline System proposes to construct the pipeline from the North Slope petroleum fields south to the ice-free port of Valdez, a distance of about 1300 kilometers (800 miles). The oil flowing through this pipeline will have a temperature of between $50°$ C. and $80°$ C. ($150°$ and $180°$ F.). Obviously where this pipeline is buried in areas of permafrost, difficult problems will arise from melting. The natural equilibrium would be permanently disrupted with disastrous consequences for the ecology in the form of mudflows and other forms of erosion. In addition, differential settling of earth materials would produce stresses on the pipeline itself which might lead to rupture and oil spill problems that would rival in seriousness those that occur at sea. It is vital, therefore, that all the potential problems must be identified *before* the pipeline is constructed, so that the system can be properly designed.

Temperature Changes

In textbooks of a generation ago much was made of the supposed disintegration of rocks resulting from alternate expansion and contraction induced by severe temperature changes. The favorite locale for such a performance was the desert. There, according to most versions, rocks expanded drastically under the noonday sun and contracted sharply with the fall of temperature at night. Presumably these dimensional changes would be greatest on the surface of a rock and least in its interior, because rocks are such notoriously poor conductors of heat. Such a process, called *exfoliation* from the Latin *exfoliatus*— stripped of leaves—was believed adequate to explain the onion-layered appearance of many rocks.

There is no doubt about the existence of exfoliated rocks; their number truly is legion, but there is uncertainty about the way in which they are formed. Peeling off of concentric rings by differential expansion of the heated surface layers from the cooler interior is an appealing solution, but there are a number of difficulties. For example, in deserts such as the Sahara and Arabia, stone monuments and buildings have survived for 4000 years with scarcely any blurring of their inscriptions.

Perhaps the most conclusive evidence that temperature changes alone are incapable of disrupting rocks comes from an interesting experiment made by D. T. Griggs. He placed a highly polished block of granite under a heat source so arranged that for 5 minutes the heat was on and then for 10 minutes the surface was air cooled by a fan. This ran the surface of the granite through a temperature range of $110°$ C., and the process was repeated 89,400 times, or the equivalent of 244 years of weathering, should each of these 15-minute cycles be considered a day. In even the fiercest desert the diurnal range is far less than $100°$ C. ($212°$ F.); if we were to consider the experiment in more realistic terms, 1000 years of actual weathering were more closely approximated.

What happened as a result of the punishment to which the rock was subjected? Nothing. The surface remained unblemished, retaining its original bright polish throughout the entire ordeal.

Griggs then introduced a little rain, as it were, into the environment by having a fine spray of water turned on during the cooling cycle. This was done for only 10 days, the equivalent of 2½ years of weathering, and in this brief time a number of notable changes occurred. The granite lost its polish, feldspar crystals clouded up with a film of clayey material on their surface, and exfoliation cracks started to appear. All this occurred in the equivalent of 2½ years, as compared with the absence of visible results after the laboratory equivalent of 10 centuries of total aridity and extreme temperature ranges.

This isolated experiment supports the observations made in Egypt by an American geologist, Barton. He noticed that almost no discernible change was visible on granite inscriptions that faced the sun, while those that were in the shade and thus remained relatively damp showed much more spalling of rock surfaces and hieroglyphs.

The conclusion made from the experiment, as well as from evidence in arid regions all over the world, is that temperature changes by themselves are incapable of making a rock exfoliate, but that water plays an important role in the process. A plausible explanation is that in many deserts, no matter how arid they may seem, a little water may be available from sporadic showers or from nocturnal dew. When this water enters into chemical combination with the more susceptible minerals in a rock they swell. It is this increase in volume resulting from *hydration* that gives the needed shove to lift off the outer layers of a rock in concentric shells. To summarize, exfoliation appears to be essentially a mechanical or disintegrative process, accomplished, however, by a chemical means—hydration.

Chemical Weathering

Chemical changes dominate in hot and humid lands where temperatures are high, a large amount of water is available, and vegetation flourishes. Organic acids, which are potent agents of rock decay, are readily generated and rocks that can stand up to their onslaught are rare. Carbonic acid (H_2CO_3) is a common acid of this type, and it results from a union of water (H_2O) and carbon dioxide (CO_2).

Of the manifold processes involved in chemical weathering, four of the more important are: (1) solution, (2) oxidation, (3) hydration, and (4) carbonation.

Solution

Among the multitude of rocks exposed on the earth's surface, few are more susceptible to chemical attack than limestone; slow disappearance through dissolving into solution is more likely than not to be its fate. This simple equation illustrates the process:

$$CaCO_3 \quad + \quad H_2CO_3 \quad = \quad Ca(HCO_3)_2$$

| calcite | carbonic acid | calcium bicarbonate |

Calcium bicarbonate is soluble and, once introduced into water, above or below ground, is carried away in solution. Thus where a layer of limestone may once have been, there may well remain nothing, because its chief constituent, calcite, is now dissolved away. This explains the profusion of caverns, underground channels, and disappearing rivers in limestone regions.

Oxidation

Rusting is a process familiar to most of us. In anything but the most severe climates, such as the Antarctic Ice Sheet, all unprotected objects made of iron rust away within a lifetime. In a rainy tropical climate the struggle to maintain steel bridges, ships, rails, and automobiles is a relentless one.

The preponderance of rocks contain some iron-bearing minerals. When these are exposed to atmospheric attack, like an old Model T frame in an automobile graveyard, they rust. Such iron-rich rocks lose their original gray, and are stained a wide variety of colors, such as red, yellow, orange, red-brown. The equation expressing this change is:

$$4FeO \quad + \quad O_2 \quad = \quad 2Fe_2O_3$$

ferrous iron oxide   oxygen   ferric iron oxide

(gray-green)   (rust-colored)

Geologists of a generation ago were prone to equate the origin of red soil and rock colors with arid regions. Of course, red-colored rocks are dominant in such landscapes as Monument Valley, Grand Canyon, and the Painted Desert, and these are unquestionably desert areas. The arid climate tends to preserve the iron oxides in the rocks, and the lack of moisture and vegetation means that organic compounds do not accumulate and thus carbon is not present to reduce (deoxygenate) the ferric iron oxide (Fe_2O_3) and change it back to drab, grayish-green ferrous iron oxide (FeO).

However, the red color of these sediments depends on their origin, not the accident of their preservation in a dry climate. Many of the red or reddish-brown sediments that are accumulating today are in tropical lands, especially those with pronounced wet and dry seasons, so that the ground is alternately thoroughly wetted and then dried out. We shall see the importance of this when we come to the discussion of lateritic soils.

Carbonation and Hydration

The two processes of chemical combination of minerals with carbon dioxide (CO_2) and water (H_2O), or carbonation and hydration respectively, are combined in this one reaction:

$$2KAlSi_3O_8 + 2H_2O + CO_2 =$$

feldspar   water   carbon

dioxide

$$H_4Al_2Si_2O_9 + 4SiO_2 + K_2CO_3$$

clay   silica   potassium

carbonate

This is one of the important reactions in nature since it involves the most abundant of all minerals, feldspar, and such standard reagents as water and carbon dioxide. The end-products of the reaction are interesting because of their role in soil formation, as well as the part they play in agriculture and life processes in general. The equation shows that the chief result of the decomposition of feldspar is the formation of a clay mineral (kaolin). Clay is about as common a soil element as there is, and the formula for this particular variety shows it to be a hydrous aluminum silicate. An interesting feature of its compo-

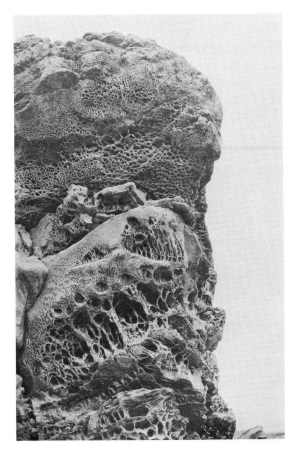

Fig. 6-12 Chemical weathering of a limy sandstone. Yellowstone River, Montana. (Photograph by C. D. Walcott, U.S. Geological Survey.)

sition is the presence of aluminum, a most abundant metal in the earth's crust. Why, then, is it not less expensive and also more commonly used? The answer lies in its link with silica in the clay. Since this bond is one of the more difficult in nature to break, an immense amount of energy is required to achieve such a separation. Most aluminum actually comes from clay which has been subjected to weathering so intense that the silica has been leached out and only aluminum oxide (Al_2O_3), or alumina, is left. This is the mineral bauxite, and many of the world's commercial deposits seem to have originated in tropical lands which have about the same climatic regimen as is responsible for the iron-bearing soil known as laterite, described further on.

Along with clay, the carbonation and hydration of feldspar frees silica which may remain dispersed throughout the deposit. Since the potassium carbonate is soluble, most of it is carried away in solution. Not all the potassium is removed from the site, however, because some of it is taken up by plants and some probably enters into combination with the clay.

The decomposition of feldspar yields a wholly different product from the original—in this case a sandy clay, quite unlike the strongly crystalline, rock-forming mineral from which it is derived.

Fig. 6-13 Disintegration of granite in the desert region of Joshua Tree National Monument, southeastern California. A coarse granitic sand composed of individual mineral grains accumulates at the base of the rocks in this dry region because chemical weathering is greatly reduced and the effects of solution, carbonation, and hydration are largely minimized. (Photograph by Ansel Adams.)

Decomposition of Granite

The way in which granite decomposes is a good illustration of the way chemical weathering affects rocks. Granite, as we learned in Chapter 4, is an intrusive igneous rock with a coarse-grained texture and consists of quartz, feldspar, and ferromagnesian minerals, such as hornblende and biotite.

Typically the feldspars decompose to sandy clay and soluble potassium-, sodium-, or calcium-carbonate, depending on their composition. The quartz survives largely unscathed, except under truly severe conditions. The ferromagnesian minerals, including biotite, break down to a rusty clay containing varying amounts of iron oxide, potassium carbonate, magnesium bicarbonate, and silica, dependent somewhat upon the composition of the original mineral.

Thus, in a warm, rainy climate, granite boulders, monuments, and buildings molder away in time to a mass of rust-stained sandy clay, or *loam*. Conversely, granite in a desert (Fig. 6-13) is more prone to disintegrate into separate sand grains, which consist chiefly of feldspar, since this is the dominant mineral; such a feldspathic sand if later cemented together to make a sandstone is called an *arkose*. In short, the same rock, granite, if placed in unlike climatic environments yields wholly unlike products.

Soils

The result of the decomposition and disintegration of rocks and rock-forming minerals is the formation of soil. Because soil is so fundamental to life, it has been studied intensively for more than a century. Unfortunately, perhaps, the problems of the origin, nature, and use of soil have been attacked piecemeal by specialists with widely differing interests, with each discipline using a wholly different terminology. The geologist's concern with soil is primarily with its origin, and his interest is likely to be directed toward establishing the relationship between parent material and soil. For example, what kind of soil results from the weathering of granite in a dry climate, of limestone in a warm humid one, etc.?

Agriculturally, an emphasis in soil classification has been to relate the various types to climatic regions. This appraoch was begun in Russia around 1870 by V. V. Dokuchaiev, K. D. Glinka, and their successors. A climatically dominated origin for soils is an understandable emphasis in such a land as Russia with its endless sweep of open steppe. In other words, the latitudinal zoning of climates across such broad plains is of greater significance than local variations in topography or rock type. The Russian concept of the ascendancy of climate as a controlling factor in soil evolution was introduced into this country in the early 1920's by C. F. Marbut, then Chief of the Division of Soil Survey in the U.S. Department of Agriculture.

In a typical *mature soil*, according to Marbut and his associates, three distinct layers should be present in a complete soil *profile*. These layers are called horizons, and in order from top to bottom are the A-, B-, and C-horizons (Fig. 6-14). The surface layer, or A-horizon, has had its soluble constituents very largely leached out or washed out, and is the so-called *eluvial zone*, from the Latin word *eluere*, which means just that—to wash out. Although soluble material very largely has been removed from the A-horizon, organic matter has been added in its place, and in some climates this is quite a bit, especially where soils are black and humus-laden.

Some of the material dissolved from minerals of the A-horizon is carried by percolating water down into the B-horizon, where it may then accumulate. Largely for this reason the B-horizon is sometimes termed the *illuvial zone*, or the layer into which material is washed. Iron oxide is especially prone to collect in the B-horizon, as is finer solid material such as clay. Should enough iron

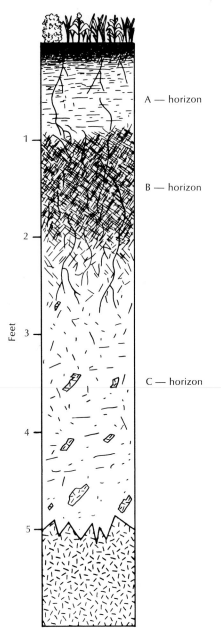

A — horizon

B — horizon

C — horizon

Feet

1

2

3

4

5

Fig. 6-14 A mature soil shows three distinct layers or horizons: A, B, and C. The A-horizon has its soluble constituents leached out, and organic matter is added. The B-horizon is often a zone of accumulation of materials leached downward from the A-horizon, and may form a hardpan composed of iron oxide or carbonate calcium (caliche). The C-horizon is a transitional zone to the parent rock below.

oxide be precipitated, it may form a cemented layer called *hardpan* by farmers, which is hard enough to turn aside an ordinary plow. In some arid regions, as in the Llano Estacado of Texas, a limy encrustation, *caliche*, may build up in the soil. In part it may be $CaCO_3$ dissolved out of the A-horizon and washed down into the B-horizon, and in part it may be lime brought up from below by capillarity and then precipitated in the B-horizon.

The C-horizon is essentially a transitional zone between the true soil horizons above and the unaltered parent material below. Thus it is a mixture in varying proportions of altered and unaltered rock fragments and soil particles. Unaltered bedrock is likely to be dominant near the base of the horizon and maturely weathered soil will be the leading constituent in the upper part. Crevices filled with soil particles characteristically will extend down into the bedrock, while higher up in the C-horizon unaltered fragments of bedrock may be isolated and completely free-floating like seeds in a watermelon. A deep mature soil showing all three horizons can take as much as 100,000 years to develop.

Zonal Soils

Mature soils with fully developed profiles and with characteristics that appear to be determined by the prevailing climate over a wide region are called *zonal soils*. Tropical soils, for example, are markedly unlike those of the Arctic, and even within a temperate climate there are wide differences which are controlled by local patterns of temperature, amount and distribution of rainfall, and nature of vegetation. *Intrazonal soils* are atypical and are not representative of a particular climatic zone. These are soils whose nature is determined by some local environment, such as the bottom of a bog or marsh or similar undrained area.

In a very general way the soils of the world can be placed in two great categories, the *pedocals* and *pedalfers*. Each word is derived from the Greek, *pedo(n)* for ground, plus an abbreviation for calcium in the first of these soil types and the symbols Al and Fe for aluminum and iron in the second. Pedocals are soils which contain such soluble substances as calcium and magnesium, usually in combination with carbonates and sulphates. These are soils which are typical of much of the arid and semi-arid West Beyond the 100th Meridian, and in extreme cases they are the alkali soil which is so lethal for agricultural plants.

Pedalfers are more thoroughly leached soils than pedocals, and occur in the United States mostly east of the 100th Meridian. These are soils in which soluble material does not accumulate, and in extreme examples they may be so degraded that only such insoluble residues as aluminum oxide (Al_2O_3) and iron oxide (Fe_2O_3) survive.

Soils and Economic Geology

Even though soil characteristics are determined more by climate than by the parent rock, some aspects of *pedology*, the study of soils, are definitely relevant to geology.

In 1807, Buchanan Hamilton, an observant Scotsman, was traveling in India, then a virtual monopoly of The Honorable The East India Company. He was greatly impressed by the sight of Hindu laborers excavating the red-brown

Fig. 6-15   The building blocks of laterite that compose this temple at Angkor Wat, Cambodia, are highly resistant because they are composed chiefly of residual iron oxide, all the other minerals having been removed by chemical weathering. There is a notable difference in weathering of the sandstone columns and statues, which are composed of minerals that are being removed by solution, carbonation, hydration, and oxidation. (Photograph by Leonard Palmer.)

tropical clay, shaping it into bricks, and using these for building material after case-hardening in the sun. In this respect, these clay blocks were much like the sun-dried bricks, or adobe, of the arid Southwest, although they differed by being taken directly from the ground and not manufactured out of mud which had been shaped in molds.

Because this tropical clay could be used so readily as a construction material, Hamilton called it *laterite* from the Latin word *latere*, or brick. Its origin was a source of wonderment to him, and has remained a puzzle to soil scientists ever since. As a construction material, laterite has served to build enduring monuments, since, among others, much of the long-forgotten city of Angkor Wat in Cambodia is built of laterite (Fig. 6-15). This alone is an indication that laterite consists of material which is virtually insoluble.

Laterite might be regarded as the skeleton of a soil, since the soluble elements, such as calcium, sodium, and potassium, are leached out, and even such a normally insoluble substance as silica (SiO_2) has been removed. This means that a typical laterite is mainly iron oxide (Fe_2O_3). Should the source rocks from which this soil is derived have a high content of aluminum, then these residual tropical clays grade into *bauxite* ($Al_2O_3 \cdot 2H_2O$), which is the chief ore of aluminum.

Laterites characteristically are found in tropical terrain that has adequate slope drainage; they do not form in continuously wet areas, such as swamps. In fact, they develop best in the monsoon or savanna tropical climate, which is one with pronounced wet and dry seasons. Essential for the formation of laterite is the alternation of heavy rainfall, which supplies the ground water to carry iron oxide in solution down from the A-horizon to be concentrated in the B-horizon, and drought, which permits the soil to aerate and allows the soil minerals to oxidize.

Laterites begin to form through the deposition of iron oxide around nuclei, such as individual mineral grains. The iron-rich nodules grow until finally they may unite to form a cement-like, concretionary layer, probably comparable to the hardpan layer of temperate-climate soils. When this happens, water percolates downward with difficulty, if at all, and the surface layers of soil, the probable equivalent of the A-horizon of temperate regions, then are readily swept away under the driving rains of the monsoon.

The result is the nearly total destruction of a region for agriculture, and in a densely populated area such as a large part of Southeast Asia, this is disastrous. Little of value can grow in a lateritic soil whose A-horizon has been stripped away so that a ferruginous clay, devoid of plant nutrients, makes up the ground surface. Such a blanket of iron-rich soil in Australia is called by an appropriate name, *duricrust*, and it spreads over an entire countryside much like an undulating pavement that covers hill crests and valley bottoms alike.

In many lateritic areas of the world a thin A-horizon is preserved only by the trees and other jungle plants which grow on it. In an effort to increase agricultural crops, this growth is sometimes stripped off, exposing the soil to the effects of weathering. The thin A-horizon is soon worn away, and the remaining B-horizon is quickly baked in the sun and becomes useless for agricultural purposes. Brazil and Dahomey in West Africa, among others, have learned this the hard way. Unfortunately, these lateritic conditions prevail in many of the underdeveloped countries of the world, so that expansion of agriculture in these areas in order to feed an expanding population must proceed with extreme caution.

The iron content of laterite in some parts of the tropics may be high enough for this blanket of regolith to be mined as ore. The iron mines of Cuba and of the Surigao Peninsula in Mindanao are examples. Bauxite shares a nearly common origin, except that it has been formed from the weathering of rocks that were richer in aluminum than iron. This appears to have been the story of the bauxite ores of Little Rock, Arkansas, which are in sedimentary layers deposited under what probably were climatic conditions very much like those of the tropical savanna today. In fact, most of the aluminum ore now processed in North America no longer comes from Arkansas, but comes from the high-alumina clays of Surinam, Jamaica, and other South American and Caribbean lands.

Soils are not only the source of some valuable ore minerals, as in the case of laterites, but they also play a part in prospecting for other ores. Extremely sensitive instruments have been developed for detecting very small quantities of metallic substances in soils. When these quantities are larger than those normally found, an underlying ore body is often indicated. This is one phase of a much more extensive technique called *geochemical prospecting*.

Since so much of the land surface of the earth is covered by soil, it becomes a vital consideration to engineering geologists. Any man-made structure, building, dam, bridge, must of necessity come in contact with soil. A thorough study of the physical properties of these soils is of the utmost importance in safe designing and construction.

Paleosols

Since the Pleistocene Epoch is the closest to us in time, it is the one about which we can learn the most. One way of deciphering the events of its history is through the study of paleosols, or fossil soils. These ancient soils may be completely buried by later deposits, in which case they are not too difficult to recognize in roadcuts and stream banks, for example. When they have never been buried, or when they have been buried and then exposed again by erosion of the covering material, they merge into modern soils, and it is hard to distinguish them.

Paleosols of the same age have been traced over great distances in this country and have been used extensively in determining and separating the events of the Great Ice Age which occurred in the Pleistocene. An example of their wide geographic distribution is the use of one paleosol to correlate events that took place simultaneously in ancient Lake Bonneville in Utah, ancient Lake Lahontan in Nevada, and in the Sierra Nevada.

Because climate in large part controls the physical properties of soils, paleosols can be used to determine the climate and environmental conditions that prevailed at the time of their formation. Care must be exercised, however, in distinguishing effects that might have occurred since the formation of the paleosol from those that were part of its original condition.

Future of Pedology

The following quotation will help to show that the future of "dirt geology" is assured.

"Well before the end of this century, making even minimal allowance for some increase in the general standard of living, the engineers and architects of the world will have to build at least as many structures as exist in the world today—buildings, dams, roads, and all allied structures. And every single one of these structures must be built in contact with the ground, thus involving geology and usually excavation as well. In the years immediately ahead, therefore, there are going to be opportunities for joint study of the soil such as have never yet been imagined—joint opportunities and joint challenges, since even today's phenomenal world rate of construction must soon be far more than doubled" (Legget, 1967).

Selected References

Kellogg, C. E., 1950, Soil, Scientific American, vol. 183, no. 6, pp. 30–39.

Knight, H. G., and others, 1938, Soils and men, U.S. Department of Agriculture, Yearbook of Agriculture, Washington, D.C.

Legget, Robert F., 1967, Soil: its geology and use, Geological Society of America Bulletin, vol. 78, no. 12, pp. 1433–60.

McNeil, Mary, 1964, Lateritic soils, Scientific American, vol. 211, no. 5, pp. 96–102.

Pewe, T. L., 1957, Permafrost and its effect on life in the north, 18th Biology Colloquium, Corvallis, Oregon.

Ruhe, Robert V., 1965, Quaternary paleopedology, in The Quaternary of the United States, H. E. Wright, Jr. and David G. Frey, eds., Princeton University Press, Princeton, N.J. pp. 755–64.

Reiche, Parry, 1950, A survey of weathering processes and products, University of New Mexico, Publications in Geology, Albuquerque, N.M.

U.S. Department of Agriculture, 1957, Soil: yearbook for 1957, Washington, U.S. Government Printing Office.

Fig. 7-1 Slumping in poorly consolidated material on a ridge crest. Madison County, Montana. (Photograph by J. R. Stacy, U.S. Geological Survey.)

7

Gravity Movements and Related Geologic Hazards

On the night of October 9, 1963, a torrent of water, mud, and rocks plunged down a narrow gorge in Italy, shot out across the wide bed of the Piave River and up the mountain slope on the opposite side, completely demolishing the town of Longarone and 2000 of its inhabitants (Fig. 7-2). It has been called history's greatest dam disaster, but when it was over, the Vaiont Dam in that narrow gorge was still intact. What could have caused the water in that reservoir, which was not even half-filled, to rise up over the dam and proceed on its destructive course? One shoulder of the dam was supported by Monte Toc, nicknamed *la montagna che cammina*—"the mountain that walks"—by the local inhabitants. Despite assurances by engineers regarding its safety and the expensive efforts that had been made to stabilize its slopes, Monte Toc not only walked that night in October, it galloped. About 600 million tons of mountainside slid instantaneously into the lake behind the dam. The water reached 244 meters (800 feet) above its previous level, and one great wave rose 91 meters (300 feet) above the dam and dropped into the gorge below. There, constricted by the narrowness of the gorge, the water increased in speed tremendously, and, snatching up tons of mud and rocks, raced on its destructive path.

The breakdown of rocks at the earth's surface through decomposition and disintegration yields a mass of unstable material that may shift downslope in response to gravity, especially where hillsides are steep. Such a transfer,

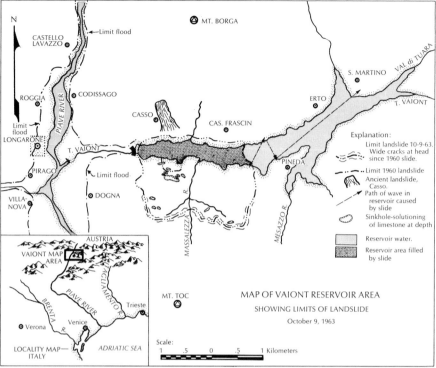

Fig. 7-2 (After George A. Kiersch, *The Vaiont Reservoir Disaster,* Mineral Information Service, Vol. 18, No. 7, 1965.)

sometimes involving both the regolith and the bedrock, may be so slow as to be imperceptible. Sometimes it may be as catastrophically violent as the landslide that plunged down Monte Toc to cause the destruction of Longarone.

This mass movement of surface material adds enormously to the effectiveness of streams in shaping the surface of the earth. Mass movement is responsible for the downslope transfer of material to rivers which then act as continuously moving conveyor belts to carry it away. The way in which the walls of the Grand Canyon flare outward from the Colorado River is largely the result of transfer by gravity of rock fragments and mineral grains downslope to where they can be carried out of the entire Colorado Plateau region by the river and its intricate network of tributaries.

How much or how little material will be shifted downhill by gravity and how rapidly or how slowly it will move are a consequence of a multiplicity of factors such as (1) climate, (2) nature of the bedrock and its attitude, (3) vegetation, (4) the principal types of weathering that are active, (5) steepness of slopes, and (6) local relief, to name some of the more obvious. Few landscapes consist solely of bare rock, and most hill slopes show some degree of rounding. Soil cover serves to alleviate the starkness of a rock-dominated landscape, and soil-blanketed and smoothed slopes are an indication that the regolith is not stationary but has a motion of its own.

This gravity-induced, downslope transfer of material is difficult to segregate into neatly compartmented little packages. The transition from gradual, virtually imperceptible movement at one end of the scale, to a free-falling mass of rock avalanching down a mountain face and filling all the valleys with thunderous echoes at the other, is blurred. In fact, these distinctions are nonexistent, since this is an entirely gradational series. A practical solution to the problem of classification is to divide mass movement into two obvious categories: slow and rapid. This is a highly subjective division, because no two people are likely to agree on their meaning. We are constantly confronted, for example, with the problem of their definition in occasional little controversies involving interpretations of traffic speeds.

Slow Movement

Creep

This descriptive word is used for the slow, glacier-like movement of the soil mantle downslope. We are likely to be quite oblivious of its existence, except to observe with dismay that building foundations may be thrown out of line, power and telephone poles tilted, and sidewalks and retaining walls cracked.

Should roadcuts be made in such shifting ground, they may reveal strata that are turned back on themselves, or thrown into a maze of convolutions where they have been dragged downslope. Soil creeping down a hill very often will drag with it fragments of the underlying rock. These commonly will decrease in number with increasing distance downslope from the outcrop, but may also form a continuous, although attenuated, line when seen in a roadcut. This so-called *stone line* generally will be at the boundary between the undisturbed ground below and the moving regolith above. This very often is a conspicuous feature in areas where creep is prevalent—look carefully for it should you ever think of building a hillside house.

Creep is an especially drastic problem in cold climates where water freezes in the ground. Here soil layers and particles may be lifted up by the expansion of freezing water, and then shifted a little farther downslope when their support is removed through melting.

Solifluction

Solifluction is an extreme sort of creep that reaches a maximum development in cold climates. Hilly terrain underlain by permafrost exemplifies it best, for while the surface layers freeze and thaw, the permafrost table remains constant. Surface water cannot sink into the permafrost, so that water which would normally percolate far down into the ground is concentrated in the active layer. The active layer, then, is far more susceptible to creep than similar terrain would be in a more temperate climate because (1) the opposing forces of ice crystallization and melting of ground ice are most active here, (2) the

Fig. 7-3   Hillside creep distorts railroad tracks along a tributary of the Yukon River, Alaska. (Photograph by W. W. Atwood, U.S. Geological Survey.)

active layer holds more water than it would under a similar precipitation regime elsewhere, and (3) this water-saturated, unstable ground rests on a frozen base over whose surface it can slide readily.

Active solifluction produces a landscape that bears some resemblance to the wrinkled hide of an aged elephant. Different parts of the water-saturated surface layer creep downslope at different rates, so that hillsides where solifluction is active are festooned with soil lobes or tongues, some of which advance rapidly and some slowly.

Another bizarre manifestation of ice-churned ground in the Arctic is curiously regular patterned ground. Sometimes from the air the tundra looks like a gigantic tiled floor. These geometrically shaped polygonal areas are thought to be the result of frost heave, which is much more effective in fine-grained soils than in coarse. When this process of frost heaving is applied repetitively over many years to a soil of mixed composition, the coarse material, such as boulders and gravel, is gradually shoved radially outward from the central area, and the finer materials lag behind and become concentrated.

Rock Glaciers

If a large mass of frost-broken rock accumulates on a moderately steep slope and contains enough interstitial ice, it may move downhill as a ponderously advancing lobe with a wrinkled, corrugated surface and steep sides, in which case it is known as a *rock glacier*. Such slowly moving rock streams are especially well developed in the southern part of the Rocky Mountains.

Earthflows

This type of earth movement is transitional between slow and rapid varieties. It is a more visible form of movement than creep, yet slower than a mudflow or a landslide.

Earthflows are characteristic of grass-covered, soil-blanketed hills (Fig. 7-6). Although they are commonly minor features, some may be quite large and cover many acres. Earthflows usually have a spoon-shaped sliding surface with a crescent-shaped cliff at the upper end and a tongue-shaped bulge at the lower. They involve the soil mantle and are most likely to occur when the ground is saturated with water. This interstitial water not only increases the weight of the mantle but drastically reduces its stability by lowering its resistance to shear.

Fig. 7-4 Soil creep in vertically foliated schist. Brazil. (Photograph by A. M. Bassett.)

Fig. 7-5   Tree trunks attempting to overcome the effects of soil creep. (Photograph by G. K. Gilbert, U.S. Geological Survey.)

Rapid Movement

Mudflows

With increasing velocity this type of gravity movement grades into ordinary stream flow, and with decreasing velocity it merges with earthflows. A typical mudflow is a streaming mass of mud and water moving down the floor of a stream channel, such as a desert arroyo. Such a viscous mass, with a specific gravity much higher than clear water, very often carries along in it a tumbling mass of boulders and rocks, some of which may be as large as automobiles.

Mudflows are a most impressive feature of many of the world's deserts. In arid lands the normally empty stream courses occasionally may fill almost at once with a racing torrent of chocolate-colored mud, following a cloudburst. Where arroyos are shallow, the mudflow may exceed the channel's capacity and spill out over the desert surface.

Mudflows, because of their greater density, are more efficient transporters of large rocks for short distances than normal streams are. They carry the large blocks and boulders often found on the floor of desert basins far beyond the base of a bordering mountain range. There they linger, long after the enclosing mud which once rafted them out beyond the mountains has been eroded away.

Mudflows not only are capable of transporting large natural objects, such as house-size boulders, but may sweep along trucks, buses, or even locomotives trapped in such a debris flow. Houses inundated by such mud streams may be buried all the way up to the eaves. When the mud has dried out, they look for all the world like so many doll houses scattered about by children and pushed by heedless hands deep into the ground.

Mudflows are by no means restricted to arid or semi-arid lands. They are characteristic of alpine regions, too, and are likely to be exceptionally destructive where a combination of steep slopes, a large volume of water freed by melting snow, and a great mass of loose debris are all available (Fig. 7-7). These mud avalanches are especially impressive when they overwhelm a forest. A lava-like mud stream may cut a swath through the forest smashing and splintering trees and branches, which then are stewn about like so much straw in an adobe brick.

Rockfalls and Rockslides

Many attempts have been made to develop workable classifications or acceptable definitions of various kinds of gravity movement. Because of the complex

Fig. 7-6   Active slump-earthflows (1) and new scarplets (2) about one foot high within an old landslide area (3) on the west side of Pleitito Canyon, San Emigdio Mountains, California. The large, hummocky area of the old slide (3) is one mile long and about 1200 feet in altitude from head to toe. (Photograph by John T. McGill.)

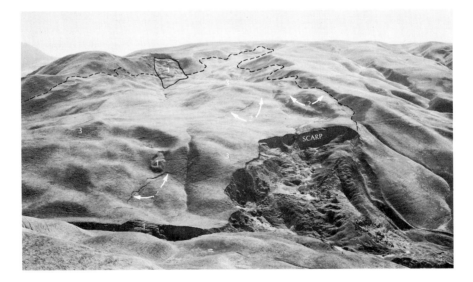

Fig. 7-7   A small village in the St. Moritz area, Switzerland, clustered at the foot of an old mudflow (left foreground) and with a few buildings on the mudflow. Additional building on the mudflow could lead to further instability, causing damage in the central part of the village.

nature of most landslides—few have a single cause, or even a single aspect— and because the terminology used to describe them comes from ordinary, everyday words, and thus is susceptible to widely differing interpretations by different users, none are wholly satisfactory.

A good solution for our purpose is to utilize in simplified form the terminology employed by Varnes (1958), in a classification designed primarily to aid highway engineers.

In this classification, material that drops at very nearly the velocity of free fall is called either a *rockfall* or *soilfall*, depending on its composition. Rockfalls can range in size from the dropping of individual blocks on a mountain slope to the failure of masses weighing hundreds of thousands of tons and avalanching down a mountain face (Fig. 7-8). In the first example, such individual blocks commonly come to rest in a loose pile of angular blocks, or a talus, at the base of a cliff. Should large blocks of rock drop into a standing body of water, such as a lake or fjord, immensely destructive waves may be set in motion with no warning at all. This is a hazard particularly feared in Norway where small deltas may provide the only available flat land at sea level. Should such a rockfall-induced wave burst through the village streets and houses, destruction is likely to be as complete as it is sudden, since these waves may range all the way from 6 to 91 meters (20 to 300 feet) high.

A much greater mass of rock that starts as a freely falling body high on a mountain face may avalanche on downslope and sweep completely across a valley floor with a velocity as great as 200 kilometers (130 miles) per hour. At such high speeds the shattered mass of fallen rock flows much as a liquid would. With entrained air acting as cushion and thus reducing internal friction, such a mass of debris behaves a great deal like the glowing, gas-charged *nuée ardentes* of Mount Pelee. One of the best examples of this type is the prehistoric Blackhawk, California, slide (Fig. 7-9). The rockfall started near the crest of Blackhawk Mountain and in its fall hit a protruding rock ledge. The mass of debris arched out into the air, much of which was trapped under the solid material when it fell back toward the ground. This cushion enabled the slide to flow a great distance over practically flat ground.

The most celebrated example of such a rockfall in North America is one that occurred at Frank, Alberta, in 1903. There, a mass of strongly jointed limestone blocks at the crest of Turtle Mountain, possibly undermined by coal mining carried on below the thrust fault at the base of the mountain, broke loose and plunged down the steep escarpment. Something like 26 to 30 million cubic meters (35 to 40 million cubic yards) fell, and then washed in one gigantic wave through the little coal-mining town of Frank—killing 70 people on the way—and swept to a high point 122 meters (400 feet) above the valley floor on the slope facing toward the source.

The great rockfall-rockslide at Gohna, India, in 1893 remains as one of the more impressive examples of modern times. There a stupendous mass of rock,

Fig. 7-8 This rockfall from the Flimserstein, Switzerland, occurred on April 10, 1939. It buried not only forests and arable land but also a building and eleven persons, all in a moment of time. (Photograph by Swissair-Photo AG Zurich.)

Fig. 7-9   The Blackhawk slide, on the north slope of the San Bernardino Mountains, California. This slide, which moved five miles from the scarp in the middle distance, started as a rockfall in the mountains. (Photograph by A. M. Bassett.)

loosened by the driving monsoon rains, dropped 1200 meters (4000 feet) into one of the narrow Himalayan gorges. A great natural dam was formed by this mass of detritus—perhaps 274 meters (900 feet) high, 914 meters (3000 feet) across the gorge at the crest, and extending for 3350 meters (11,000 feet) up and down stream. This pile of broken rock, involving about 3.6 billion cubic meters (5 billion cubic yards), impounded a lake 237 meters (777 feet) deep when the waters of the river were dammed.

The British engineers, then in India, proved to be a remarkably foresighted lot. They predicted the date of the dam's failure within ten days of the time that it actually occurred, over the two years of its span of life. All bridges were removed downstream while the dam was in existence, the river channel was cleared of obstacles, a telegraphic warning network was set up, and everything was prepared for the imminent flood. When it came it set a world record. Around 7600 million cubic meters (10,000 million cubic feet) of water were discharged in four hours and made a flood whose crest was 73 meters (240 feet) high. Interestingly enough, after the flood was over, the river channel close to the dam, instead of being deepened, was raised 71.3 meters (234 feet) by the sand and gravel deposited after the flood crest had passed and the river flow had returned to normal.

Landslides

A multiplicity of downslope movements is included in this broad general term, and no useful purpose is served here by reviewing the many schemes for classification that have been proposed by geologists, engineers, and other specialists. That so many people are concerned indicates in itself the important role landslides play in everyday life. The difficulties they make for us—and they are considerable, as well as being expensive—are very largely our own doing. Without our disturbance of natural slopes, landslides would be important chiefly in remote mountainous terrain or on hillslopes underlain by notably unstable rocks. Today, landslides are a problem of increasing magnitude as a result of growing urbanization and the mounting demand for high-capacity freeways. Both trends require larger excavations for building foundations and deeper cuts and higher fills for highways. Oversteepening of slopes is a likely cause to start the ground moving.

Landslides may involve the bedrock alone, or they may be limited to the overlying soil mantle, especially if it is deep and water-saturated. More

Fig. 7-10   A landslide in the Palos Verdes Hills, near Los Angeles, California. (Photograph by George Cleveland.)

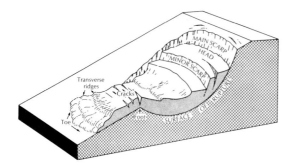

Fig. 7-11   The principal parts of a slump-type landslide. (From Highway Research Board Spec. Rept., 1958. By permission.)

typically, perhaps, they involve both soil and rock. Varnes recognizes two major categories of landslides: (1) glides and (2) slumps. In the first, the slippage is dominantly planar; that is, a large mass of rock may become separated from its fellows and glide outward and downward along the surface of an inclined bedding plane. In slumps, the motion is rotational—usually along a concave-upward slip-plane, so that the upper part of the landslide is dropped down below the normal ground level and the lower part is bulged above it.

The diagram (Fig. 7-11) shows the principal parts of such a slump-type slide, and since by far the greater number of landslides are variants of this form, some of its details are worth noting.

Most of these slides start abruptly with a crescent-shaped *scarp*, or cliff, at their head, sometimes known as a breakaway scarp. Downslope from this may be a number of lesser scarps, and in plan these are almost always concave downhill. Between the individual scarps the surface of the slide customarily is tilted or rotated backward against the original slope of the ground. This causes more instability since these backward-rotated wedges on the surface of the slide make collecting basins in which small lakes or ponds can form. With the slip surface leading downward from the marginal scarp, a channel is available for water to seep into the slide, greatly increasing its instability.

The concave-upward slip-plane down which the jumbled mass of soil and rocks moves may approximate a cross section of a cylinder whose axis parallels the ground surface, if the slide is sufficiently broad. Otherwise, the pattern of the surface of rupture is likely to be spoon shaped.

The slide may advance downstream from the point where the surface of rupture intersects the ground (Fig. 7-11), as a glacier-like lobe of jumbled debris whose surface typically is a chaotic pattern of hummocks and undrained depressions. If the slide moves down a forested slope it often presents a desolate scene of broken trunks and trees. The inexorable thrust of the foot of such a slide against man-made structures almost invariably leads to their collapse, and it is this part of the slide that commonly is responsible for blocking canals, highways, railroads, and engulfing other types of excavations.

Among such slides, the immense ones that closed the Panama Canal at Culebra Cut shortly after it was opened in 1914 and that kept it closed more or less continuously until 1920 are impressive examples. Of the 128 million cubic meters (168 million cubic yards) excavated in the Gaillard Cut, landslides made necessary the removal of at least 55.5 million cubic meters (73 million cubic yards). Great masses of loose, unstable volcanic ash, shale, and sandstone slid on a gently inclined rupture surface toward the canal excavation. One unexpected result was that the bottom of the canal was heaved upward —once as much as nine meters (thirty feet)—until what had been the canal bottom appeared as an island in mid-channel.

All landslides are the result of the forces of gravity acting upon earth materials which are in an unstable condition. Although the movement itself may be extremely rapid (or almost imperceptibly slow), landslide conditions do not suddenly spring into being, but develop gradually until one element of the situation reaches a point which triggers the others. Instability may occur naturally or may be man-made, man often acting as the trigger in an unstable area, as he disrupts a precarious balance. The natural process of erosion can create an unstable condition, for example, when a cliff is undercut, and the material above breaks off and causes a slide. When erosion wears away the toe of an existing landslide or other slope of unconsolidated material, the rest is unsupported and will then move until it is stable once again. If layers of rock slant down toward a road cut or railroad cut, the rocks may slide along the bedding planes between the layers. A normally stable area may become unstable when a heavy load, such as a building or excavated material, is placed upon it. This last was one of the causes of the extensive slides into the Panama Canal during its construction.

Another important cause of instability is the presence of treacherous materials, of which clay is the most notorious and widespread example. Clay minerals when wet have a tendency to expand, so that layers of sedimentary rock which have a high percentage of clay are especially susceptible to movement under the right conditions. In fact, a substance called *quick clay*, is responsible for a number of disastrous landslides.

Quick clay is composed primarily of flakes of clay minerals less than two microns in diameter arranged in very fine layers, and it has a water content which can often exceed 50 per cent by weight. It is commonly part of the debris left behind by retreating glaciers, so it is not surprising that Norway, Sweden, and parts of eastern Canada have several such landslides a year. Quick clay has a most amazing and treacherous quality: ordinarily it is a solid, capable of supporting 952 kilograms (2100 pounds) per square foot of surface, but the slightest jarring motion immediately turns it into a liquid. When the clay layers were originally deposited, generally in salt water, they naturally contained sodium ions which somehow keep the fine clay particles together. If such clays are then exposed to weathering in the atmosphere, the sodium ions are leached out by rainwater and their cohesive effect is lost. Any sudden shock can produce liquefaction. In one slide in Sweden the trigger is believed to have been a pile driver, whose hammering started the movement, with the result that 32.3 million cubic meters (106 million cubic feet) of soil and gravel slid down into the nearby river, carrying with it thirty-one houses as well as a paved highway and a railroad which it picked up on the way. One person was killed, fifty injured, and three hundred homes destroyed in less than three minutes. As we shall see in the discussion of earthquakes, quick clay was responsible for much of the damage to Anchorage, Alaska, when its movement was set off by the 1964 quake.

Water is the hidden devil in the ground in many landslides. When natural drainage conditions are changed, either by natural processes or by man, the ground-water situation can be changed to a potentially dangerous one. Probably the most frequent trigger of landslides is excessive rainfall, a soaking rain that continues relentlessly day after day until the ground is completely saturated. This unusually high water content makes the soil or rock much heavier, and makes it difficult for those soils and rocks which contain clay to stick together. In addition, it acts as a lubricant along potential sliding planes.

Prediction, Prevention, and Control

Because the conditions favorable for gravity mass movements build up slowly, it should be possible to determine ahead of time which areas could become troublesome. As the population continues to increase, and construction follows this growth, such prediction becomes ever more important for the prevention of human anguish. Obviously, the more we know about the geology of a region, the easier it will be to point out those regions where earth movements are likely to occur. This knowledge is gathered by geologists and recorded on geologic maps which show the distribution of the different types of rocks on the surface, and in cross sections which are informed guesses of the distribution of the rocks below the surface. When such work is related to geologic hazards —landslides, earthquakes, earth subsidence, and the like—it is called "urban geology" or "environmental geology," and it will become increasingly important as a direct application of science and technology to the humanizing of our surroundings.

Civil engineers have devised many ingenious methods of controlling unstable slopes, although not enough persons are aware that it is easier, cheaper, and safer to apply those methods before the gravity movements have taken place rather than after. Consolidation of unstable materials is one method of control, and an unusual example of this was used in the construction of Grand Coulee Dam in Washington. Freezing temporarily stabilized a landslide which threatened to move again. The frozen "arch dam" effectively blocked further movement so that construction could proceed. An example of slope treatment and drainage is provided by the history of the large slides that interrupt the hilly terrain of the Ventura Avenue oil field in southern California. Some of these landslides cover 64.7 hectares (160 acres), and in the rainy winter of 1940–41 one block of 24 hectares (60 acres) slid as a single unit for a distance of approximately 30 meters (100 feet).

Since these slides sheared through oil wells when they moved (in this particular episode twenty-three of them were cut off at depths as much as 30 meters (100 feet) below the ground surface), unusually extensive and expensive efforts were made. Continued slide movement was partially curbed by covering the surface with tar and by drilling horizontal drainage holes into the slides —64 kilometers (40 miles) of them. Vertical wells were also drilled through the slides to a porous layer of sandstone beneath it. This sandstone then served as a conduit to carry water away from the slides and into adjacent solid ground where the excess flow could then be pumped out.

Not only is geologic knowledge of hazardous areas vital, but this knowledge must be known to those persons in authority who can and will use it properly. In the Vaiont Reservoir disaster, with which we opened this chapter, there were a number of danger signs preceding the slide, signs which were misinterpreted or unheeded. Even while the area was being considered as a dam site, it became obvious that Monte Toc was unstable, both from its local nickname and from its past landslide history which includes a slide in historic time and a prehistoric one which blocked the valley. In 1960, after the construction of the dam, a small slide occurred, and a pattern of cracks developed which approximately marked the subsequent large slide. A by-pass tunnel was constructed to handle the projected overflow from possible future small slides, and instruments were set up to measure the creep and accompanying small tremors. In addition, holes in the fractured area were drilled to try to find the slide plane, but no evidence of one was discovered. As it turned out, the holes did not go deep enough. It was not until the day before the disaster that the engineers realized that the whole mass was moving as a single element. Even though the bypass was open and the water was being lowered, by then it was too late.

While not all gravity mass movements can be prevented, co-operation among geologists, engineers, city planners and managers, and the public can prevent many of them, and predict the possibility of others, thus ensuring proper land usage as land itself becomes more and more a rare and valuable commodity. M. R. Hill says: "We must focus our efforts on converting the unforeseen to the predictable, transforming the predictable into the preventable, preventing the preventable, and restraining the foolish."

Selected References

Kiersch, George A., 1965, The Vaiont Reservoir disaster, Mineral Information Service, vol. 18, no. 7, California Division of Mines and Geology, Sacramento, Calif.

Kerr, Paul F., 1963, Quick clay, Scientific American, vol. 209, no. 5, pp. 132–142.

Shreve, Ronald L., 1968, The Blackhawk landslide, Geological Society of America Special Paper No. 108.

Terzaghi, Karl, 1950, Mechanism of landslides, Geological Society of America, Berkey Volume, pp. 83–123.

Varnes, D. J., 1958, Landslide types and processes, Chap. 3 in Landslides and engineering practice, Highway Research Board, Spec. Rept. 29.

Fig. 8-1   A stream in the Sierra Nevada, California. (From the Cedric Wright Collection, courtesy of the Sierra Club.)

8

Stream Transportation
and Erosion

Few natural phenomena are more intimately involved with human affairs than rivers. In centuries past such streams as the Nile, the Tigris, and the Euphrates literally were the givers of life as they threaded their way across a weary desert land (Fig. 8-2). The ancient civilizations depended on this water for irrigation; it was through this communal enterprise that many of the attributes of modern urbanized society arose. The beginnings of mathematics, surveying, and hydraulics developed in the designing of dams and canals. One of the earliest was a long dike built about 3200 B.C. on the west bank of the Nile with cross dikes and canals to carry flood waters into basins adjacent to the river.

Boundary disputes and ownership problems logically led to a system of codes and usages that evolved into a pattern of laws and courts much like ours today. The Code of Hammurabi (c. 1900 B.C.) includes a provision that if a landowner damages his neighbor's land through neglect of a portion of a canal that was his responsibility, he was liable for all the damage.

Rivers have long played a role as natural barriers—two that were of decisive importance in Roman times were the Rhine and the Danube. The crossing of the Danube by the barbarians commonly is cited as one of the events heralding the fall of the Roman Empire.

Contrasting with their role as barriers is the function that rivers serve as

Fig. 8-2   A river meanders across the Iranian desert in the land of Elam and flows past the ancient city of Susa, where Esther was chosen queen. (Photograph by Aerofilms and Aero Pictorial Ltd.)

communication routes. The Mississippi packet boat, with its flashing wheels and double columns of smoke, is gone forever, to linger on only in memory, or in the pseudoreality of Disneyland. Its place is usurped by the vastly more powerful diesel-propelled towboat (which actually pushes its load), and its broad acreage of heavily laden barges driving against the current. The endless parade of diesel-powered barges that surges up and down the Rhine is an impressive sight to the European traveler.

Rivers from time immemorial have been routes from the sea to the interior. Explorers have followed them; most of the world's leading cities are built on their banks. They are identified indissolubly with the history and national aspirations of almost all the lands that border them. It would be difficult to conceive of a Germany without the Rhine, a Vienna without the the Danube, or a Russia without the Volga or the Don.

Stream Flow

Although streams play so vital a role in our lives, and their control had engaged the efforts of men for centuries, many aspects of their behavior remain as mysterious today as they have always been. Great impetus has been given in recent years to the study of stream flow because of its importance in the design of high-head hydroelectric plants, dams and spillways, and increasingly complex irrigation systems. Every leading nation is actively engaged in research into the nature of stream flow, and the majority of them maintain large and well-equipped hydraulics laboratories. The largest in this country is the U.S. Waterways Experiment Station, operated by the Corps of Engineers, U.S. Army, at Vicksburg, Mississippi. Here, elaborate models of the Mississippi River have been constructed (Fig. 8-3), and an immense amount of data collected and analyzed in order to find ways to bring this unruly river and its tributaries under control.

From laboratory studies and field investigations around the world, a number of observations can be made about the nature of stream flow. Water appears to move principally in two ways: by *laminar flow* and by *turbulent flow*.

Turbulent flow is of vastly greater importance in the motion of water in natural streams. Individual water particles, instead of gliding past one another as in a beautifully arranged ballet, thrash about in the most irregular fashion imaginable (Fig. 8-4). Familiar examples of turbulent flow are the tumultuous rush of water through Niagara Gorge downstream from the plunge pool at the base of the Falls, or the surging maelstrom of white water at the bottom of the spillway of the Grand Coulee Dam.

In turbulent flow, the main component of motion of the water is forward, downslope in the direction that the stream is flowing, but in addition there is a vast amount of purely random movement of water particles. Sometimes they swirl upward like autumn leaves, or like dust devils in the desert—at other times they descend just as violently in the vortices of whirlpools and eddies. In

Fig. 8-3   A section of the Mississippi Basin Model at the U.S. Waterways Experiment Station, Vicksburg, Mississippi. (Courtesy of U.S. Waterways Experiment Station.)

Fig. 8-4 Turbulent flow patterns in the Colorado River. (Photograph by Martin Litton.)

part, it is this erratic flow pattern that makes the actual velocity of a stream such a difficult parameter to measure.

Velocity

In very general terms, the velocity of a stream can be defined as the direction and magnitude of displacement of a point per unit of time. When we speak of speed, we really mean only the magnitude of the displacement. Customarily we measure this in kilometers per hour. Few streams, however, attain velocities in excess of 16 kilometers per hour (10 miles per hour), and velocities of less than 8 are more likely to be the rule. The velocity as much as any single factor is responsible for determining the size of particles that a stream can transport, as well as the way in which it carries its load.

Where is such a thing as an average velocity likely to be located within a stream? Different parts of the water in a stream advance at different rates. The velocity front is essentially curved from the stream surface down to a point a short distance above the bed. Then the velocity gradient becomes very steep from this inflection point down to where it becomes zero at the stream bottom. The sharp break in the stream profile is considered to be at the boundary between the zones of laminar and turbulent flow (Fig. 8-5). This is the critical point on the curve, and the rather inappropriate name of *bed velocity* is given to it. The inappropriateness stems from the fact mentioned immediately above—that the velocity is zero directly on the bottom of the stream. The so-called bed velocity is of the greatest importance in determining the size of sediment particles a stream picks up in the process of eroding its channel, as we shall see later. The average velocity of a stream depends on such factors as the volume of water, the gradient (or *forward* slope), the cross-sectional shape of the channel, the roughness of the channel surface, and the quantity of sediment the stream is carrying.

Increased gradient obviously speeds up a stream's flow. Where its gradient becomes zero, as when a river empties into another river or a lake, then its velocity quickly becomes zero. Where slopes are vertical, as in a waterfall, the velocity approaches that of free fall. So it is in principle with almost all streams. Where the gradient is low, a stream lazes along; where the gradient is high, the water leaps and quickens in a headlong dash to the sea.

The cross-sectional shape of the channel has an effect on the velocity, too. If this approaches a semicircle, then the stream has achieved the shape that includes the maximum area within the shortest perimeter.

A stream flowing in a channel with an extremely narrow cross section tends to erode the banks and increase the width until a more equable balance is reached between velocity and channel width. In a flattened channel, the current slows near the banks where the velocity is least, and as a consequence deposition occurs there. The result is a narrowing of the channel until the best hydraulic fit is achieved between channel cross section and discharge.

The effect of roughness on velocity is obvious. A smooth, clay-lined channel will promote an even, uniform flow, while an irregular, bumpy, boulder-filled one induces enough turbulence within a stream to significantly retard its velocity.

An increase in the sediment load, and a corresponding decrease in the per-

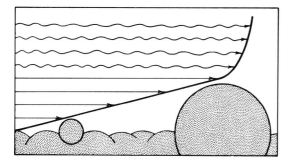

Fig. 8-5   Velocity gradient near a stream bed. The swift and turbulent flow slows and becomes laminar near the bed.

centage of water, has a strong braking effect on the velocity. This is readily understandable because the more sediment is mixed in with the stream, the muddier it becomes, until finally it may become a mudflow. Such a muddy, viscous stream may come to a halt when the ratio of solids to fluid becomes so high that the viscosity is so great that the stream can no longer flow. This is sometimes demonstrated by the ephemeral streams produced by short-lived thunder showers in the desert.

Discharge

Another important property of streams is their discharge. This is a measure of the quantity of water that passes a point in a given interval of time. The discharge of most rivers is far from constant. The flow of northern rivers fluctuates with the melting of snow and ice—those that drain northward to the Arctic are likely to have especially serious problems because their lower courses are still frozen when the headwaters have thawed out and are in flood. Tropical rivers, too, especially those in monsoonal regions, show large seasonal variations, determined by the rhythm of wet and dry seasons throughout the year. Streams of the arid southwestern United States show as great a range as any. Throughout much of the year they may have no surface flow at all. During a sudden cloudburst they may be converted into raging torrents filling the channels from bank to bank.

Stream Transportation

The roiled cloud of sediment brought down by the Mississippi River stains the waters of the Gulf of Mexico far seaward of the river mouth. In the Southwest, within a generation some fairly large reservoirs have silted up completely, and the lakes once backed up behind the dams are converted into dreary expanses of muddy or dusty silt, depending on the season. After a heavy rain, many streets and sidewalks are slippery with a coating of mud, or have a scattering of stones over their surface. Scores of other examples from everyday life are enough to convince an observant person that the land is inevitably wasting away, and that much of its substance is being swept to the sea.

This annual wastage can be an imposing amount and is demonstrated by a single example, the Mississippi, which every day carries around 2,000,000 tons of sediment to the Gulf of Mexico. When the river is in flood, this may rise to as much as 4,000,000 tons. This colossal drain on the central lowlands of the United States has resulted in the construction within the past million years or so of a broad platform of sand, silt, and clay that covers an area of around 31,080,000 square kilometers (12,000 square miles) at the river's mouth, with a central thickness of at least a mile.

Most people know that streams carry a heavy burden, but they may be uncertain as to how it is moved. Even experts in hydraulics are uncertain as to the actual way in which a stream moves its load. The attempt to answer this problem justifies a vast amount of research and the publication of scores of papers.

There is, nevertheless, a broad spectrum of agreement that a river moves its load in three major ways—in part by solution, in part by suspension, and in part by rolling, sliding, and moving it bodily along the bottom of the channel.

Solution

This is the load of dissolved material in a river that has been supplied largely through the leaching out of soluble minerals in rocks. It is the invisible dissolved load which gives some river water its distinctive taste. This is especially true of western rivers which cross arid or semi-arid regions. When such water evaporates it leaves behind a white residue that blankets the entire surface of the ground; from a distance extensive patches of it may look like fields of snow. Such accumulations of alkali are fatal to nearly all crop plants, and are the bane of many irrigation districts. Keeping these soluble residues from accumulating in the soil is an unceasing struggle, and one that has been lost in a number of reclaimed areas.

The solution load in a stream, although it is invisible, is imposing. A plausible estimate is that rivers carry around 2,500,000,000 tons of dissolved material to the sea each year. This loss of soluble chemicals is so great that the lands of the world would be lowered by about .3 meter (1 foot) in 30,000 years through solution alone.

Suspension

The contrast between a limpid trout stream in the high mountains and the muddy, roiled water sluicing through an arroyo as the aftermath of a desert cloudburst is largely a function of the suspended load. This is the visible cloud of sediment suspended in the water above the stream bottom. If the sediment is so finely subdivided that it approaches colloidal dimensions, it may remain buoyed up in the moving water almost indefinitely. This is likely to be true for such particles as clay flakes; it is not so true for silt or sand, or for larger particles. How long such fragmental material keeps afloat depends on many factors: (1) the size, (2) the shape, (3) the specific gravity of the sediment grains, (4) the velocity of the current, and (5) the degree of turbulence.

The effect of these factors is obvious. Flat mineral grains, such as mica flakes, will sift down through the water much like confetti, when compared with the more direct way in which nearly spherical grains settle out. Specific gravity is important, too, because denser substances such as gold nuggets, with a specific gravity of 16-19, are deposited far more rapidly than feldspar grains of the same dimensions, but with a specific gravity of about 2.7.

One of the more important factors in keeping particles in suspension is turbulence. If a particle is about to settle out and then is caught in an upward swirl of water, it may be whisked up suddenly in much the same way that tumble-weeds spiral upward in a swirling wind eddy in the desert. This means it is most unlikely in a stream for a grain of sediment to settle out at a uniform velocity all the way down from the top to the bottom of a river. Rather, such a particle follows a complex path. In part, it drifts forward down the river with the moving current, and in part it swirls erratically up or down much like a sheet of paper caught in a vagrant wind.

Typically, not all of a stream's suspended load is distributed uniformly throughout a stream. Uniform distribution is likely to be true of the finer grain sizes, such as silt or clay, but the greatest concentration of larger grains, such as sand, is closer to the bottom. A stronger current is required to keep sand in suspension, more turbulence is needed to get it up off the bottom, and sand-sized grains settle out more rapidly than do clay-sized particles. In fact, this basal concentration of sand grains may be in suspension briefly, only to sink back to the bottom of the channel, whereupon it becomes part of the so-called bed load.

Bed Load

Part of the river's burden is transported by rolling or sliding, either as individual particles or collectively along the bottom. This is the bed load. In terms of work accomplished by a stream in cutting down or widening its channel, this is the part that does the lion's share. The bombardment of sedimentary particles against the sides and bottom of the channel wears it away as effectively as though it had been worked over by an abrasive such as carborundum.

The bed load is a virtually impossible quantity to measure in a natural stream as compared to the dissolved or suspended load. Both of these are diffused through the main body of the river. The bed load not only is moving along the most inaccessible part of a stream—the bottom—but it is not measurable by any ordinary sampling devices. Enough observational evidence has been collected to suggest that, in great floods, the whole body of sediment on the floor of the Colorado flows bodily downstream. Evidence from other streams, too, is convincing that much of the bed load is moved during floods. Then the channel bottom may be scoured to roughly the same depth that the river surface crests above the normal level. Although the way in which the bed load moves is not known, some knowledge of what may be happening is gained from laboratory studies with controlled conditions of velocity, channel shape and cross section, and amount and size of sedimentary particles.

Individual particles may move by sliding, rolling, or saltation (from the Latin word, *saltare*, to jump). This last process is somewhat analogous to playing leapfrog. A sand grain may be rolling along the bottom, or even may be stationary, when it is caught up by a swirling eddy of turbulence. Then it bounds or leaps through the water in an arching path. Should the velocity be great enough, it may be swept upward to become part of the stream's suspended load; if not, the particle sinks again, either to remain stationary, or perhaps to continue its downstream progress by leaps and bounds.

Competence and Capacity

The size of sedimentary particles a stream can transport, or its *competence*, depends primarily upon the velocity. At low velocities a typical stream may run clear, and sediment grains on the floor may rest relatively undisturbed. With rising velocity the water becomes increasingly roiled and larger and larger particles are set to moving. Finally, even such impressive loads as railroad locomotives may be swept along—several were rafted away, out of the roundhouse and into oblivion, during the Johnstown Flood of 1889. In a cloudburst in the Tehachapi Valley of California in 1933, a far larger behemoth of the rails, a Santa Fe steam freight locomotive and its fully loaded tender, was swept several hundred yards downstream from the tracks and completely buried in the stream gravels.

The most stupendous example in the United States of the transporting power of running water probably was provided by the failure of the San Francisquito dam in California in 1928. When this 62.5-meter-high (205-foot-high) concrete structure collapsed in the darkness, a wall of water 38 meters (125 feet) high swept down the canyon with a velocity of perhaps 80 kilometers (50 miles) per hour. This was enough to raft individual blocks of concrete weighing as much as 10,000 tons half a mile downstream. Partially buried in the gravel and boulders swept down river by the same torrent, they much resembled gray icebergs stranded on an arctic beach.

Capacity is a term for the potential load that a stream can carry and, like competence, it is dependent, in part, upon the velocity with which current is flowing. A sluggish stream meandering across a swamp is capable of moving very little in comparison with a boulder-rolling mountain torrent. Capacity is also dependent upon discharge. Obviously, a rivulet of water moving with the same velocity as the mile-wide Mississippi can move but a fraction of that river's immense burden.

Grade

If we remember that *capacity* is a measure of what a stream theoretically can do and *load* is a measure of what a stream actually is doing, then we can keep these two terms straight. For example, a Lackawanna boxcar may have such a figure as "Capacity 100,000" stenciled on its side, yet actually be carrying a load of 80,000 pounds of automobile frames.

A stream whose load is greater than its capacity is not likely to be spurred on by excess of zeal into carrying more than is expected of it. Instead, the overload is dropped as abruptly as a Peruvian llama deposits its burden if it is convinced that its carrying capacity of around 45 kilograms (100 pounds) has been ex-

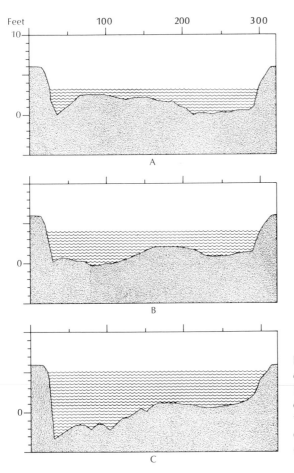

Fig. 8-6 Changes in the shape of the channel of the Rio Grande, near Bernalillo, New Mexico, during the progress of a flood in the spring of 1948. (A) May 15, discharge 1540 c.f.s. (B) May 22, discharge 4920 c.f.s. (C) May 28, discharge 12,400 c.f.s.

ceeded. When a stream deposits its excess load we say that it is *aggrading* its channel and this may happen when too much sediment is supplied or when the particle size exceeds the competence. Conversely, an underloaded stream— one whose capacity is greater than its load—is likely to pick up an additional quantity of material and thus erode its channel, or *degrade* it.

When a stream is balanced between these extremes, and has achieved equilibrium, so that its slope and discharge give it a current which is balanced for its load, it is at *grade*. A *graded stream* is defined by Mackin (1948), and his statement includes all the essential elements:

> A graded stream is one in which, over a period of years, slope is delicately adjusted to provide, with available discharge and with prevailing channel characteristics, just the velocity required for transportation of the load supplied from the drainage basin. The graded stream is a system in equilibrium; its diagnostic characteristic is that any change in any of the controlling factors will cause a displacement of the equilibrium in a direction that will tend to absorb the effect of the change.

A word often confused with grade is *gradient*, which means slope. The two are interrelated because a graded stream will have developed a gradient whereby it will be able to maintain this delicate balance between velocity, load, channel cross section, and size and amount of load. Such a gradient constitutes the so-called *profile of equilibrium*, and ideally it would be a concave-upward curve — nearly horizontal at the river mouth and steeper near the head. The reason for this is that there is a progressive change in conditions downstream, such as increasing discharge and volume of water as successive tributary streams make their contribution, a decrease in size of sedimentary particles constituting the stream's load as they undergo the wear and tear that is a necessary accompaniment of their journey to the sea, as well as a decrease in the proportion of actual load to discharge, and commonly a greater erodibility of the channel where a stream is flowing on its own deposits near the mouth than where it is carving its way through resistant bedrock in the headwaters zone.

Today virtually everyone recognizes that valleys, even such imposing ones as Hells Canyon in Idaho, Grand Canyon in Arizona, or Kings Canyon in California — all of them more than a mile deep — were cut by the narrow ribbon of

Fig. 8-7   A field of boulders swept by flash torrents from the distant canyon. Near Manzanar, California. (Photograph by Ansel Adams.)

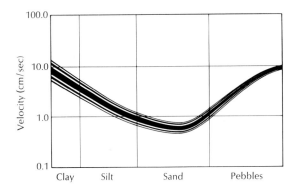

Fig. 8-8 Graph showing how swiftly a current must flow to erode material of various sizes.

turbid water in the stream at the bottom, barely visible thousands of feet below the canyon rim (Fig. 8-9).

This belief, so obvious now, was disputed by our predecessors, and was not fully acceptable even to such illustrious figures as Sir Charles Lyell and Charles Darwin. As late as 1880 many geologists were content to believe that while streams were capable of some downcutting, nonetheless the deeper and more impressive gorges, such as Grand Canyon, resulted from a violent sundering of the earth's crust. The presence of such a river as the Colorado was purely fortuitous—instead of cutting the canyon it simply followed the course it does because that was predetermined as the lowest and easiest route to follow.

Although a belief in a cataclysmic origin unquestionably has greater appeal to the imagination, it simply means that many geologists of a century ago had overlooked the succinct and eloquent statement made in 1802 by a Scottish mathematician and amateur geologist, John Playfair. Although this essay of his is now widely quoted, it is worth repeating because of the clarity of his style— an attribute not always characteristic of scientific writing today—and because his choice of essentially simple, straightforward words made the point so long ago that streams do in fact excavate the valleys they occupy.

> If indeed a river consisted of a single stream, without branches, running in a straight valley, it might be supposed that some great concussion, or some powerful torrent, had opened at once the channel by which its waters are conducted to the ocean; but when the usual form of a river is considered, the trunk divided into many branches, which rise at a great distance from one another, and these again subdivided into an infinity of smaller ramifications, it becomes strongly impressed upon the mind, that all these channels have been cut by the waters themselves; that they have been slowly dug out by the washing and erosion of the land; and that it is by the repeated touches of the same instrument that this curious assemblage of lines has been engraved so deeply on the surface of the globe.

We saw in Chapter 7 that in such a chasm as the Grand Canyon, the Colorado River is responsible for cutting its channel to a depth of a mile below the plateau, but is not responsible for the excavation of the full 21-kilometer-wide (13-mile-wide) trough. The widely flaring portion of the valley, outside of the narrow slot actually cut by the Colorado, is the result of mass movement downslope, of weathering, and of slope wash. The role of the river has been to serve as a gigantic conveyor belt, running without cessation, sweeping away most of the debris supplied to it, and cutting downward to maintain the gradient of the sides.

Valley Widening

The long-term goal of all rivers is to wear the land down to a nearly featureless plain virtually at sea level. Stream erosion for all practical purposes ceases at sea level. This was called the *base level of erosion* about seventy-five years

Fig. 8-9 The Snake River, here merely a mile below us, has cut Hells Canyon, Idaho. (Courtesy of Oregon State Highway Dept.)

Fig. 8-10 The successive cross-profiles of a valley, beginning with a narrow canyon and progressing toward a peneplain. On the right, the slopes are diminished by downwasting; on the left, the slope angles are parallel and the valley walls retreat by backwasting.

ago by Major John Wesley Powell, pioneering geologist of the far western United States and leader of the first party to explore the Grand Canyon of the Colorado. In his thinking this included much more than merely carving a narrow, water-filled stream channel down to sea level; it also involved the reduction of all the interstream areas until an entire region was worn down nearly to sea level.

The effect of the slow wasting away of the area between streams is shown in the accompanying diagram (Fig. 8-10). A once broad and nearly level upland, trenched by narrow canyons, may over many years decline until ultimately it is reduced to a nearly level plain not very far above sea level. To such a broad, nearly featureless plain the term *peneplain* is applied, from the Latin, *paene*, meaning almost, and the English, *plain*.

A peneplain, since it is a product of widespread degradation by streams and mass wasting, can never have a perfectly level surface, although it may come very close to it. Since it is a product of erosion rather than deposition, the surface of the plain truncates the underlying bedrock. This bedrock surface may not be worn down everywhere to the same monotonous level; portions underlain by more resistant material may stand higher than their surroundings. Such isolated, residual hills, or even mountains, are called *monadnocks*, after Mount Monadnock in New Hampshire.

Few incontestable examples of peneplains have been described from around the world, although many partial peneplains, or partially convincing examples, have been described. Many expanses of nearly level plains of barren rock, such as the Hudson Bay region of Canada or much of the Scandinavian Peninsula, have complex histories, in each of these cases involving widespread glacial stripping.

Actually, there is less than general agreement among geologists about the way in which a combination of stream erosion and mass wasting operates to produce these broad plains. One way that it may be achieved is by *downwasting*, a process through which the steepness of slopes adjacent to a stream valley gradually diminishes through soil creep, gravitative transfer, and the decomposition of rocks. This process might be regarded as a point of view based on the work of W. M. Davis, a pioneer American scientist. A different solution was advocated by an Austrian, Walther Penck, and according to him would yield the same end product, a nearly level, stripped plain, but produced by *backwasting*.

For this process to operate, an equilibrium slope has first to be established on the canyon walls, as in Figure 8-10, and once the proper inclination of this slope has been reached for a particular climate, dominant sort of weathering, vegetative cover, and kind of bedrock, then the slope will continue to retreat essentially parallel to itself. As it recedes, a graded, stripped rock surface develops at the base, and as the slope above it retreats, the platform widens. Ultimately, all land above the level of the widening platform will be stripped off as separate retreating equilibrium slopes meet, and then an entire region will be worn down to a base level of erosion with a very gentle gradient toward the sea.

A final choice between these two explanations would be premature in view of the state of knowledge today. The answer to the problem of the wearing away of the land between the rivers, just as to many problems in the real world, is complex, and rather than being all one or all the other, involves elements of each. If any differentiation can be made, the possibility seems strong that backwasting may be a leading process in the more arid lands of the world, and downwasting may be paramount in lands where vegetation blankets slopes, soil is deep, and mass movement plays a strong supporting role.

Regardless of the details of the mechanism of its accomplishment, geologists today are in essential agreement that the running water of streams—given time —can wear away the highest mountain made of the most resistant rocks, until it is reduced to a nearly featureless plain. That so much of the land surface of the world fails to conform to this drab description is a testimonial in itself of the recency and continuing nature of the forces of deformation affecting the rocks of the earth's crust.

In other words, seldom does a portion of the crust remain stationary long enough for erosion to run its course. More often than not the earth's surface may be re-elevated, as in the Grand Canyon region, and a plains area, once near sea level, may then become a plateau hoisted thousands of feet above its former base level. This process probably has been repeated again and again through the long antiquity of the earth. The forces of crustal deformation are ranged in never-ceasing opposition to those of erosion. When erosion prevails, extensive degraded rock plains develop. With a reversal of the forces, such a peneplain, broadly uplifted, may survive after many millennia only in tattered remnants of upland surfaces, or accordant ridge crests, high in the alpine zone of some lofty mountain range.

Stream Deposition

The broad plains bordering many of the large rivers of the world have been tempting sites for settlement since the beginning of history. Both the Egyptian and Babylonian civilizations were essentially riparian, and the life of their people was bound to the river, be it the Nile or the Euphrates, the giver of life in an arid land.

Across the wide expanse of lowland bordering a river, a stream such as

Fig. 8-11   The gauge that measures the height of the Nile at Elephantine Island, Egypt. The records of floods along this river extend back thousands of years. (Photograph by Jean B. Thorpe.)

the Nile is free to spread its waters in time of flood. In fact, the annual flood was an event of such importance in the survival of Egypt that a whole pantheon of deities centered their activities around whatever divine force was responsible for this phenomenon (Fig. 8-11).

A plain, such as the one flanking the lower Nile, quite appropriately is called a *flood plain*. In the United States the region adjacent to the lower Mississippi River is a superb example. This is probably the most thoroughly surveyed of the world's largest rivers, both with regard to the configuration of the river itself—its bends and channel patterns—and with respect to the thickness and character of its deposits as well.

Characteristic Flood Plain Features

From the nature of the term itself, we should logically expect the surface of a flood plain to be covered with deposits made by the river in time of flood. This is true for the lower Mississippi River, whose channel from a short distance below the junction with the Ohio down to the Gulf of Mexico lies wholly upon its own alluvium.

Such a flood plain normally is bordered by low bluffs (Fig. 8-12) which mark the outer limits of the band of swampy ground that the river has been free to wander across. Most such rivers, and the lower Mississippi is typical, are confined by low embankments, or *natural levees*, which slope gently away from the river. A section of low-lying ground, the *backswamp*, may be found between the bluff and the natural levee. This may be a swampy, ill-drained, boggy section whose surface waters cannot flow back into the river because the slope of the natural levee is against them.

Natural levees are built by the river when it overtops its banks during a flood. Rather than surging violently over its banks at a single outlet, the typical flood moves away from the river toward the backswamp as a sullen, tawny, inexorably spreading tide of muddy water. The greatest check to the velocity of the flood waters comes when they leave the relatively compact hydraulic configuration of the stream channel and first encounter the lake-like expanse of flood waters that have inundated the backswamp. The bulk of the suspended sediment is deposited forthwith where the sudden velocity drop occurs at the channel edge. The result, then, is the gradual building up of a narrow embankment, chiefly of fine sand and silt on either side of the lower Mississippi River. Because their water content is less than the immediately adjacent, ill-drained backswamp, and the grain size of their sediment is larger than that of the swamp muck, the natural levees provide the only solid ground in such a saturated region as the lower Mississippi flood plain, and for this reason, roads, settlements, and farms are clustered along the higher and firmer ground of the levee immediately adjacent to the river. The backswamp with its intricate pattern of bayous, branching channels, and lakes is virtually uninhabited, with the exception perhaps of birds and muskrats and their hunters.

A bird's-eye view of a flood plain is likely to look like the aerial photograph of Figure 8-13, which shows the river swinging in broad curving bends, or *meanders*. The word comes from an actual river, the Menderes, in western Turkey, and the derivation of the word is from the Greek *maendere*, to wander.

Before speculating about the origin of these beautiful but perplexing sinuosities in a river's course we need to know something of their geometry, as well as their nomenclature. Fortunately, the names we give their various parts are taken from the American vernacular; they are the same ones that were in use along the Mississippi by the river men of a century or so ago.

The common name for a broadly curving part of a river is a *bend*. The convex bank in such a curve is a *point*. A straight stretch of the river is a *reach*. A now abandoned and partially filled river channel—ordinarily sited just inside a point—is called a *chute*. The level of the stagnant backwater might rise during a flood, and in the days of steamboat racing a chute would provide

Fig. 8-12   Diagram of a flood plain.

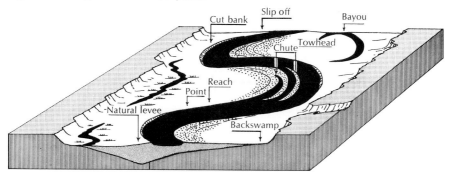

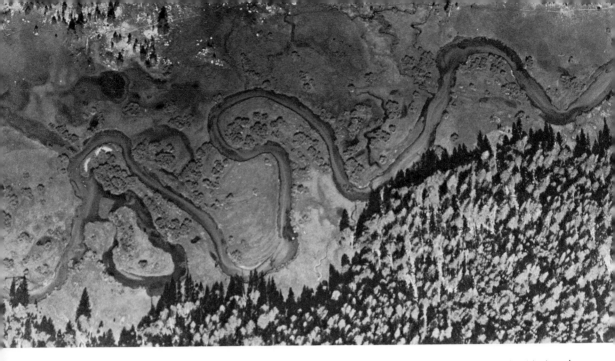

Fig. 8-13   Meanders on the Lyell Fork of the Tuolumne River, in Yosemite National Park, California. The river flows toward the right. (Photograph by Hal Roth.)

a short cut for a daring pilot trying for a fast upriver passage. Here the danger for a packet boat was impaling the hull on the roots of a waterlogged tree, or snag.

Oddly, there is no common term for the path followed by the thread of maximum velocity of a stream. In general, the main current clings to the outer side of a bend where it is shunted by centrifugal force. In such a bend, where the current is strong, the bottom of the channel is scoured more deeply than elsewhere, and this section of deeper water close to the bank is given the old-fashioned name of *pool*.

When the current leaves a bend, it normally does so on a tangent and may occupy a variety of positions while in a reach. On entering the next bend downstream it crosses over to the opposite bank, and such a place was known as a *crossing* to the steamboat pilots. They disliked crossings because here the current was diffused, its strength dissipated when compared to where it ran deep and strong along the outer curvature of a bend. A crossing, then, was beset by shoals and shallows, or sand bars, and at low water it was necessary to take a steamboat across under slow bell, sounding all the way. The leadsman's cry could be heard echoing along the river through the stillness of a summer noon, "By the mark five, by the deep four, by the mark three, Mark Twain!" The last is a depth of two fathoms (12 feet).

At a truly low stage of the river some of the bars appear above the water surface as low, tawny sand islands, appropriately known as *towheads*.

If we draw two profiles across the channel of such a river as the Mississippi, the first in a reach, where the current is approximately in midstream, and the second in a bend, the two channel cross sections will be quite unlike. In the reach, the Mississippi flows in a broad, shallow, nearly flat-floored trough.

Typical of such a pattern are the river's dimensions in the delta just before it breaks up into separate distributary channels. There the river is around 1200 meters (4000 feet) wide and roughly 15 meters (50 feet) deep, with an approximately trapezohedral shape; that is, a flat bottom and comparatively steep, flaring sides.

The form is quite different in a bend. The channel nearest the concave, or outer, side of the bend is deep and nearest the point it is shallow. In short, the cross-sectional pattern is wedge-shaped, with the deep part of the triangle being the pool. The parts of the profile that stand above the water surface have distinctive names, too. The concave side is the steeper, since it is continuously being undermined by the river, and commonly it is known as the *cut bank*. The gentler, or convex, point is a site of deposition and has been built up gradually through the accumulation of sand and silt. It is called the *slip-off slope*. In a metaphorical sense the river appears to have slipped away from it as the diameter of the meander has enlarged through undermining of the cut bank.

How meanders enlarge is a question that has been long debated. The most generally accepted belief is that the cut bank is undermined by the river, especially when in flood, by deep scouring in the pool, thus over-steepening the cut bank and slicing away its foundations. Commonly, the bank fails by slumping into the stream through this removal of support at the base, rather than being sawed horizontally by the river.

How the slip-off slope grew — so that additions to it just about keep pace

Fig. 8-14 Entrenched meanders. The Goose Necks of the San Juan River, Utah. (Photograph by A. M. Bassett.)

with the retreat of the cut bank, with the result that the river always maintains about the same width—puzzled investigators for many years. Where does the sand come from to keep the slip-off slope advancing as the cut bank recedes? At one time it was believed that there was a cross-channel transfer of material from the cut bank to the slip-off side—which in a sense maintained a balance between cutting on one side of the river and filling on the other.

An impressive series of large-scale laboratory experiments was made by the U.S. Waterways Experiment Station in Vicksburg. These demonstrate very convincingly, by using models with river lengths ranging from 15 to 46 meters (50 to 150 feet) and widths of from .3 to 1.5 meters (1 to 5 feet), that much of the contribution of sand and silt to a slip-off slope comes from the erosion of the cut bank next upstream. This is a fully expectable process since the current is at maximum efficiency in the pool, and when material from a cut bank slumps into the river through undermining, the sedimentary material is delivered at the point where it can be most effectively transported. As a result, most of it is swept away from the concave part of the bend in short order.

A careful study of the course taken by the current shows that its erosive effectiveness is directed against the outside of the curve, the bends tending to enlarge in diameter rather than contract. The neck of land between adjacent bends narrows correspondingly. This narrowing may continue until finally the neck is cut through, the former channel becomes a crescentic lake, and the land area of the point is converted to an island.

The new and more direct channel of the river is known as a *cut-off*, and the abandoned channel, whose entrances soon silt up, becomes a crescent-shaped lake, or *bayou*, as it is known along the Gulf Coast. All told there have been something like twenty naturally occurring cut-offs on the lower Mississippi since 1765. About fifteen have been made artificially by the Mississippi River Commission since 1932 to straighten out the river's course, thereby increasing the gradient and thus the velocity, and as a consequence diminishing the flood hazard by improving the hydraulic efficiency of the channel.

Historically, one of the more interesting cut-offs of the Mississippi occurred at Vicksburg in 1876. Before that date the river made a broadly sweeping curve past the city, now isolated from the main stream by the cut-off that created Centennial Lake (Fig. 8-15).

The river then achieved by itself what General U.S. Grant failed to do in 1863 when the Union Army made an abortive attempt to dig a short canal across the neck to divert the river and thus bypass Vicksburg. It is too bad that the effort failed—perhaps because Grant was not fully persuaded of its merit—because had it succeeded it would not only have made Admiral Porter's lot much easier by not having to run his fleet of Union gunboats past the Confederate batteries, but such an artificial diversion would have been a signal achievement for the branch of the science recognized today as military geology.

Meanders seem to develop best when river banks are eroded with relative

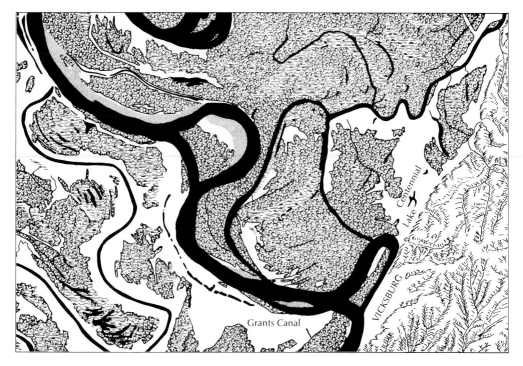

Fig. 8-15   Before 1876 the Mississippi River flowed around the meander that is now Centennial Lake.

ease; if they are too resistant, the stream cannot establish a rhythmically swinging pattern. There also needs to be some degree of inhomogeneity in the composition of the stream banks, such as a mixture of sand and silt. If the material is too uniform in size, and relatively resistant to erosion—such as clay—then the channel very likely will remain essentially straight, as the Mississippi does below New Orleans.

If the banks consist of readily erodible material, there is a tendency for the channel to widen appreciably, and perhaps locally to develop a *braided* pattern (Fig. 8-16). This is a channel pattern where the stream instead of flowing in a single channel divides and subdivides in a complex fashion. The actual size of the river is not a factor in determining whether or not a stream meanders or braids—the lower Ganges and the Amazon, for example, have braided channels—and the same stream may show both types of patterns at different places along its length and at different times in its history. In a general way, braids seem to be characteristic of those reaches of a stream with steeper gradients, while meanders apparently typify those with gentler slopes.

Deltas

Herodotus, in the fifth century B.C., impressed by the branching pattern of the distributaries of the Nile, compared the form of the watery, muddy region be-

Fig. 8-16 Dry stream beds in Death Valley, California, show a braided pattern of small channels. The width of the channel at lower right is about 75 feet. (Photograph by William Garnett.)

tween Cairo and Alexandria to the Greek letter *delta*, Δ. The comparison is so apt that it has won acceptance in most of the languages of western Europe from that day to this.

The Nile delta is a nearly ideal example of this particular landform, so much so that few others measure up to its perfection. The map shows how the main channel of the Nile separates into a host of branching lesser arms, called quite appropriately *distributaries* (Fig. 8-17). Another interesting feature of the Nile delta is the bordering bays and lakes, of which one, Abu Qir Bay, is a good example. This was the site of the Battle of the Nile where the French Fleet was destroyed by Nelson, thus ending Napoleon's hopes for an Eastern empire. Similar *delta-flank depressions*, as such water bodies are called, border many of the other deltas of the world. Well-known ones are Lake Ponchartrain by the Mississippi delta, the Zuider Zee (Ijsselmeer) and marshes of Zeeland adjacent to the Rhine, and the lagoon surrounding Venice at the mouth of the Po.

Not all the world's rivers have deltas where they enter the sea. Two large ones that do not are the St. Lawrence and the Columbia—the St. Lawrence because it has little chance to pick up much sediment in the short run between Lake Ontario and the Gulf; the Columbia because it discharges directly into the open sea whose powerful waves and currents quickly redistribute the river's burden of sand and silt.

The best-developed deltas are most likely to be constructed where a river moves a large load of sediment into a relatively undisturbed body of water. Examples of such impressive accumulations of riverine deposits are the great deltas at the mouths of the Ganges-Brahmaputra in India and East Pakistan, the Indus in West Pakistan, the Tigris-Euphrates in Iraq, the Niger in Nigeria, the Yangtze Kiang and Hwang Ho in China, the Mississippi in the United States, the Danube in Romania, and the Volga in the Soviet Union. One of the most picturesque of the deltaic worlds is the Camargue, the land of cowboys and semi-wild cattle at the mouth of the Rhone in southern France.

In America, the Mississippi is by far the best known geologically, not only because of the long record of channel changes but because something like 90,000 oil wells have been drilled in its sediments, and because repeated geophysical surveys have been made up and down its length. All this information combined gives us a uniquely detailed, three-dimensional picture, not only of the 31,080,000 square kilometers (12,000 square miles) of delta surface but of the mile-deep thickness of sediments below the waters of the Gulf of Mexico as well.

The great delta of the Mississippi is not a single simple structure, but is a compound feature built up through a complexly overlapping pattern of so-called *subdeltas*. The relative succession in which these appeared is shown on the accompanying map (Fig. 8-18) which also gives the estimated dates at which the river held the positions shown. From these it can be seen that the river established its present course only very briefly before the arrival of the

Fig. 8-17   The delta of the Nile.

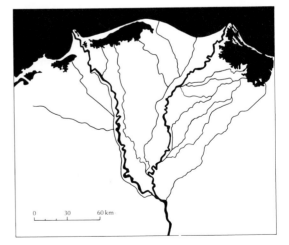

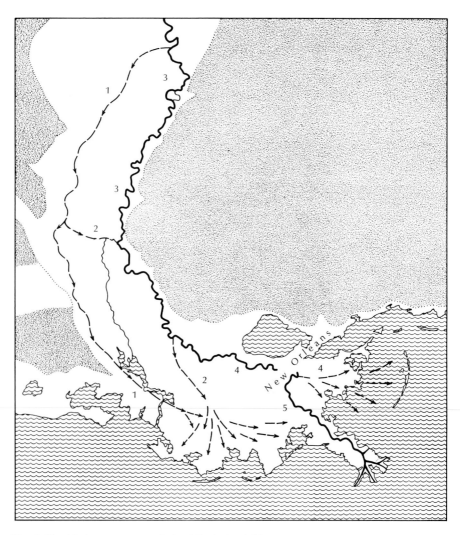

Fig. 8-18 Various courses of the Mississippi River.

first white men, Cabeza de Vaca in 1528(?) or de Soto's followers in 1544. Natural levees border the channels of distributaries that are growing outward into the Gulf. The height of the natural levees increases gradually inland from the Gulf until they are around 4 meters (13 feet) high at New Orleans, a distance of 165.7 kilometers (103 miles) from Head of the Passes.

Head of the Passes is where the present-day river breaks up into three major and a number of minor distributaries (Fig. 8-19), and here the river depth is around 12 meters (40 feet). The slope of the channel bottom in these distributaries is upstream against the gradient of the river surface, because most deposition occurs at the mouth of the distributaries, and here, without the modern interference of dredging and jetty construction, the river builds up a shallow sand bar where it enters the Gulf of Mexico. Thus, the normal low water depth at the mouth of a distributary was as little as 3 to 4.5 meters (10 to 15 feet)—before improvement—and in some of the minor, less frequently used channels, as little as .9 to 1.5 meters (3 to 5 feet).

This strange deltaic world—half water and half land, the abode of water birds of great variety, with a continuously changing pattern of lakes, swamps, marshes, and constantly shifting streams—is one of the more important environments for man on Earth. From the beginning of historic time their fertile soil, their network of waterways, and their position as a meeting ground between seafarers and those who make their living along the rivers of the world have made deltas tempting sites for coastal cities. Such a bewildering maze of channels, large and small, was made to order for piracy, as witness the successful operations for many years of the brothers Lafitte at Barataria in the Mississippi delta.

Quite a price is exacted, however, from a delta city in return for its communication advantages. Such a city is under constant threat of inundation by flood, building foundations are insecure—a visit to Venice is an impressive illustration of what differential settling can do to structures built on delta mud. Even such a shallow excavation as a grave may fill with water (the vaulted sepulchers of New Orleans are an answer to this problem), and the development of an unpolluted local water supply is a difficult task. Overriding these many adversities is the greatest threat of all, that the river may change its course completely or the channel may silt up. A good example of this latter fate is the ancient city of Ravenna; in the days of Justinian and Theodora it was a leading seaport on the Adriatic, now it is something like 9.6 kilometers (6 miles) from the coast.

The low country bordering the mouths of the Rhine provides an impressive

Fig. 8-19 At Head of the Passes the Mississippi River divides into several distributaries which lead out to the Gulf of Mexico. (Courtesy of Humble Oil and Refining Company.)

example of the problems besetting the inhabitants of a delta. This is one of the more densely inhabited regions of the world, and one where an immense volume of seaborne and riverborne trade moves through such ports as Rotterdam and Amsterdam. These cities, already below sea level, throughout their entire existence have not only had to endure attacks overland—a hazard to which delta cities are peculiarly vulnerable—but have had to labor valiantly to hold back the sea. The explanations of all the difficulties faced by the people of the Rhine delta are complex, but they certainly include the currently rising level of the sea, the compaction of the water-logged clays on which these cities stand, and the apparently geologically active subsidence of this part of the European coast.

Many of these same difficulties plague New Orleans, not the least being the problem of subsidence. Indications of a relative lowering of the land with respect to the sea are everywhere. Among these are the presence of what were once Indian settlements far out on the bottom of Lake Ponchartrain, drowned cypress tress and inundated farmland around the lake margin, and the sunken streets and graves of the deserted settlement of Balize near Head of Passes. The now silent, shell-paved streets are buried under the marsh about four feet below sea level.

There is little disagreement over the evidence of subsidence in many of the world's deltas, but there are strong differences of opinion as to its cause. There are those who believe the addition of perhaps as much as 2,000,000 tons of sediment a day, as in the Mississippi delta, makes for an excess load on the earth's crust, which bows down as a consequence. Opponents of this belief point out that the excess load is not so great as might appear at first since the weight of the displaced sea water has to be considered, too. Thus, the material added to the crust may have a net density of about 1.8, or so, and how this lighter material displaces heavier subcrustal material with a specific gravity of perhaps 3.3 is a tricky question to answer.

The answer probably is not a simple one, but certainly involves many of the factors operating in the Netherlands. The addition of an extra burden of sediment, as in the Mississippi delta, may not be adequate in itself to bring about a broad crustal downwarping, but if a load such as the Mississippi's accumulates in a region where subsidence is the dominant geological process, this additional weight certainly is not going to work in an opposite direction.

Rational Use of Rivers

In the early days of our country's history, before the transcontinental railroad was completed and before the present web of highways was constructed, rivers provided the principle path of transportation for both goods and men. Now, of course, they are no longer used in this way to any great extent, but they are certainly used in other ways. One of the most important is as a supply of water for both individuals and industry, the natural outcome of which is the use of rivers as a means of disposal of our monumental amount of waste products and sewage.

In man's attempts to harness the forces of nature for his own benefit, he has built and is proposing to build a great many dams across our rivers. The water stored behind the dams is being used for a number of purposes: production of electric power, maintainance of a steady supply of water for irrigation purposes, and flood control, among others. At the same time, the number of rivers included in National Parks, Monuments, Forests, and Wilderness Areas, implies that they are also valuable for recreational uses and aesthetic values.

That there is a mighty conflict among these uses is clear from even a minimum aquaintance with the mass media. Choices must be made, and too often they are made by a small group of persons who do not necessarily represent the views of the majority of the persons affected. This may not be the result of any deliberate intent to deceive, but rather of neglect in informing the community and the country at large so that their views may be known.

Even when plans for development of a river are disclosed, the proposers have often been remarkably stubborn in refusing to listen to the opposition's arguments or to consider their alternatives. One prominent hydrologist believes that a reason for this is that the proposed benefits can be quantitatively stated, while the "nonmonetary values are described either in emotion-laden words or else are mentioned and thence forgotten" (Leopold, 1969). In an effort to remedy this situation he has devised a chart to evaluate quantitatively some of the aesthetic factors of river sites. Disclaiming any personal bias, he applied his method to twelve river sites in Idaho in the vicinity of Hells Canyon of the Snake River, an area where the Federal Power Commission wants to construct one or more additional hydroelectric dams.

After comparing the Hells Canyon site with the others which are also capable of hydropower development, he found that this site is the most worthy of preservation. He then compared it to other river areas which are known to be deserving of preservation because they lie within national parks: (1) Merced River in Yosemite National Park, (2) Grand Canyon of the Colorado River, (3) Yellowstone River near Yellowstone Falls, and (4) Snake River in Grand Teton National Park. The outcome of this second comparison was that "Hells Canyon is clearly unique and comparable only to Grand Canyon of the Colorado River in these features" (Leopold, 1969). It remains to be seen whether this method will be a valuable tool in determining the rational use of our river resources.

The Last Word

As we have seen, a stream, from its beginnings as a spring or as rain to its end at sea level, and including all its tributaries and distributaries, is an enormous system of interdependent and interacting forces trying to achieve equilibrium. In this case it is a struggle which, by erosion, transportation, and deposition molds the landscape over a considerable part of the earth's surface.

Let Mark Twain, in his role of steamship pilot, have the last word on streams. In *Life on the Mississippi* he writes about river meanders and how the river is shortened when it cuts through the narrow neck of a meander. He grossly misuses the principle of uniformitarianism and is not very kind to science in general, but who will argue.

Therefore, the Mississippi between Cairo and New Orleans was 1215 miles long 176 years ago. It was 1180 after the cutoff of 1722. It was 1040 after the American Bend cutoff. It has lost 67 miles since. Consequently its length is only 973 miles at present.

Now if I wanted to be one of those ponderous scientific people, and "let on" to prove what had occurred in the remote past by what had occurred in a given time in the recent past, or what will occur in the far future by what has occurred in late years, what an opportunity is here! Geology never had such a chance, nor such exact data to argue from! . . . Please observe:

In the space of 176 years the Lower Mississippi has shortened itself 242 miles. That is an average of a trifle over one mile and a third per year. Therefore, any calm person, who is not blind or idiotic, can see that in the Old

Fig. 8-20   "The Army Corps of *who* put us up to this?"

Oölitic Silurian Period, just a million years ago next November, the Lower Mississippi River was upwards of 1,300,000 miles long, and stuck out over the Gulf of Mexico like a fishing rod. And by the same token any person can see that 742 years from now the Lower Mississippi will be only a mile and three-quarters long, and Cairo and New Orleans will have joined their streets together, and be plodding comfortably along under a single mayor and a mutual board of aldermen. There is something fascinating about science. One gets such wholesale returns of conjecture out of such a trifling investment of fact.

Selected References

Garner, H. F., 1967, Rivers in the making, Scientific American, vol. 216, no. 4, pp. 84–94.

Leopold, Luna B., 1962, Rivers, American Scientist, vol. 50, pp. 511–37.

———, 1969, Quantitative comparison of some aesthetic factors among rivers, U.S. Geological Survey Circular No. 620, Washington, D.C.

———, Davis, Kenneth S., and the Editors of Life, 1966, Water, Time Inc., New York.

———, and Langbein, W. B., 1966, River meanders, Scientific American, vol. 214, no. 6, pp. 60–70.

———, and Wolman, M. G., 1957, River channel patterns: braided, meandering, and straight, U.S. Geological Survey Professional Paper 282–B.

Mackin, J. Hoover, 1948, Concept of the graded river, Geological Society of America Bulletin, vol. 59, pp. 561–88.

———, 1963, Rational and empirical methods of investigation in geology, in The Fabric of Geology, Claude C. Albritton, Jr., ed., Addison-Wesley Publishing Co., Reading, Mass.

Fig. 9-1   Lettuce field, Salinas Valley, California. (Photograph by Ansel Adams.)

9
Ground Water

In Xanadu did Kubla Khan
A stately pleasure-dome decree
Where Alph, the sacred river, ran
Through caverns measureless to man
  Down to a sunless sea.

Coleridge's verse quoted above reflects a remarkable image of most people's picture of the nature of water within the earth. Many of us are prone to speak glibly of underground rivers flowing for miles beneath the parched surface of some of the world's most absolute deserts, and to many of us springs are nearly as mysterious as they were to men of long ago. Springs played a leading role in Greek and Roman mythology and inspired legends which were certainly picturesque and more colorful than the prosaic opinions we now hold.

Generally there were two leading schools of thought. One held that springs drew their water from the sea—how the salt was eliminated and how the water was elevated to the great heights it reached in mountain springs remained unanswered questions. The other belief was that springs and streams had their origin within subterranean caverns, large enough perhaps to have atmospheres of their own from which water condensed as a sort of rain within the earth to feed them.

It was the middle of the seventeenth century before the key to the problem, that there was a relationship between rainfall and the discharge of springs, was demonstrated. According to Meinzer (1939):

> Perrault made measurements of the rainfall during three years, and he roughly estimated the area of the drainage basin of the Seine River above a point in Burgundy and of the run-off from this same basin. Thus he computed that the quantity of water that fell on the basin as rain or snow was about six times the quantity discharged by the river. Crude as was his work, he nevertheless demonstrated the fallacy of the age-old assumption of the inadequacy of the rainfall to account for the discharge of springs and streams. Perrault also exposed water and other liquids to evaporation and made observations on the relative amount of water thus lost. He also made investigations of capillarity, established the approximate limits of capillarity in sand, and showed that water absorbed by capillarity cannot form accumulations of free water at higher levels.
>
> Mariotte computed the discharge of the Seine at Paris by measuring its width, depth, and velocity at approximately its mean stage, making the velocity measurements by the float method. He essentially verified Perrault's results. In his publications, which appeared after his death in 1684, he defended vigorously the infiltration theory and created much of the modern thought on the subject . . . he maintained that the water derived from rain and snow penetrates into the pores of the earth and accumulates in wells; that this water percolates downward till it reaches impermeable rock and thence percolates laterally; and that it is sufficient in quantity to supply the springs. He demonstrated that the rain water penetrates into the earth, and used for this purpose the cellar of the Paris Observatory, the percolation through the cover of which compared with the amount of rainfall. He also showed that the flow of springs increases in rainy weather and diminishes in time of drought, and explained that the more constant springs are supplied from the larger underground reservoirs.

Although little was known by our ancestors of the reasons for water being in the ground, there was a considerable use made of it in ancient times. The well was the center of village life for centuries and still is over much of the world. Indeed, the office drinking fountain is no substitute for it as a communication center. Not only was the well the focal point of village life, but it was an absolute essential to the survival of a walled city or castle. It would be a foolhardy baron who would attempt to hold off a siege without an intramural source of water.

The demand for water created by large concentrations of people in urban centers has resulted in a more extensive development of underground sources of water supply than most people realize. Almost everyone knows of the heroic measures the Romans took to conduct water to their cities by building imposing, valley-spanning aqueducts. Oddly enough, the Romans knew little of the nature of water in the ground. They placed their chief dependence on springs and streams, with the result that they went prodigious distances to the Apennines for water when they had a perfectly adequate supply almost directly underfoot had they dug for it.

Others did, and their underground pursuit of water led to the construction of some of the more remarkable burrow-like excavations known. The chief example are the *kanats* of ancient Persia, now Iran. The kanats center largely around Teheran, and for the most part are dug in the gravels of the great apron of alluvial fans at the base of the Elburz Mountains. The kanats are long, mole-like burrows that serve as collecting galleries in the porous gravels of the fan. Some are 24 kilometers (15 miles) long, and individual tunnels may be as much as 152 meters (500 feet) deep. In the old days they were truly multipurpose structures because they served as a source of drinking water and as a means of sewage disposal. In general a kanat follows a water-bearing layer of sand or gravel within the fan, and every few hundred yards is connected with the surface by a shaft sunk during construction.

A remarkable water-collecting tunnel of antiquity was built in Egypt around 500 B.C. This tunnel in the Nubian sandstone gathers water which has probably been introduced into the rock as seepage from the Nile. All told, the tunnel system has a length of a hundred miles or so, although no one can say with certainty because it is almost entirely caved in. Enough water still escapes from the tunnel entrance that it was once thought to be a spring. Actually, the construction of this extraordinary enterprise was recognized as being of such importance that the temple of Ammon was dedicated in its honor and the Egyptians were reconciled to acknowledging the Persian, Darius I, as their Pharaoh.

Not only did men of long ago drive tunnels to intercept water in the ground, but they drilled wells to surprising depths. An outstanding achievement among dug wells was one at Orvieto, in Italy, which was sunk to a depth of 61 meters (200 feet) in 1540. It had two spiral staircases inside the walls, one above the other, with one being used by descending, the other by climbing, water-bearing donkeys.

Drilled wells, once the spring-pole method came into use, went to great depths, as much as 1500 meters (5000 feet) in China, for example. Deep wells drilled at Artois in France in the twelfth century and Modena in the Po Valley of Italy flowed water at the surface. These excited great interest since they were the first true artesian wells of medieval times.

Origin of Ground Water

Nearly all the water in the ground comes from precipitation that has soaked into the earth. Additionally, some water is included with marine sediments when they are deposited, and some (called *juvenile*) reaches the upper levels of the crust when it is carried there by igneous intrusions, volcanoes, hot springs, etc. In practical terms, these two are minor parts of the total budget of usable water in the ground.

Many things happen to water that falls as rain or snow. Much of it evaporates and goes directly back to the atmosphere. Some of it is picked up by plants and returned to the air by their process of transpiration, which is about the same for them as sweating is for us. Then, as we know from ordinary observation, a good deal of the rainfall runs off over the surface of the ground in rills and streams. Finally, some part of it sinks underground and becomes the ground water responsible for springs, and caves, and wells.

No one actually knows how water is divided among these various destinations proportionately, but Leopold and Langbein (1960) estimate it as being somewhat like the following.

Over the United States an *average* of about 76.2 centimeters (30 inches) of rain falls per year. Of this amount, approximately 53.3 centimeters (21 inches) are returned directly to the atmosphere by evaporation and transpiration. Only 22.9 centimeters (9 inches) runs off in streams directly to the sea, and of the total runoff nearly 40 per cent escapes by the Mississippi River—an impressive fraction of the continental supply.

Where does ground water come from, then, if the budget balances as closely as 53.5 + 22.9 centimeters accounted for out of the 76.2 centimeters that fall? The answer is that although the amount of water entering the ground by infiltration is slight—perhaps as little as .25 millimeter (0.10 inch) per year in some places, more in others—with the passage of many millennia great quantities of water slowly accumulate in the ground. It is this vast reservoir built up gradually in the thousands of years since the end of the ice age that we draw on today—unfortunately in some areas more rapidly than it is replenished.

Occurrence of Ground Water

Probably the most familiar aspect of ground water is to see it standing—very often green, scummy, and unappetizing—in a shallow well. If we simply bail the water out with a bucket or a hand pump, very little happens to its level. If we install an electric pump and run it wide open for a while, the water level drops. If this process is continued actively for a long time in a number of wells, the water level continues to drop. This has happened in scores of irrigation districts, especially in the arid and semi-arid West, where, in a sense, farmers are mining their water supply more rapidly than it is being replenished.

To return, however, to the undisturbed water level in the well. If we were to determine its altitude and then to compare it with the level at which water stands in nearby wells, then in many regions we should quickly discover that the water surface is nearly level. This surface at which water stands in wells is called the *water table*. All the voids, or openings, in rocks below its surface are filled with water, or are *saturated*. Above the water table the pore spaces in the ground may be any combination from completely dry to partially full, and these openings are said to be in the *zone of aeration*.

Actually, the water table very rarely is dead level. Instead it is more likely to be a blurred replica of the ground surface (Fig. 9-2), rising under hills and sinking under valleys. It intersects the surface at lakes (Fig. 9-3) and streams, and also at springs. Sometimes a stream has water added to it from the water

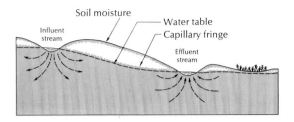

Fig. 9-2   Occurrence and flow of ground water.

table, especially if the stream is at a lower level (Fig. 9-2). Then it is called an *effluent stream*. If the stream is above the water table, and thus adds its contribution to the supply of water in the ground, it is an *influent stream* (Fig. 9-2).

The diagram (Fig. 9-2) shows that there are three elements in the occurrence of water in the uppermost layers of the earth. These are (1) the belt of soil moisture, (2) the intermediate belt with its variable amounts of water or air filling the voids between sand grains, and (3) the water table and the saturated ground below it.

The belt of soil moisture is the portion of the profile with which we are likely to be familiar. This is the ground layer that becomes wet after a rain or a lawn watering. Sometimes this layer may become completely saturated and be converted into a quagmire of mud and water; at other times it may be bone dry and dusty from the top of the ground down.

Commonly there is a lower margin to this surface belt of soil moisture. It may be only a few centimeters down or it may be several meters. When we dig downward the ground generally becomes drier, until perhaps most of the soil moisture seemingly may have disappeared—as Seneca believed it did. Typically, though, in this intermediate belt the water has percolated slowly downward through soil openings until it reaches the water table. How well or how rapidly it percolates is dependent largely upon the porosity and permeability of the ground. What these two terms mean will be described in a few paragraphs ahead.

Extending a short distance upward from the water table is the *capillary fringe*. This is a band of thread-like extensions of water which has migrated upwards in the minute passageways between the individual soil grains. This movement is achieved in about the same way that kerosene climbs in the wick of a kerosene-burning lamp, or water in the confines of a narrow glass tube in a chemical laboratory.

Figure 9-2 shows average conditions over much of the world. However, in such places as swamplands and marshy ground, the water table is either at or very close to the surface and the intermediate zone is lacking. Elsewhere, as in desert lands, the water table may be scores, or even hundreds, of meters underground. Contrary to what many people think, the zone of high water content under the water table does not continue downward indefinitely into the earth. In other words, drilling a well to great depths will not necessarily increase the flow of water. With increasing depth the pore spaces in the rocks close up, and with this decrease in size the water-bearing capacity diminishes

Fig. 9-3   A small volcanic explosion pit on Paoha Island in Mono Lake, California. The level of water in the explosion pit is controlled not by the drainage area of the pit, but by the level of the ground-water table which in turn is controlled by the surface elevation of Mono Lake itself. Note the matching sets of ancient beach lines on the shores of Mono Lake and in the explosion pit resulting from changes in hydrologic conditions. (Photograph by Hal Roth.)

until the rocks may be completely dry. In deep mines the upper level may be flooded and require constant pumping to keep them operational while the lower levels are dry and water has to be brought down for use in drilling.

Porosity

This property is of the greatest importance in controlling the movement of water in the ground. We are familiar with the general meaning of the word when we think of a porous substance as one that contains many holes. Actually, porosity is the percentage of the total volume of the rock that is occupied by openings. If half the available space were to be in openings, the material would have a porosity of 50 per cent; if only one quarter, then 25 per cent, and so on.

Many factors determine the porosity of a rock, and chief among them is the arrangement of the constituent particles in such material, for example, as sand. Should the sand grains be arranged as in Figure 9-4a, the porosity is around 47 per cent; if they are packed more densely, as in Figure 9-4b, the porosity drops to approximately 26 per cent.

An important point is that the *size* of the spherical grains in this illustration remains the same in both cases; it is the arrangement that is different. It does not matter at all if the grains are the size of BB's or basketballs. Porosity is a wholly relative matter. In fact, relatively fine-grained materials, such as silt, may have higher porosities than such seemingly open material as gravel. Part of the reason for this is the pore spaces between the individual gravel particles may be occupied by sand, or even finer fragments, and this sharply reduces the size of the pores. The porosity is also drastically decreased if any natural cement is introduced into the interstices. In other words, the porosity of sandstone will be less than that of sand.

Among the higher porosities are those of newly deposited muds, such as those of the Mississippi Delta. These may actually reach the incredible value of 80–90 per cent—which means they are *dilated*. This is to say that they contain so much water, as a quicksand does, that the individual particles scarcely touch one another. This situation is the exception, and more normal porosities are less than 15–20 per cent. In tough, dense, homogeneous rocks, such as obsidian, basalt, or granite, the porosity may be as low as a few hundredths of a per cent.

Permeability

This property of a rock is a measure of its capability of having a liquid transmitted through it. Therefore, the actual size of the openings in a rock is of vastly greater importance than the percentage that is open space. For example, a silt or clay may have a higher porosity than a gravel, but the permeability is less. Along with the size of the openings—large ones obviously being more permeable than small—is the matter of connections between the openings. If the pore spaces in a rock are not joined together, then water will not flow, no matter how large they may be.

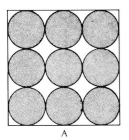

A

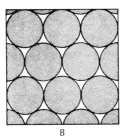

B

Fig. 9-4 Porosity (a), spherical grains packed in such a way that intergrain voids make up approximately 47 per cent of the volume; (b), spherical grains packed in such a way that intergrain voids make up approximately 26 per cent of the volume.

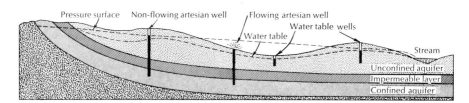

Fig. 9-5   Aquifers and water wells.

Aquifers

Not all rocks are equally permeable, nor do they all have equal water-holding capacities. A layer such as a permeable, highly porous sandstone not only may be able to hold much more water than its enclosing rocks but also may provide a route along which ground water moves with relative freedom. Such a favorable layer that readily yields water to a well is called an *aquifer*. One that is too impermeable or tight to accept water is called appropriately an *aquifuge*.

There are many different kinds of aquifers, but two of the common types are shown in Fig. 9-5. The first, or *unconfined aquifer*, is the simpler. This may involve little more than a relatively permeable layer of sand under perhaps a cover of clay or loam. The water table, as we have already seen, is very likely to be a subdued replica of the ground surface, and for this reason the water level in any two wells may not stand at quite the same altitude. Furthermore, the two water table wells shown on the diagram are non-flowing and thus will have to be pumped.

Figure 9-5 also shows a *confined aquifer*. Here a layer of permeable sandstone is enclosed between layers of impermeable shale. Typically, a sandstone aquifer may crop out in a band paralleling a mountain front and then dip below the adjacent plain. This is the relationship east of the Rocky Mountains and under the Great Plains portion of the Dakotas and Colorado. The aquifer there is the Dakota sandstone, and since the first wells were drilled into it in the 1880's it has yielded a prodigious quantity of water. Wells when they were first drilled flowed at the surface of the ground over much of the area underlain by this productive, water-bearing layer. With the loss of pressure through the years, many of the wells are now on the pump.

Artesian Wells

Wells that flow at the surface of the ground, such as the early ones drilled to the Dakota sandstone, were called *artesian wells* from the Roman province of Artesium, now Artois in sourthern France. For almost everyone the term artesian means a well that flows freely. In practice, though, the term has a more restricted use, and is now applied to a well in which the water is under pressure because a confined aquifer has been penetrated (Figs. 9-5, 9-6).

Whether or not the water reaches the top of the ground depends on the relationship of the *pressure surface* and the shape of the terrain (Figs. 9-5). In a flowing artesian well the pressure surface is above the ground and in a nonflowing well it is below the ground level.

The pressure surface is the level to which water rises in a confined or unconfined aquifer. Theoretically, in a confined system it could be equal to the highest point on the aquifer which is shown on the diagram projected as a level line from the water table out into the air (Fig. 9-5). The pressure surface does not coincide with this line because of the frictional loss of energy of the water as it moves through the aquifer.

Should a large number of wells tap an artesian reservoir, the pressure drops and the flow will ultimately diminish. Such was the case with the Dakota sandstone and the other aquifers associated with it. Where forty to seventy years ago water poured out of the ground under high enough pressure in some places to operate waterwheels, now, after the drilling of around 10,000 wells, the pressure has dropped to the point where many have to be pumped, and in flowing wells the yield is only a few gallons per minute.

Fig. 9-6   Artesian well. Gallatin County, Montana. (Photograph by J. R. Stacy, U.S. Geological Survey.)

Pumping Wells

Those who have lived on a ranch dependent on a well for irrigation water are fully conscious of the fact that when the well is pumped the water level in it drops. A short time after pumping ceases, the water rises, although not always to the level it may have had before should the drawdown be exceptionally severe and continued.

How much of an effect does a single well have on the water table of an entire district? Does the water level in all the adjacent wells rise or fall in concert? The answers to these questions have been established through observation in many localities over many years. If the well is pumped heavily, and water is taken out of the ground faster than it can be replenished, then the water table is pulled down in the form of an inverted cone centering on the well, and this is known as a *cone of depression*. Obviously the water level in nearby wells will be affected more drastically than in more distant ones. Studies show that the effect of an individual well may be felt by others over distances of as much as a quarter of a mile away.

Hard pumping by many wells serves to make the rims of individual cones overlap until the water level of an entire basin may be lowered. A striking example in the United States is the southern part of the San Joaquin Valley in California. This central valley of the state has long depended on pumped water for its agricultural survival. Something like 40,000 wells lift around 1,234,000 liters (326,000 gallons) a year out of its underground reserves, which is approximately 25 per cent of all the ground-water yield of the United States. This quantity of pumped water is 264,978 more liters (70,000 gallons) than are supplied by infiltration from streams of the nearby Sierra Nevada, from rainfall, and from other sources such as leakage from irrigation ditches. The result has been a lowering of the water table throughout the years; in some places as much as 76.2 meters (250 feet) since 1905. Close to the western margin of the Sierra Nevada and about the mid-point of the valley, the water level in wells in one district dropped from 16.7 meters (55 feet) in 1921 to about 45.7 meters (150 feet) today. No wonder, then, that California is engaged in one of the most prodigious efforts in the history of mankind to transfer water from the northern, more generously endowed parts of the state to the parched southern counties.

Extensive withdrawal of ground water may have other unfortunate effects. One of these is subsidence which occurs in areas underlain by unconsolidated sediments. These sediments become compacted as their water content is removed and the land surface subsides. Sinking of the Santa Clara Valley, California, was first noticed as long ago as 1920, and since then six million dollars have been spent building levees to keep San Francisco Bay from flooding this fertile valley. The only remedy would be to recharge the aquifer with water brought it from elsewhere which would stop the subsidence.

Geologic Role of Ground Water

Water in the ground does work of geologic significance comparable in many ways to the more visible achievements of rivers, glaciers, lakes, and the sea on the earth's surface. Among its more significant accomplishments is providing the means by which the various natural cements, such as calcite ($CaCo_3$), silica (SiO_2), and iron oxide (Fe_2O_3), are introduced into the pore spaces of unconsolidated sediments. These are reasonably soluble substances, and they may be dissolved from rock or soil layers by water when it starts its journey underground. Later, when the saturation is sufficiently high, and temperature and pressure relationships are right, these substances may come out of solution. Gradually, as they are deposited on the surface of individual grains, much like scale is deposited on the inside of a kettle or hot-water heater, pore spaces become drastically reduced until finally they may become sealed off almost completely, with an accompanying fall-off in permeability.

Limestone Caverns

Ground water plays a unique role where rocks are readily soluble in water. Among these are limestone, marble, gypsum, salt, and other evaporite deposits. The dissolving of these rocks and thus the slow wasting away of their substance underground, may be compared to erosion by surface streams, just as the process of cementation described in the preceding paragraph in a sense is equivalent to deposition.

The most widely known effect of the solution of rocks by ground water is the formation of caves, such as the Carlsbad Caverns, Mammoth Cave, and the ones decorated by Stone Age Man in Europe. In addition to these examples, there are scores of others in many parts of the world. Their dark, silent recesses have intrigued explorers since the beginning of time, and even today there are few states without active speleological groups within their borders.

The origin of limestone caves has long been debated, and is far from being settled. The crux of the debate is whether or not the solution responsible for the removal of thousands upon thousands of cubic yards in some of the larger caverns occurred above or below the water table. The problems here are: (1) in those parts of caverns today that are above the water table the leading process appears to be deposition rather than solution. At least this is the process responsible for making stalactites and stalagmites. The alternative (2) that solution occurred below the water table encounters difficulties because the water very often is already saturated with lime; it cannot pick up any more and thus solution stops. To get around this dilemma a continuing supply of circulating water is called for, and this also has to have a low content of lime in solution—a difficult feat to achieve in a region dominantly underlain by limestone.

Fig. 9-7   Lehman Caves, Nevada. (Photograph by Hal Roth.)

A further complicating factor is that many caverns include deposits of clay, silt, and even gravel. This leads some geologists to conclude that many such caves were eroded, at least in part by subterranean streams. Such rivers are fairly common in limestone terranes; there are quite a number in Indiana and Kentucky.

A theory that appears to be applicable to the Carlsbad Caverns is that caves were formed by solution at a time when the water table stood higher than it does now. As a result of canyon cutting by nearby streams the water table was lowered and passageways made by solution along joints and bedding planes were then opened to the air. Following this it was possible for such distinctive features of the cave world as stalactites, stalagmites, columns, and ribbons and sheets of travertine to grow.

Few geologic phenomena arose more curiosity than the strange, in fact, eerie patterns made by dripping water in the timeless darkness of caverns underground. Most familiar of these to visitors to the great number of national, state, and privately controlled caves are the iciclelike pendants of travertine hanging down from the cave roof (Fig. 9-7). These are *stalactites*, and they normally form where dripping water seeps from the rocks above the cave. When this water reaches the air some of the CO_2 contained in solution escapes, thus increasing the saturation of the water to the point where $CaCO_3$ is deposited. Also, if some of the water evaporates, a residue of lime is deposited. Since the drop of water that hangs suspended momentarily from the cave roof is likely always to be about the same size, the tiny ring of travertine left by it will nearly always have the same diameter. Gradually, these successive rings pile up to form an icicle-like pendant, customarily with a narrow tube extending for its length. Seldom, though, is such perfection actually achieved or long maintained. The tube may become plugged, the amount of water may vary, or new holes may break out along the sides rather than at the tip. All these vagaries lead to the great variation in form that stalactites show.

Stalagmites are deposits built upward from the cave floor, and characteristically they grow below a stalactite. When a drop of water falls from the tip of a stalactite it may lose some of its CO_2 content on landing; or its water may become concentrated through evaporation, with the result that more lime is deposited. Thus a counterpart accumulation of lime gradually builds up from the floor of the cave to oppose the stalactite growing downward from the roof. Stalagmites, unlike stalactites, do not have a central tube, and since they are built up by the water that spatters over their surface they usually are thicker and have more diversified shapes. With two such structures growing toward one another, if the initial distance separating them is not too great, and if an adequate rate of growth can be maintained, stalactites and stalagmites eventually meet and fuse. The resulting column of travertine is called a *pillar*.

Other cave deposits may take on a wide variety of shapes, and these fluted, or columnar, or sheet-like masses are the ones that guides or operators of caves use in achieving imposing feats of indirect lighting effects.

Karsts

The landscape that may develop in a region underlain by limestone differs in a multitude of ways from one characteristic of less soluble rocks. A striking and well known region of this type is the *karst*, the portion of Yugoslavia bordering the Adriatic, the Dalmatian Coast. This is one of the picturesque coasts of the world, with the sea penetrating far inland in long, fjord-like inlets. These are separated by barren, whitish limestone ridges and islands which contrast vividly with the wine-dark waters of the sea—to use the 2700-year-old imagery of Homer.

This is one of the historic coasts of Europe. The once-forested slopes of the now barren hills of what was then known as Illyricum provided timbers for the galleys of Rome, and later for the wide-ranging vessels of the Venetian Maritime Republic. Today this is a harsh, stony land, and it is difficult to visualize the widespread forest that once mantled its slope before destruction by overcutting and overgrazing.

This region has one of the heavier rainfalls of Europe, yet it is strikingly devoid of surface streams. Limestone is so permeable that rainwater sinks rapidly into the ground, especially if joints and other fractures abound. Streams flow for short distances, disappear underground, and then reappear several kilometers away as a river emerging full-born from a giant spring.

Such a limestone terrane as this is pocked with large numbers of closed depressions, some large, some small. Commonly such depressions are floored with clay, and this thin accumulation of reddish soil (characteristically called *terra rossa* in the Mediterranean world) is likely to be all that is available for agriculture. In Yugoslavia the larger depressions may be several kilometers across—large enough at any rate to shelter a village and its surrounding patchwork of fields. The origin of these large depressions is uncertain. While they are partly due to removal of material by solution, part of their origin also appears to be the result of the folding and faulting of the underlying limestone.

Smaller closed depressions in Yugoslavia are almost certainly caused by solution. Some of them extend downward into the earth by near-vertical shafts which commonly lead into deep caverns. In America, such solution pits are called *sink holes*, and they may be open to the air, or they may be closed and floored with clay, or they may hold small lakes. These lakes are an impressive sight from the air, especially when sunlight glances from their surfaces. Unlike lakes in other kinds of ground, the water surface stands at different levels because the lake is floored with virtually impervious clay. Should this clay seal be broken, then the lake will drain away through solution channels into the underlying limestone.

Sometimes sink holes serve as natural wells. Their steep sides extend downward for scores or even hundreds of feet, until they intercept the water table which stands as a pool of water, somber and green, at the bottom. Renowned

Fig. 9-8   Home lost in the collapse of a sinkhole in Bartow, Florida. (U.S. Geological Survey.)

of such occurrences are the *cenotes* of Yucatan. The Yucatan Peninsula is a nearly level limestone plain, devoid of surface streams because the rainwater sinks almost immediately into the ground. When the peninsula was the site of the Mayan Empire the dense agglomerations of people at such cities as Chichén-Itzá were dependent upon so slender a supply of water as the dank fluid at the bottom of a limestone sink. No wonder that, to preserve this tenuous link with survival, a maiden burdened down with bangles and ornaments was ceremoniously hurled into the cenote each year in order to assure a continued supply.

Sinkhole-prone areas around Bartow, Florida, have been under study using airborne remote sensing devices. This is a vexing geologic hazard common to many areas in Florida and in other parts of the country where it causes damage to homes (Fig. 9-8) and building foundations and makes the maintenance of stable road beds for highways very difficult. Data obtained from instruments in airplanes reveals thermal and vegetation patterns that may indicate the presence of caverns subject to collapse.

Prospecting for Ground Water

"Prospecting" for water is not so strange as it may sound, for ground water is essentially a mineable resource. Radiometric dating has shown that thousands of years may be required to accumulate water in underground reservoirs, while the permanent lowering of water tables with continued pumping indicates that annual rainfall is not sufficient in most cases to replenish the amount of water removed. As the demand for ground water increases, because of the population explosion and because of the increasing pollution of surface waters, prospecting for new water supplies becomes an important task for the hydrologist.

The search for ground water has produced its colorful prospectors, just as has the search for gold, although they may not be as picturesque as the grizzled old man with his burro and pick and gold pan. It has long been customary to use "water witching" to determine the location of a new well. The water witch walks back and forth over the land, holding two ends of a forked stick, or "dowsing rod," and keeping the stick horizontal. When the stick, through some magical power, dips sharply downward of its own accord, the dowser announces that this is the place to dig. More often than not this patently unscientific method brings in a good well, at least often enough to perpetuate the belief in water witching, even among those who should know better. One reason it is successful is that in many places in the world there is a plentiful supply of ground water no matter where a well happens to be drilled. Another is that some water witches are very observant. They have noticed that the water table is closer to the surface in valleys than on hilltops, or that certain types of vegetation are indicators of water.

Much more accurate methods, however, are now required to find the vast quantities of water we will need. This is especially true of the underdeveloped nations, many of which are in arid regions where water must be found before agricultural production can be increased.

There are several direct methods of searching for water. Mapping the rock units present in an area will show where potential aquifers may be expected, as will the well "logs," or records, of existing wells. Indirectly, underground rock structures are indicated by man-made earthquakes. This method, called seismic refraction, depends upon the fact that earthquake waves travel at different speeds through different rocks, according to whether they are solid or porous and whether or not they contain water. Much the same information can be determined by measuring the resistance of rock formations to an electrical current. Infra-red aerial photography can show the temperature differences between areas underlain by water-bearing formations and those that are not.

Because ground water is in a sense an unrenewable resource, its proper exploitation and management is of the utmost importance. The Sahara Desert, for example, is underlain by seven basins, each of which has a tremendous supply of water waiting to be tapped. These basins, however, do not coincide with national boundaries, and co-operation among the thirteen countries involved is imperative for the wise and efficient development of this resource. In such cases as this a computer model can be useful. Computer models for an entire ground-water system can simulate the flow of the ground water, the soil conditions at each well, the amount of water being removed, and the actual water table level at any time. It can also predict the effects of future withdrawals and help plan the distribution of wells and the timing of water removals. In some areas efficient management may require the collection and storage of rainfall run-off in huge underground reservoirs to prevent excessive loss through surface evaporation. In others it may be economically feasible to recharge a ground-water system by pumping into it water brought from a distance.

Geysers and Hot Springs

By far the most spectacular manifestation of ground water is its appearance at the surface in the form of geysers and hot springs. Certainly they are the leading attraction of Yellowstone National Park, and it is a rare household that does not include a member who has seen Old Faithful run through its repertoire. Yellowstone is not the only geyser area in the world; in fact, the extensive one of Iceland gives its name to this sort of aqueous outburst, since all are named for a large Icelandic spring, *geysir*. Another large and touristically attractive geyser region is the Rotorua region of North Island in New Zealand. This is currently being developed as a source of thermal power.

Although the actual process that goes on in an erupting geyser is something of a mystery, enough is known of the physical laws operating so that a plausible explanation can be offered. Incidentally, its general elements are much the same as the one advanced by the German chemist, Bunsen, whose burner is known—sometimes by direct personal contact—to every student of chemistry.

A generally held view is that ground water percolating downward in a geyser area comes in contact at depth with a source of heat. This may be cooling volcanic rocks, or steam, or other gases given off by magma. Even though the water at the bottom of a tube may be heated to 100° C. (212° F.) it does not boil because the boiling point is raised with an increase in pressure. We are more familiar with the opposite effect—the lowering of the boiling temperature in the thin air of high mountain tops to the point where potatoes, for example, do not cook through.

Thus in a column of water standing in a tube-like opening in the ground the temperature at the bottom of the column may rise above the boiling point at normal atmospheric pressure. However, nothing is likely to happen until all the water is heated to the top of the column, perhaps to the point where it begins to surge, or spill over the rim. Should enough drain off, then the pressure throughout the column is reduced, with the result that superheated water near the bottom flashes over into steam. This is enough to propel the whole column of water upward, and since a similar pressure reduction and near-instantaneous conversion to steam occurs throughout its length, a mixture of hot water and steam is hurled skyward—in Old Faithful for

Fig. 9-9   Old Faithful geyser, Yellowstone National Park, Wyoming. (Photograph by Ansel Adams.)

something like 45.7 meters (150 feet). It is on the details of how the subterranean geyser reservoir is filled after being blown clear, and on how some geysers achieve their remarkable periodicity that much of the debate centers.

The castellated rims, platforms, and parti-colored deposits surrounding the geysers and hot springs of Yellowstone are especially interesting features for the park visitor. In general, there are two kinds of hot water deposits. Those deposited directly from mineral-rich geyser water often are composed of silica—supplied in part from the underlying volcanic rock—and these deposits are called *siliceous sinter*. They are likely to be grayish colored and to consist of amorphous silica, very much like opal. Limy deposits, made by calcareous algae that can survive in the temperatures of hot springs and pools, are called *travertine* (Fig. 9-10).

Hot springs are more widely distributed over the face of the earth than geysers. There are over one thousand in the United States, and most of them are located in the montane parts of the Far West. Fundamentally, hot springs are a consequence of bringing ground water into contact with a source of heat in the earth's crust. Typically, this source may be volcanic rocks that have not yet lost all their initial heat. Or it may be juvenile water, freed by igneous bodies at depth, which has cooled to something less than the boiling temperature by the time it has reached the surface.

Ground water may also make its way by means of an aquifer down to a level where the temperature of the water is raised by the temperature increase of around $1°$ F. for every 18 to 30 meters (60 to 100 feet) of depth. If this heated water can be returned to the surface quickly, without too great a reduction in temperature, a hot spring results.

Geothermal Development

Naturally-occurring areas of hot water and steam are known as geothermal regions, and man has cast his speculative eye on them and asked, "Are they of any use?" Although the development of such areas is still in its infancy, the answer is "Yes." Italy has been using volcanic steam for the production of electricity since 1904. Steam from a deep magma body rises to the surface through rifts in the rocks. It is also obtained by drilling wells, and a typical one will produce 220,000 kilograms (485,000 pounds) of steam an hour at $204°$ C. $(400°$ F.).

Japan has recently started an intensive program for producing electricity from its widespread geothermal areas. Such an energy source is particularly appealing to nations like Japan where oil and coal are almost nonexistent and hydroelectric power is limited. Since Japan must import about 70 per cent of its fuels, its geothermal potential is of great importance.

In the United States the first and the only successful large-scale exploitation of geothermal energy is at The Geysers in northern California. In spite of its name, this area has no geysers, but steam rises from hot springs, wells,

Fig. 9-10 Jupiter Terrace, Yellowstone National Park, Wyoming. (Photograph by Ansel Adams.)

and fumaroles, as well as steaming ground. Steam from wells as deep as 2700 meters (9000 feet) is piped to turbines which can consume one and two-thirds million pounds of steam per hour and produce 82,000 kilowatts of electricity, enough for a city of 90,000 persons. Continued expansion of this facility is possible and is being planned.

In Iceland volcanic steam is used to heat buildings, and steam pipes placed in the fields warm the soil so that crops that ordinarily would not survive in that climate can be grown. Steam for heating and for the generation of power should be relatively pure. Some hot springs, however, are highly charged with chemicals, and their water can quickly corrode pipes or deposit a heavy scale in them. While these pose handling problems, they can also be the source of valuable chemical industries. Such an area occurs in the Salton-Mexicali Trough which straddles the California-Mexico border east of San Diego. Here 20 per cent of the briny water that comes to the surface changes naturally and instantly to steam which is used for generating electric power. The remainder of the water is pumped into ponds where solar evaporation concentrates potassium, sodium, and calcium chlorides. The highly concentrated brine which is left presents a serious disposal problem, for even though it is rich in elements such as lithium, lead, copper, and iron, among many others, these cannot be economically extracted at present.

Selected References

Ambroggi, Robert P., 1966, Water under the Sahara, Scientific American, vol. 214, no. 5, pp. 21–29.

Leopold, L. B., and Langbein, W. B., 1960, A primer on water, U.S. Geological Survey, Washington, D.C.

Marsden, Sullivan S., Jr., and Davis, Stanley M., 1967, Geological subsidence, Scientific American, vol. 216, no. 6, pp. 93–100.

McNitt, James R., 1960, Geothermal power, Mineral Information Service, vol. 13, no. 3 pp. 1-8, California Division of Mines and Geology, Sacramento, Calif.

Meinzer, Oscar E., 1939, Ground water in the United States; a summary, U.S. Geological Survey, Water-Supply Paper 836-D.

Stefferud, Alfred, and others, 1955, Water: yearbook of agriculture, U.S. Department of Agriculture, Washington, D.C.

Fig. 10-1   Sea stack. Point Loma, California. (Photograph by William Estavillo.)

10

Ocean-Land Interfaces

One of the most visible and dramatic interfaces on the earth occurs where the land and the ocean confront each other. It is an area, unique both physically and biologically, that has fascinated man for thousands of years. Even before the bikini and the surfboard, the coastal regions of the world have been foremost among man's playgrounds. When an almost irresistable force meets an almost immovable object, the resulting conflict is bound to be worth observing and investigating.

Waves

Waves are intriguing and can be almost hypnotically fascinating. Although one wave may look like any other, no two are ever the same. Their rhythmic beat depends not only upon the local wind for the shorter, steeper waves, but also upon far distant fiercer winds that have set the long, even-spaced ridges of the ground swell moving outward from a storm center half a world away.

As we watch the endless procession of waves, it is difficult to believe that it is the form of the wave that moves forward through the water and not the water itself. This statement may not appear to make sense at first, but watch a bottle bobbing on the surface of a bay. Waves pass under it repeatedly, but

Fig. 10-2   The sea attacks a rocky coast at Shore Acres State Park, Oregon. (Photograph by Ray Atkeson.)

other than a slow drifting with the current the bottle holds its position remarkably well. An analogy for waves in the sea is the waves the wind makes when it blows across a field of grain. Ripples follow one another across the stalks of wheat, and yet the wheat does not pile up in a heap on the far side of the field. Instead, the wave motion in the grain results from the up and down nodding of the individial stalks each time a wave passes through them.

As long ago as 1802 it was known that water particles within a wave do not move forward with the advancing wave itself but follow a circular path (Fig. 10-3). Detailed studies have been made in the years since, but the basic principles of water motion in a *wave of oscillation* are the same as those discovered then. The diagram is intended to show (ideally) the paths followed by water particles within a wave of oscillation. The same diagram also shows how rapidly wave motion diminishes with depth. For practical purposes wave motion ceases to be effective when the water depth is approximately equal to one-half the wave length.

Thus far a few terms have been introduced that need some clarification. *Wave length* is the horizontal distance separating two equivalent wave phases, such as two crests or two troughs. The *velocity* is the distance traveled by a wave in a unit of time, and this can be related to its other physical properties by a number of simple relationships. The *period* of a wave is the length of time required for two crests or two troughs to pass a fixed point. The *frequency* is the number of periods that occur within a set interval of time—say a minute.

There appear to be finite limits to the size that waves can reach. Among the larger waves whose dimensions have been reasonably well established was one that was 34 meters (112 feet) high when it was sighted off the stern of the U.S.S. *Ramapo* in 1933 during a gale in the North Pacific. Wave lengths are likely to be less than most people imagine, but they are impressively large at times. One of the largest swells ever reported had a wave length of 792 meters (2600 feet), which would give it a period of 22.5 seconds and a velocity of 125.5 kilometers (78 miles) per hour. These are formidable figures when one considers the enormous masses of water involved.

How such volumes of water are set in motion is a fair question. Almost everyone knows that the wind driving across the surface of the sea is the primary cause. If this is true, though, how is it that on completely windless days a tremendous surf may belabor some exposed coast? Or that in a violent gale the wind may hammer the sea flat into a wildly turbulent mass of dark, malevolent-looking water streaked to the horizon with foam?

For the formation of large waves in deep water the following requirements must be met. First, there must be a strong wind in order to set large masses of water to moving. Second, it must have a fairly long duration—more than just a sudden gust of wind is needed. Third, the water must be deep, at least deep enough to round out the full circular pattern—9-meter (30-foot) high waves are not likely to grow in a water basin only 3 meters (10 feet) deep. Fourth, the distance wind friction can operate on waves—called their *fetch*—is important for their growth, too. When waves have a long uninterrupted run, they have an opportunity to reinforce one another. The ripples crossing a small pond are a good example. They are small on the upwind side of the pond, yet they may have grown to fair dimensions by the time they reach the downwind shore.

It is not surprising, then, that some of the largest seas are those driven before

Fig. 10-3 Cross section of ocean wave traveling from left to right, showing the wavelength, the paths of individual water particles, and the diminution in size with depth.

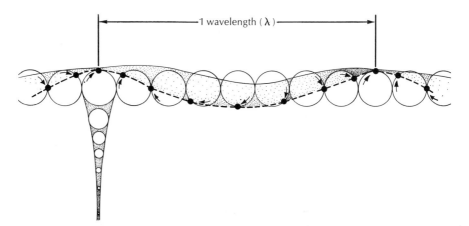

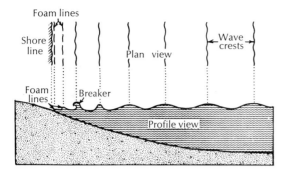

Fig. 10-4 How waves shorten, steepen, and break as they advance into shallow water.

the strong westerly winds which rule south of Cape Horn. Around the margins of Antarctica is an unbroken sweep of ocean which encircles the earth; in other words, these waves have unlimited fetch.

To return to an earlier question: how, if waves are formed by the wind, can they travel shoreward in an endlessly rhythmically advancing succession on a dead calm day? The answer is that such waves, to which the name of *swell* is given, may have originated in gales thousands of miles away. Waves that break on the exposed coast of Cornwall may have had their start in the far-distant reaches of the South Atlantic. Waves crashing on the west coast of the United States may have been born in the Antarctic, while waves that the surfers ride in Hawaii may have come from the Arctic.

The swell is made up of *long-period waves*, and these outrun the more mixed-up, randomly distributed, *short-period waves* which are characteristic of a storm. These short-period waves are left behind, and the more uniformly spaced swell far out-distances the gale winds localized around some cyclonic center.

Wave Refraction and Surf

As waves move from deep water shoreward they begin to feel bottom when the depth of water becomes about equivalent to one-half the wave length. As these waves move into shallower water near shore, their length is shortened and their height is increased relatively (Fig. 10-4). Their velocity is also reduced, and if one looks at their crests along a pier or breakwater, they do indeed seem to rise up out of the sea. This shallowing of the water at the shore has two major effects on waves of oscillation as they move landward; first, they may be *refracted*, and second, when they reach shoal water close to the beach, they break.

The effect of refraction is shown in the accompanying diagram (Fig. 10-5). Two things are especially important. First, waves approaching an irregular coast adjust themselves to the irregularities until they achieve a near-parallelism to the shore throughout all of its indentations. Second, wave attack is concentrated on the headlands, and thus these promontories are gradually beaten back. The fact that wave energy is focused on the headlands is demonstrated by the lines on the diagram, which are drawn so that they are everywhere at right angles to the advancing wave fronts. We may consider their equal spacing on the outermost wave front as indicating equal amounts of available wave energy. Then it is apparent on the diagram that most of the energy in an advancing wave is concentrated on the projecting salients along a coast and is diffused in re-entrants, such as a small bay.

As we have seen, when waves approach the shore and encounter shallower water, their length is shortened and their height increased. Finally, they become unstable and break. Without breakers, most coasts of the world would lack their most picturesque element.

The formation of surf is a complex phenomenon. The endlessly changing pattern of breaking waves—and their variations with the tide, with wind and calm, with storm, and with the lulls between—have been an inspiration to generations of painters, photographers, writers, and ordinary daydreamers. Few manifestations of the natural world are more dynamic, or make us more conscious of the force of moving water, than standing in the surging mass of a strongly running surf.

Although much remains to be learned about how waves of oscillation break, a good deal is known as a result of observations, model studies, and the application of hydrodynamic theory. In a general way, waves of oscillation break when the stillwater depth is roughly equal to 1.3 times the height of the wave. At least two major causes appear to lead to the formation of breakers. The first of these is a speed-up of the circular velocity of particles in a wave that has moved into shallowing water, which makes the velocity of such particles at the wave crest exceed the decreasing velocity of the wave form. The second is that in the shallow depths near shore the amount of water needed to complete the wave form is not available. This is especially conspicuous in the so-called *plunging breaker*, in which a wave curls over in a beautifully molded half cylinder, topples, and crashes with a thunderous roar. The entrained air is violently compressed, and in its efforts to escape converts the entire roller into a froth of foam-whitened water.

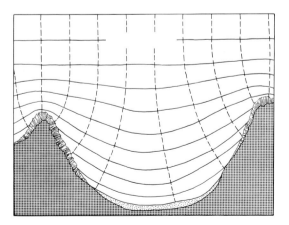

Fig. 10-5 How waves are refracted. The solid lines are wave crests, which crowd together as they move into the shallow water around the rocky headlands but stay more widely spaced in the deeper waters of the bay. Wave energy, equally distributed between the parallel dashed orthogonal lines in deep water, is concentrated where the orthogonals converge on the headlands, and is diminished in the bay.

Fig. 10-6   The coast Guard Cutter *Pontchartrain* wallows in the trough of a following sea in the North Atlantic. (Official U.S. Coast Guard photograph.)

Spilling breakers (Fig. 10-7) are those in which the crest foams over and cascades down the front of the advancing wave without actually toppling over. Such breakers ordinarily diminish in height as they move landward, and they may advance in rows simultaneously over a broad stretch of beach.

Wave Erosion

The shore zone, where breakers are able to bring their full force to bear against the land, is one of the areas where the erosive work of the sea is concentrated. Since there is an upper and lower limit to such attack, the work of waves might be likened to that of an enormous horizontal saw. The upper limit of their effectiveness is the maximum height that waves can reach at high tide during a storm. This will differ among coasts, depending on how boldly the land faces the sea and how strongly gales drive waves before them. It also depends upon how long the generating wind continues to blow in one direction, and, of course, upon the fetch. On the more rugged coasts, storm-driven waves have been known to reach heights of 60 meters (200 feet) or so.

A *tsunami*, an earthquake-caused wave, at Hilo, Hawaii, on May 23, 1960, provides an excellent example of the power concentrated in a moving mass of sea water. There, in a series of waves, the largest reaching a maximum height of 10.6 meters (35 feet), devastated the waterfront section of the town, taking sixty-one lives and causing $20 million worth of damage. Geologically, an impressive result was a demonstration of the transporting ability of moving water. Automobiles were swept for whole blocks, and in some cases were piled three-high; in others they were wrapped like limp fish around the bases of coco palms. These waves, which were not far from storm waves in their dimensions, moved generators and sugar mill rolls and ripped up whole slabs of asphalt paving. Boulders from the sea wall, weighing twenty-two tons or so, were transported inland 152 to 182 meters (500 to 600 feet).

Many of the waves that caused the damage were approximately 6 meters (20 feet) high and, according to wave theory, moved through the shallow water of the harbor at 72 kilometers (45 miles) per hour. This means that they were capable of exerting pressures of nearly two tons to the square foot against objects placed in their path. Other large waves have had measured forces of up to seven tons per square foot.

The lower limit of wave erosion is much less certainly known. Estimates vary widely, possibly reflecting the prejudices of the estimator. Observational evidence seems to indicate that most waves can generate currents capable of moving sand at depths of not much more than 12 meters (40 feet)—although

Fig. 10-7 Spilling breakers. (Courtesy of Humble Oil and Refining Company.)

Fig. 10-8   Wave erosion undercuts a sea cliff. Baja California, Mexico. (Photograph by William Estavillo.)

ripples and other signs of disturbance of the sea floor are often seen at depths of 106.6 meters (350 feet) below sea level. Gravel and cobbles are not moved by wave currents much below 6 meters (20 feet). The lower effective limit of wave transportation and erosion is called *wave base;* obviously its depth differs on different coasts, just as the upper limit does.

Waves generally accomplish most of their erosion by abrasion, just as streams do. In times of calm, very little erosion takes place; in fact, if there is a source of sediment, deposition commonly occurs. During storms, however, when the waves are highest and most capable of carrying abrasive agents such as sand, gravel, and even cobbles, their erosive power is great. Since waves cannot erode above their maximum height, their action is restricted to a narrow zone of the shore. On a cliffed shoreline the base of the cliff is undercut, and if the rocks are weak because of jointing or poor consolidation, landslides will occur. In such cases landslides are actually the eroding agent of the upper parts of the cliff.

The shore zone is one of the most dynamically active erosional theatres on earth. During high tide, it is submerged. When the tide is low, and especially if the tidal range is great, then a broad expanse of the shore zone may be exposed to the air where it can be acted upon by agencies such as the rain and the wind, and in the Arctic it may be modified by shore ice. In limestone regions, the alternation of wet and dry periods is important in erosion also.

Shoreline Classification

Though the problem of classifying shorelines may appear simple, it is complex enough that no system devised in the last century has won universal acclaim. Even the proponents of some of the proposed classifications have grown disenchanted with their own creations.

One classification that was quite widely accepted was based upon whether the shoreline had been submerged by the ocean, or whether it was recently emerged sea floor. Local crustal instability has caused some coastal areas to be uplifted or depressed relative to sea level. In addition, there have been actual changes in sea level itself. By their very nature these latter will be world-wide in their effect, and to them the name of *eustatic* change of sea level is given.

That sea level is inconstant may surprise some people (Fig. 10-9); it has certainly fluctuated widely through quite a vertical and horizontal distance in postglacial times, or within the last 17,000 years. During the latter part

Fig. 10-9   Recent changes in sea level.

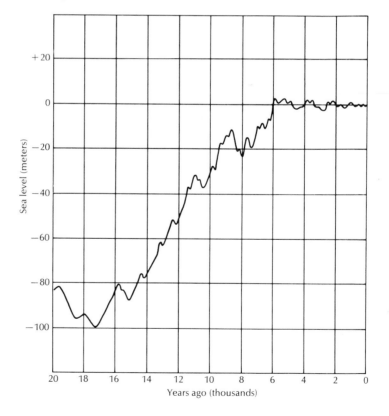

Fig. 10-10   Elevated marine terraces. Monterey County, California. (Photograph by B. Willis, U.S. Geological Survey.)

of the ice age, sea level was as much as 183 meters (600 feet) lower than it is today. Then sea level apparently rose in a series of steps up to about 4200 years ago. In these last 4000 years it has remained fairly constant, with a very slight rise. Interestingly enough, this world-wide relative stability coincides roughly with the appearance of the maritime civilizations around the shores of the Mediterranean Sea and the Persian Gulf, so that ancient harbors of the Egyptians, Persians, Phoenicians, and Minoans correspond roughly to the sea level of today.

Today's sea level is far from stationary, however. Beginning around 1850 it started to rise, as shown by a careful comparison of tide gauges from seaports all over the world. This rise, which now amounts to about 11.4 centimeters (4.5 inches) per century, almost certainly correlates with the virtually world-wide recession of glaciers and is a result of the return of their water to the sea.

On some coasts the effects of this eustatic change are augmented or diminished by movements of the earth's crust. In areas such as northern North America and Scandinavia, which were burdened under a load of ice in the

Pleistocene, the land is now rebounding. Perhaps the greatest rate measured at the present time is along the shores of the Gulf of Bothnia where the rise amounts to 11 millimeters per year (slightly less than 0.5 inch).

Other coasts, such as those of the Netherlands which are in a deltaic area of rapid subsidence, are sinking, in this case at the rate of around 20.3 centimeters (8 inches) per century—about half from subsidence and half from rising sea level—so that, beginning with the great floods of medieval times, the Dutch have been compelled to resort to the construction of an extraordinary complex of dikes and coastal defenses upon which their national survival depends.

The significant thing to remember is that the last major postglacial change to affect the shorelines of the world has been the rise of sea level by something like 183 meters (600 feet), thus making nearly all the coasts of the world show the effects of submergence. This may take the form of long incursions of the sea into former river valleys, such as those of the Delaware, the Hudson, and the Potomac, or the invasion of coastal valleys, such as those forming the arms of San Francisco and Sydney harbors, or the flooding of ice-deepened valleys, such as the fjords of Norway, Scotland, Labrador, Alaska, or New Zealand.

It is a rare coast where recent uplift has overcome the 183 meters of submergence, so that the effects of elevation are dominant over those of subsidence. The coast of southern California is one of these, however, as is shown by a series of elevated marine terraces stepping up and back from the shore. Even in those parts of Scandinavia where postglacial uplift amounts to nearly 305 meters (1000 feet), this has not been enough to drive the sea from the fjords, where the water extends 80 kilometers (50 miles) or more inland. In this case, though, the bottoms of the fjords have been overdeepened by glacial erosion and are probably below sea level even in interglacial times. Nevertheless, the complicated effects of these emergences and submergences mean that almost no shorelines can be classified as purely one or the other. Most are complex combinations of the two.

Other classifications have been based on whether a shoreline is erosional or depositional. This can be useful for some coasts, but in other areas erosional and depositional periods are cyclical. Winter storms may remove all the sand from a beach, sand that is redeposited during the calmer summers. Occasionally there is an even shorter cycle of approximately two weeks which is related to the lunar tidal cycle.

One factor which makes classification difficult is the fact that shorelines range through all the climatic environments of the world, from Antarctica,

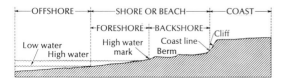

Fig. 10-11 Subdivisions of the shore zone.

Fig. 10-12 The embayed coastline north of San Francisco, California, showing the Pacific Ocean at the left and part of San Francisco Bay at the right. (Photograph by Aero Photographers.)

where for thousands of miles the shore is an unbroken cliff of ice, to coasts in the tropics which may be fronted by mile-wide forests of sea-dwelling trees such as the mangrove. In addition, the many kinds of rocks that crop out, their attitudes, and their relative resistance to erosion control to some degree the pattern of landforms that may develop along a particular coast.

Examples of Coastal Development

Recognizing, then, the extraordinarily complicated nature of the coastal environment and the lack of any satisfactory classification of the world's shorelines, the discussion here is limited to only two representative types out of the myriad possibilities. The two selected, the *embayed coasts* and *plains coasts*, are typical of many of the coasts of the temperate parts of North America and Europe. The same principles that are developed in the description of these two types can be applied to other shorelines in other lands and in other environments.

Embayed Coasts

Typically these are coasts in which the sea extends inland, sometimes for long distances, in embayments such as those shown in Figure 10-12. Should these indentations have been shaped by stream erosion before their invasion by salt water, such an embayed coast is known as a *ria coast*, from the name applied to the southern shore of the Bay of Biscay. The Gulf of Maine and the southern coast of Brazil are other examples.

How a particular coast may become embayed may be difficult to determine. Perhaps the land subsided, in which case the sea invaded or "drowned" pre-existing river valleys. This is not likely, however, as subsidence usually occurs where there is rapid deposition, a condition not typical of ria coasts. More probably the land remained stationary and the postglacial rise of sea level flooded low-lying parts of it. In either case the effect is the same.

Something of the progressive changes that the relatively commonplace ria coast is likely to undergo as a result of its modification by the waves and currents of the sea are shown in the accompanying set of diagrams (Fig. 10-13).

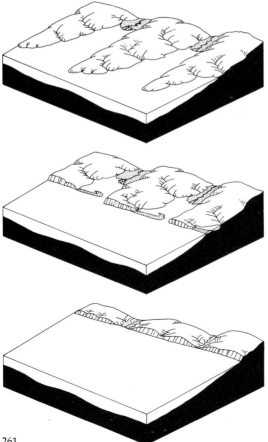

Fig. 10-13 Development of ria coast. In the top diagram, the embayments are as yet unmodified by waves and currents. In the center block, cliffs front the headlands and spits partly block the bays, into which streams are building deltas. In the lower diagram, the coast has been straightened to a line of cliffs.

261

In the first stage the sea has come to rest upon a land mass whose surface has been shaped by stream erosion. Former ridges now extend seaward as headlands, and the sea may reach inland as an embayment or estuary, perhaps much as Chesapeake Bay does.

The second diagram shows some of the changes that might be anticipated with time. Since most of the erosional energy of the sea is concentrated on the headlands, they are soon made to terminate in sea cliffs. If there is a longshore current in the sea, then the debris supplied by the rivers is deposited in a submarine embankment called a *spit*, which may partially enclose the entrance to the bay. Streams entering the bay deposit their sediment when their velocity is abruptly checked and thus build deltas out into the relatively still water.

As a sea cliff recedes before the onslaught of the waves, a planed-off rock bench is cut at its base, called a *wave-cut platform*. Sometimes it will be bare, abraded rock, interrupted, perhaps, by tide pools, and occasionally by unreduced remnants of the cliff, known as *stacks*. Where the platform is mantled with sand, it is the *beach*. Seaward of the platform there may be an accumulation of wave- and current-transported material which constitutes the *marine-built terrace*. Whether this last feature will exist or not depends in part on the strength of waves and currents, especially the longshore current. Should the current be strong enough, it may sweep the sediment into the mouth of the next bay down the coast instead of accumulating it at the base of the cliff from which it came.

In the later history of such a coast the headlands may retreat inland until a time comes when an equilibrium is formed between sea erosion and deposition of sediments from terrestrial erosion. If there is very little sediment being added, as in southern California, for example, the headlands may recede as far as the innermost bay head. Then the coast will have lost its original indented character and will be cliffed throughout much of its length. What irregularities there are will very largely reflect the relative resistance of the rocks cropping out along the cliff face.

Plains Coasts

Some coasts of the world are fringed by long narrow sand bars roughly parallel to the coast but separated from it by a narrow body of water. These are called *barrier beaches*, *barrier islands*, or *barrier spits*, depending on their form. A map of the United States shows that such barriers make long chains, like linked sausages, extending from Long Island, New York, around Florida and the Gulf of Mexico, to the southern end of Texas. A comparable chain of sandy islands stretches along the low coasts of the Netherlands, Friesland, and Denmark around the southeastern margin of the North Sea.

In the Civil War these sandy bars and islands were the locale of scores of amphibious landings and minor naval operations. Gaining control of the off-

Fig. 10-14   A curved spit nearly closed the opening into Bolinas lagoon, north of San Francisco, California. Tidal channels meander across the mud flats. (Photograph by Aero Photographers.)

shore islands and the passes between them was of the utmost importance to the Union Navy and proved to be a formidable task. The shallow, unlighted passes (or openings) through the barrier islands, with their endlessly shifting channels, were a boon to the blockade runners whose shallow-water craft were more maneuverable than the vessels of the blockading fleet.

Unpromising as such low, and often unstable, islands may be, on occasion they may be the sites of large and flourishing communities—including such metropolitan concentrations as Galveston and Atlantic City, as well as scores of beach and resort centers.

The diagram (Fig. 10-15) of a cost bordered by these sandy islands shows that it has about the same pattern as the coast where cities such as these are built. To reach such a city one must cross a broad stretch of shallow water, a *lagoon*, separating the island from the mainland. The island itself is commonly only a few feet higher than sea level, and the most conspicuous natural objects on its surface are sand dunes, which are much overshadowed at Atlantic City by the high-density development of beach hotels.

Fig. 10-15 A plains coast, bordered by an offshore bar on which are sand dunes. The bar shelters a lagoon.

Barriers have their beginnings some 20,000 years ago, during the ice age, when sea level was much below what it is today. A great part of the continental margins which are today submerged was then dry land. The ancient beaches at the land-ocean interface had landward ridges built by storm waves, much as beaches do today, and, just as now, these ridges stood above sea level. As the great ice sheets melted and sea level rose, the beach ridges moved landward as the beaches did, but often there was a depression on the landward side. Where these depressions were below seal level they formed lagoons, and the progressing storm ridges became the barriers we see today. They were able to maintain their height above sea level because the newly submerged land areas were a source of sediment which the waves used to build the barriers ever higher.

Wayward Beaches

"Beaches are the ocean's welcome mat." They are more than that, however; they are also the land's first line of defense against wave erosion, particularly by hurricane and other storm waves. Much of the force of such storm waves is expended upon the gradual slope of the beach, and even if the beach is removed, it will usually be rebuilt by deposition in more peaceful times if there is a source of sediment. Sea cliffs do not have this renewal ability; once parts of them are eroded and removed, they are gone forever. There are protective as well as recreational reasons, then, for close observation and study of our beaches in order to preserve them.

Where do beaches come from? Beach sands are a product of weathering and erosion far inland from the ocean. These sands are brought to the coast by streams in their inexorable trip to the sea, and there they are dumped when the stream no longer has the capacity to carry them. All the sand on any given beach, however, is not necessarily derived from the nearest stream or river Longshore currents are constantly in operation shifting a river of sand parallel to the coast and predominantly in a constant direction. This *sand transport*, as it is called, provides the raw material from which the waves can build beaches between storms.

An equilibrium is built up which nicely balances the amount of sand available and the variations in wave energy. Even if some of the sand is captured by submarine valleys and shunted off to deeper parts of the ocean floor, it is replaced by sand brought down to the sea by streams, and so the balance is maintained.

Permanent loss of beaches can have a devastating effect on man and his works, and, ironically enough, it is usually caused by man and his meddling with this equilibrium. Every dam he builds upstream accumulates behind it sand that would otherwise reach the coast. Thus he intercepts the supply. In the construction of beach resorts, much of the available sand may be removed to provide flat areas for the foundations of buildings.

Probably the most disruptive activity, however, is the building of jetties and breakwaters for the purpose of improving harbors and creating marinas. This interference with the longshore current and sand transport creates all kinds of havoc, resulting in damage that is extremely expensive to correct. The longshore current will deposit sand on one side of a jetty or breakwater, which acts as a dam, and sand will accumulate until the beach is excessively wide, while on the other side, the current removes all the available sand and

Fig. 10-16 A wave-cut platform cut across Tertiary sedimentary strata, at Seal Rocks, Oregon. A sill of basalt forms the row of strata. (Photograph by Parke Snavely.)

Fig. 10-17 Elevated wave-cut platform. Point Loma, California. (Photograph by William Estavillo.)

carries it away until there is no beach at all. In addition, whatever sand does manage to get around the end of the jetty proceeds to fill up the harbor or marina the jetty was built to protect. In one case, at Port Hueneme, California, a jetty that extends seaward 914 meters (3000 feet) essentially takes the sand and leads it down Hueneme Submarine Canyon.

Unfortunately man has seldom recognized that his own structures are the undoing of the precious beaches. Instead he blames the ocean and attempts to conserve his beach property by fighting the ocean rather than working with it. In addition to building jetties and breakwaters to trap the sand, he also attempts to control erosion by building a seawall. This is seldom effective because it provides a vertical surface for the sea to batter against, rather than the gradual slope of a natural beach. The result is that excessive wave erosion not only prevents the accumulation of a new beach, but also quickly undermines the seawall itself. Commonly, too, seawalls are not built high enough to keep out the most damaging storm waves, but instead keep the water that does come over them from returning to the ocean.

Another remedy often tried is the groin, a smaller version of a breakwater. This does indeed trap the sand on its upcurrent side, but the property owner below the groin is apt to be somewhat upset when the undernourished long-shore current rapidly removes his beach. He may even erect a groin on his own property. Thus groin-building has a natural tendency to proliferate and, in fact, the New Jersey shore, for one, bristles with more than 300 groins in varying states of neglect and disrepair.

Beach engineering has progressed to the point where suitable and effective structures have been designed, but usually they cost more money than the individual property owner wishes to pay, and, in the absence of any clarification of legal responsibility, local governments are often unwilling to take on the added tax burden.

One of the remedies that has proved useful is the artificial beach, constructed by dumping a large quantity of sand brought in from elsewhere. This is tricky, however, for if the sand is not the correct size, the artificial beach may disappear faster than the original beach. If there are cobbles in the imported sand, as happened at Oceanside, California, the sand vanishes, leaving a very stony shore area. While this may not be suitable for recreational purposes, it does prevent further erosion. Off Long Branch, New Jersey, on the other hand, 459,000 cubic meters (600,000 cubic yards) of sand was dumped too far offshore. Beyond the reach of the longshore current, the sandpile remained essentially intact.

But where conditions are right, the aritificial beach works quite well, and dune fields, too, are being rebuilt by the addition of sand and stabilized with a plant cover. One of the remaining problems concerns determining the proper

Fig. 10-18   Sea caves formed by wave erosion. Adak, Aleutian Islands, Alaska. (Photograph by G. Fitzgerald, U.S. Geological Survey.)

Fig. 10-19   Wave-cut platform and sea cave, elevated above sea level since they were formed. Vancouver, British Columbia. (Photograph by W. W. Atwood, U.S. Geological Survey.)

conditions under which an artificial beach is feasible. This is a difficult research problem since measurements in a zone as active as the ocean-land interface are hard to make, although progress is being made. Sometimes it is a matter of finding enough sand. Usually, however, a suitable source can be found nearby —an upcoast harbor that is filling with unwanted sand, or an ancient beach a short distance offshore or even inland. But with man's increasing use of the shorelines of the world the search for sand will become of increasing economic interest. A proposal has been made to bring sand from the Grand Bahama Banks to replenish the eroded sand along 100 miles of the southeast coast of Florida. This would involve the design and construction of a special hopper dredge, as well as the payment of a suitable royalty to the government of the Bahamas. If we assume the average population of the cities on this coast over twenty years to be 10 million, the cost per person per year is only twenty-five cents. The company making the proposal claims that the cost of a series of stopgap measures would be at least as much—with much less satisfactory results.

The size and scope of beach erosion problems make it obvious that control cannot long be left to individual landowners. For reasons of economy alone, as well as the large-scale research, planning, and organization required, such operations will increasingly fall to local, state, and federal agencies. This situation may lead to much more public ownership of beach property and, hopefully, to its increasing availability for public recreational use.

Fig. 10-20 Wave-cut nip at base of sea cliff, now submerged. (Official U.S. Navy photograph.)

Fig. 10-21 Moriches Inlet, Long Island, New York. Currents carry great quantities of sand into the lagoon through the pass that cuts the offshore bar. (Photograph by Fairchild Aerial Surveys, Inc.)

Selected References

Bascom, Willard, 1959, Ocean waves, Scientific American, vol. 201, no. 2, pp. 74–84.

Carson, Rachel, 1950, The sea around us, Oxford University Press, New York.

Engel, Leonard, 1961, The sea, Life Nature Library, Time Inc., New York.

Shepard, Francis P., 1963, Submarine geology, 2nd edition, Harper and Row, New York.

Fig. 11-1   Sand channel in a deeply submerged reef at a depth of 200 feet off the north shore of Jamaica. Predominant organisms are large branching corals and vividly colored sponges. (Official U.S. Navy photograph.)

11

Oceans and Islands

The winch motor slows to a hum, the heavy cable creaks as it follows its tortuous path around wheels and through the tensiometer, and finally a five-foot length of large-diameter pipe breaks the surface of the water. Scientists and crew manhandle the dripping object aboard the ship, tip it upside down, and eagerly start searching through the mud and rocks the pipe dredge has brought from the bottom of the sea. For we are on the research vessel U.S.N.S. *de Steiguer*, and the pipe dredge is only one, although perhaps the simplest, of the many pieces of oceanographic equipment with which she is outfitted.

Ocean Technology

The *de Steiguer*, an AGOR, is one of many such research vessels constantly probing the oceans and the ocean floors. Because of the inaccessibility of this area of the earth's surface, it is the last to be explored. Only in the past twenty years has technology provided many of the instruments needed for advanced oceanographic study, and their development has spurred interest in that study to a phenomenal degree. Simple devices like the pipe dredge do not

Fig. 11-2 Crane swings the submersible *Deepstar* out over the stern of the research vessel. Crewmen pay out stern line and wait to disconnect the crane's quick-release hook from the craft. Attached to the brow are plastic bottles used to collect water samples. (Official U.S. Navy photograph.)

require any complex technology, but of what use are its contents, or any other information about the sea floor, if we do not know where we are? Until recently, once a ship was out of sight of land, its position could not be determined within a mile at the very best, usually not within five miles. If you imagine making a map on land with that limitation on your accuracy, you can see the difficulties of mapping the ocean floor, much less trying to interpret the forms you may have found. More accurate methods were devised, but they were too expensive to put on oceanographic vessels. Now, however, with the advent of the orbiting satellites, a ship's position can be determined within 15 meters (.01 mile).

Another development which has helped to study the sea floor is the deep-diving research submersible (Fig. 11-2). A very specialized one, the Trieste, has reached the greatest known depth of the earth, about 10,900 meters (36,000 feet). A number of smaller submersibles, each carrying two to four persons, have also been built, and they can dive 600 to 2400 meters (2000 to 8000 feet), which is extremely useful in exploring the continental shelves. They are equipped with windows for direct observation of the sea floor and manipulator arms for sampling and other maneuvers (Fig. 11-3). Such submersibles even rescue each other. When the submersible *Deep Quest* was diving off the coast of San Diego, one of her propellers got fouled in an aluminum cable so thoroughly that she was effectively held captive on the sea floor. A call for help brought the little "electronic pickle" *Nekton*, which, with the aid of some paper-clip and masking-tape technology, cut the cable and freed the *Deep Quest*.

Recent technology has also made possible drilling in deep water. It was only in 1961 that the very first deep drilling took place. That expedition drilled in water 3600 meters (12,000 feet) deep near Guadalupe Island off Mexico, twenty times deeper than had ever been drilled before. The ship maintained its position for a month without the aid of anchors and drilled 183 meters (600 feet) into the ocean floor. Improvements in the succeeding eight years led to the construction of the *Glomar Challenger* which in its first voyage drilled the sea floor in 6096 meters (20,000 feet) of water and penetrated over 835 meters (2738 feet) of ocean bottom. In only one year of operation the *Glomar Chal-*

Fig. 11-3   Sampling arm of *Deepstar* obtaining rock specimens from the 600-foot terrace off Mission Beach, California. The plants are sea anemones and are about 2 feet high. (Official U.S. Navy photograph.)

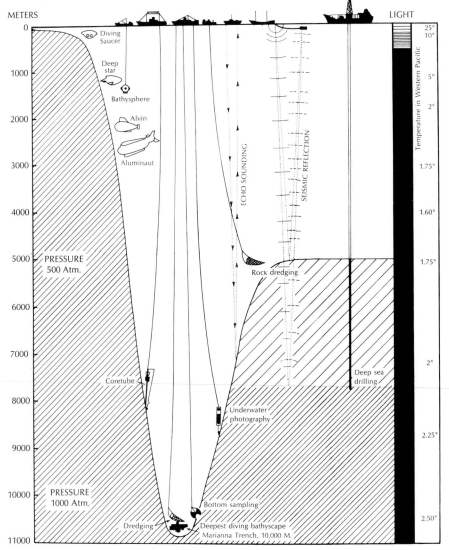

Fig. 11-4 Recent progress in man's observation of the sea floor. (From Heezen and Hollister, *Face of the Deep,* courtesy Oxford University Press.)

lenger added significant information to our knowledge of the ocean basins. It obtained samples of the oldest sediments—Jurassic, about 140 million years old—ever taken from the ocean floor, collected an almost complete geologic column from the Upper Cretaceous to the Recent, and made the first discovery of oil in deep-sea areas.

On smaller ships, too, such as the *de Steiguer*, bottom samples can be taken, but it is necessarily a much simpler operation, and the samples do not come from very deep within the bottom sediments. One method is called gravity coring because the force of gravity acting on the coring device as it is released from the ship is the only force that causes it to penetrate the bottom (Figs. 11-5, 11-6). Even though gravity coring is not very deep, it is quite adequate for many purposes. There are sampling devices other than the gravity corer and the pipe dredge. Some grab samples from the bottom and bring them to the surface in their jaws, while the box corer brings up about one square foot of undisturbed bottom sediments (Fig. 11-7).

Even small ships can "see" the ocean floor and, surprisingly, the rocks beneath the floor with acoustic devices (Fig. 11-8). Almost all ships carry one of these instruments—an echo-sounder. This creates a loud noise underwater by an explosion of gas or by an electronic impulse; the sound travels to the sea floor and is reflected back to the ship. Since the speed of sound in water is almost constant, the time it takes a sound impulse to journey to the floor and back is an accurate measure of the distance to the bottom. On most deep-sea expeditions the echo-sounder performs this operation every second, resulting in a continuous visible recording of the bottom profile as the ship steams along. Menard (1969) describes it thus:

> We cruised over hills and low mountains a mile and a half below and visible only on the echo-sounder. However, after a while the distance between the echo-sounder and the bottom is forgotten. As the marine geologist surveys a new range of undersea mountains, he senses them around him. I was once surveying with a captain new to the game. He remarked, as we headed again toward a peak on the map we were making, that he could not suppress a captain's feeling that we would hit it, even though he knew perfectly well it was a mile below the ship.

Now, if we make the sound impulse very much stronger, some of the sound waves will penetrate the ocean floor and will be reflected back from rock layers beneath (Fig. 11-9). In this way marine geologists can see not only what the shape of the bottom is, but can calculate the thicknesses and attitudes of strata below the surface. This is called sub-bottom profiling and is quite enough to make a land-based geologist sigh with envy. True, the terrestrial geologist can see the surface for miles around and take samples with a few strokes of his hammer, and he can even "see" the underlying rocks in very limited areas with some noisy and cumbersome equipment, but he cannot get a continuous sub-bottom profile as he bumps along in his jeep. Actually, the marine geologist and the land geologist proceed in quite opposite ways. On land the geologist maps the structure of rocks on the surface and then has to imagine what they do in cross section, while the marine geologist has the cross section provided for him by the sub-bottom profiler, and then he has to imagine the map of the sea floor.

Features of the Ocean Floor

The excitement generated by all this attention to the ocean floor would be incomprehensible if the floor were as flat and dull as everyone supposed for

Fig. 11-5 Preparing to lower a piston corer from the research vessel *Atlantis*. (Courtesy of Woods Hole Oceanographic Institution.)

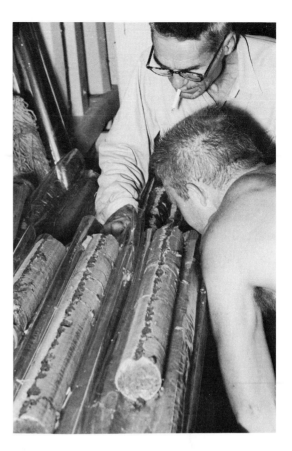

Fig. 11-6 Geologists aboard ship examine cores taken from the floor of the Pacific. (Courtesy of University of California, San Diego.)

Fig. 11-7 Box corer, used for undisturbed ocean bottom samples. (Official U.S. Navy photograph.)

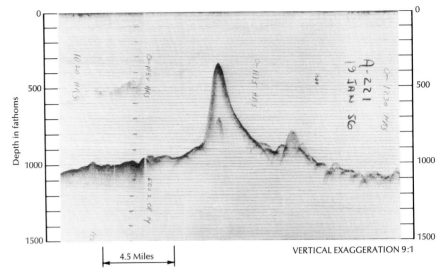

Fig. 11-8 A fathometer record of a seamount in the Caribbean. (Courtesy of Woods Hole Oceanographic Institution.)

years. It just *had* to be flat because nothing could happen down there with no erosion by wind, or ice, or even running water since the ocean floor must be the basest of all base levels. The excitement, of course, has been finding out that the ocean floors are a long way from being flat and that much has taken place down there and is still going on.

Continental Shelves

In the preceding chapter we discussed the important ocean-land interface as if it existed only at sea level. This, of course, is not true, since the entire ocean floor is an ocean-land interface, although not so dynamic a one. We barely got our feet wet before, but now we could profitably use one of those small submersibles (Fig. 11-2). As we leave the beach and proceed seaward, we find a relatively shallow platform that surrounds almost all the continents and that is known as the *continental shelf*. These shelves have an average width of 40 nautical miles but can be as wide as 320 kilometers (200 miles), as it is off the coast of Newfoundland, or as narrow as 16 kilometers (10 miles), such as the shelf off San Diego. The continental shelves, far from being smooth, gradually sloping platforms, often have an extremely irregular surface with depressions, hills, and terraces, showing as much as 10 fathoms (60 feet) of relief. In addition, in glaciated regions the inner or landward part of a shelf may actually be deeper than the outer edge. This outer edge is called the *shelf break* and is characterized by a sudden steepening of the slope. The depth of the shelf edge ranges from 20 to 550 meters (65 to 1804 feet) and averages 133 meters (436 feet).

Most continental shelves are composed of sedimentary rocks and many are given form by natural dams. Common around the Pacific Ocean are dams formed by long blocks of rocks that have been broken or folded and pushed upward. The basins thus formed behind the dams are then filled by sediments from the continents. Sometimes the dam rises above the surface of the sea and exists as islands, such as the Farallons near San Francisco and the Channel Islands off southern California. At other places, chiefly the east coast of the United States, such dams are very ancient and sediments have not only filled the basins behind them but have covered the dams as well. In such cases, and where there are no dams, the shelf sediments form a great wedge, oftentimes thickening in their seaward direction.

Off the coast in the western Gulf of Mexico, the dams are giant domes of salt which have pushed up through overlying sedimentary rocks. In tropical waters, dams are often formed by outlying reefs that grew millions of years ago and that are composed of ancient marine microscopic plants, or algae as they are called, and of more recent marine corals. Shelves off the eastern Gulf coast and southeastern United States are examples of this type of shelf.

During the Pleistocene Ice Age, when much of the ocean's water was withdrawn from the basins to lie in great ice sheets on the land, extensive parts of the continental shelves must have been exposed to subaerial and glacial ero-

Fig. 11-9 Diagram of the method of sub-bottom profiling with an actual profile. (Official U.S. Navy photograph.)

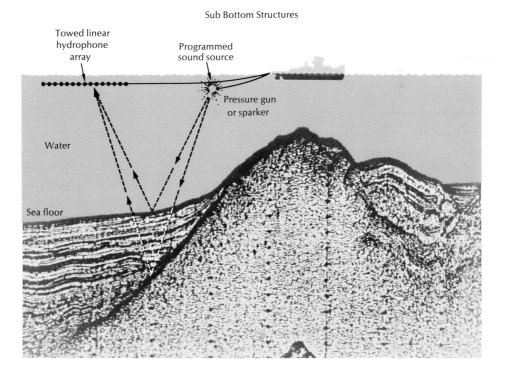

sion. In places this erosion exposed the tops of the dams, revealing igneous and metamorphic rocks which have not been covered to this day. The temperature during the Ice Age did not remain cold for the entire period, but had fluctuations between colder and warmer times. These, of course, were reflected in changing levels of the sea. When the temperature remained constant for any length of time, the waves cut cliffs and terraces in the exposed continental shelves. These are now submerged, and some of them have since been buried by new sedimentation. Others, however, still persist and can be detected by echo-sounders. They provide us with clues to the earth's recent history.

Continental Slopes

The continental slopes extend downward from the shelf break to the deep-ocean floor. It should not be thought that the gradients of the slopes and shelves are very great, a conclusion that might be drawn from looking at typical drawings of cross sections of ocean basins since these drawings have high vertical exaggeration. The average slope of the shelves is $0°07'$, while that of the continental slopes is $4°17'$ for the first 1000 fathoms (6000 feet). Like the shelves, the slopes are not gradually descending, flat expanses, but have basins, hills, valleys, and canyons. In general, the slopes are a quite rugged transition

Fig. 11-10   A sea anemone and a shrimp, living in perpetual darkness on the sea floor in about 400 fathoms of water, are momentarily illuminated for photography. (Courtesy of Woods Hole Oceanographic Institution.)

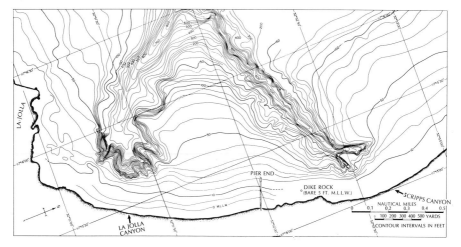

Fig. 11-11   Topographic map of Scripps and La Jolla Canyons. Each line connects all points of equal depth below sea level. (Official U.S. Navy photograph.)

zone, about 16 to 32 kilometers (10 to 20 miles) wide, which connects the two main levels of the earth's surface—sea level, or close to it, and ocean bottom, 2000 fathoms (12,000 feet) below sea level (Fig. 11-10).

Submarine Valleys

One of the most characteristic features of the continental slopes is the numerous valleys that have been cut into them. Some are caused by landslides and other gravity mass movements in the unstable materials of the slopes, and others follow breaks in the underlying rocks. Those valleys which are properly called canyons, however, are among the most impressive features of the ocean bottom and the most difficult to explain.

Although submarine canyons are world-wide in distribution, it is very difficult to pick out one as typical, because so few have been studied in detail. Probably the best known is La Jolla Submarine Canyon and Fan Valley, with its major tributary, Scripps Canyon (Fig. 11-11). These could hardly escape intensive investigation, located as they are just off the coast of La Jolla, California, and the site of Scripps Institution of Oceanography, as well as being within a few miles of the Naval Undersea Research and Development Center. These canyons are of interest also as the location of the first dives by a submersible in the United States when the Trieste explored there in 1960. In addition, they head very close to shore, making their shallower portions down to 60 meters (200 feet) easily accessible to small boats and scuba-diving scientists. Deeper portions have been studied with Jacques Cousteau's Diving Saucer.

From the beach, a flat, gently sloping, sand-covered terrace extends seaward. About 213 meters (700 feet) from shore at a depth of 12 meters (40 feet), the bottom suddenly drops away in the precipitous 24-meter (80-foot) headwall of

Fig. 11-12   Headwall of La Jolla Submarine Canyon showing exposures of upper Pleistocene lagoonal deposits, depth 125 feet. (Official U.S. Navy photograph.)

La Jolla Canyon (Fig. 11-12). At the base of the cliff is the wide, bowl-shaped head of the canyon. From here seaward, the valley widens out on each side and then narrows again until it is a rock-walled gorge whose steep sides are covered with a lush growth of marine plants and animals. The bedrock of the canyon floor shows through its normal sandy covering in some places; in others there are thick mats of seaweed and other organic material. Sometimes progress down the canyon is by giant steps or terraces which have been caused by slumping in the loose bottom sediments.

Partway down, Scripps Canyon joins La Jolla Canyon, and a glance through the submarine gloom reveals that its walls have been so undercut at their bases that they actually overhang as much as 6 meters (20 feet). Gradually the height of the walls decreases, and the valley widens until it is a little more than half a mile wide with an entrenched channel wandering across it. The walls are composed of semi-consolidated clay rather than hard rock, and there are natural levees on either side of the valley. We have entered the La Jolla Fan Valley which is cut, not in the La Jolla Terrace as was the canyon, but in its own fan-like deposits. Eventually its depth below its surroundings decreases until it merges with the sea floor. Such submarine fans, built by sediment channeled down the canyon, are not uncommon, and if a number of canyons or other submarine valleys are spilling sediment out onto the ocean floor, a marine equivalent of an alluvial plain is formed. This sediment-laden boundary of coalescent fans between the continental slope and the deep-ocean floor is called the *continental rise*.

The origin of submarine canyons has been argued by geologists for decades, and it is not necessary here to go into the many theories that have been proposed—only to be discarded. Much work remains to be done before a final solution can be found, but current thought seems to favor subaerial erosion as a starting point for the heads of many canyons. This is reasonable in view of the great lowering of sea level during the ice age. However it will not account for the erosion necessary to produce the features we see today where canyons extend as continuous features out onto the deep-sea floors. The proto-canyons must have been profoundly modified by submarine erosion as well. What the dominant agent of that erosion is remains in some doubt, and in fact it appears that there are several processes active in canyons that cause submarine erosion. We know that strong bottom currents do exist throughout the ocean, and they have been observed in canyons. These currents can transport sand-sized sediment and go both up and down the canyons. Observations of canyon floors before and after storms have shown the considerable effects of storm-generated currents. Marine organisms no doubt play their part in the mechanical and chemical disintegration of canyon walls. Density currents of heavily laden water are triggered by the slumping of oversteepened and unconsolidated material near the heads of canyons. Although these may transport very fine material great distances, their power as erosive agents is doubtful. Quite possibly, slow but steady gravity mass movements of sediment-charged mats of coarse organic material may prove to be the most important agent of erosion. Where sand is rapidly added to the head of a canyon, spectacular sand flows and falls may occur (Fig. 11-13).

The Deep-Ocean Floor

Although the waters of all the oceans of the world form a continuous surface, the configuration of the continents and submerged ridges divides the ocean floor into a series of major and minor basins. The major ones, of course, are the Atlantic, the Pacific, the Indian, the Arctic, and the Circum-Antarctic Oceans. Land barriers at the edges of the Atlantic and Pacific separate the so-called *marginal seas* from the principal basins. These marginal seas include the Mediterranean, the Black Sea, the Gulf of Mexico, and the Caribbean.

Within the basins themselves are other subdivisions, as for instance in the Pacific, where the southeastern half of the basin is considerably shallower than the northwestern half. Of considerably more importance, however, is the world-wide, system of ridges that almost encircles the globe (Fig. 11-14). Ever since ships started taking soundings in the ocean, such shoal areas have been

Fig. 11-13   Sand spilling over a "fall" about 30 feet high in the Cape San Lucas Submarine Canyon, Baja California, Mexico. (Courtesy of University of California, San Diego.)

observed, but their continuity and wide distribution were not apparent until much later. As a consequence, individual portions of ridges, or rises as they are also called (not to be confused with continental rises, however), have been named, such as the Mid-Atlantic Ridge and the East Pacific Rise. These ridges are approximately 2 to 4 kilometers (1 to 2 miles) above the sea floor, 1000 to 4000 kilometers (620 to 2485 miles) in width, and 10,000 kilometers (6200 miles) long. They are often accompanied by extensive volcanism and are also often the location of numerous earthquakes. In fact, their topography in general is that of pronounced faulting or breaking of the earth's crust.

Fractures associated with the ridges are of two types. One type runs along the crest of the ridge and appears as a trench or rift which may persist for many miles. In addition, there are fracture zones which are perpendicular to the ridges and also generally perpendicular to the continental margins. As can be seen from Figure 11-14, movement along these ruptures has been horizontal, so that the ridges are broken into small segments slightly offset from each other. These fractures are not all active, and many are active only in those portions that cross the crests of the ridges.

Along some continental margins, particularly those that bound the Pacific Ocean, are found island chains with accompanying deep trenches. These areas are the most active parts of the earth's crust today. Usually the island groups are crescent-shaped, which has led to their being called *island arcs*, and the island arcs are often, but not always, convex toward the ocean basins. The trenches are the deepest part of the ocean floor, reaching the nadir in the Marianas Trench at approximately 10,850 meters (about 35,800 feet). When we speak of islands and trenches as the most active areas of the crust, we are referring to the facts that they have the most volcanism, the largest gravity anomalies, the strongest shallow earthquakes, and almost all the earthquakes that originate at great depth.

Not all oceanic volcanoes occur associated with trenches, island arcs, and ridges. Most ocean volcanoes in fact are not high enough to rise above the surface of the water. These are known as *seamounts* (Fig. 11-8) and have the familiar shape of land volcanoes of the intermediate type, with steep sides and a small summit area. They are generally about one kilometer above the ocean floor. The number of such submerged volcanoes is almost unbelievable. Menard (1964) estimated that there are 10,000 in the Pacific Basin alone, which

Fig. 11-14 The mid-ocean ridges. (Modified from Sir Edward Bullard, "The Origin of the Oceans," *Scientific American,* Vol. 221, No. 3.)

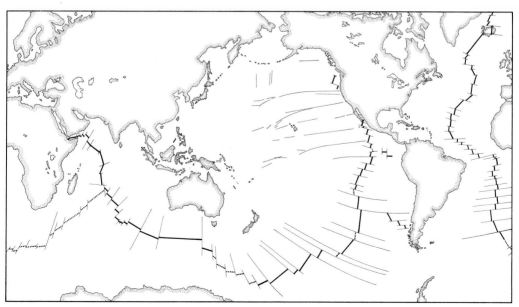

Fig. 11-15   A barrier reef encircling a volcanic island. (Office of Naval Research.)

makes them one of the ocean floor's most prominent features. Although some are isolated cones, many others occur in clusters and linear arrangements. All so far investigated are of basaltic composition.

Some of these submerged volcanoes were once at or above sea level. This is shown by the existence of *guyots* and *coral atolls*. Guyots are simply submarine volcanoes that have truncated summits, the planation having occurred by wave action when the volcano was nearer the surface than it is today.

The origin of coral atolls, ring-shaped islands made up of the skeletons of marine animals, has been another subject of intense discussion among geologists. They were first investigated by Darwin, the great naturalist, on his famous voyage aboard the *Beagle*. He put forward the theory that corals started building reefs around volcanoes when those volcanoes stood above or at sea level (Fig. 11-15). He believed that the volcano gradually sank, or the sea level rose, and the coral was able to keep building its reef upward (Fig. 11-16) until the volcano was deep under water and only the uppermost, active part of the ring-shaped reef remained visible (Fig. 11-17). The trouble was that for many years no evidence for the volcanoes could be found, and various other theories were suggested to account for the atolls. Finally, before the atomic tests at Eniwetok and Bikini, extensive studies of both atolls were undertaken, studies which included drilling through the reef structures and seismic profiling. These studies revealed the volcanic islands below and proved that Darwin was correct (Fig. 11-18). Oceanographers and marine geologists devoutly wish that other problems could be solved so simply and neatly.

The birth of a volcano is an event seldom witnessed on land and even less frequently seen at sea (Fig. 4-34). On November 14, 1963, seamen aboard a fishing vessel noticed a column of black smoke rising from the sea southwest of Iceland. Closer inspection confirmed that it was ash and steam rising from a submarine eruption. By 3 p.m. of that day the smoke column was four miles high and was visible from Reykjavik. A ridge had developed below the surface of the water, and the sea was breaking on obstacles it had never encountered before. That night the ridge grew enough to emerge above the surface of the ocean, and the island Surtsey was born.

As might be expected, the birth was attended by numerous geological midwives who, although they could not assist the delivery, were devoted observers. They kept watch from the sea, from the air, and from the island itself when it had calmed down sufficiently to permit landing. Thorarinsson (1967) describes the action thus:

Fig. 11-16   Block diagram of the succession of reef types around a sinking volcanic island, from a fringing reef (front block) to a barrier reef (middle block) to an atoll (rear block). (After W. M. Davis, *The Coral Reef Problem,* Spec. Pub. No. 9, 1928. By permission of American Geological Society.)

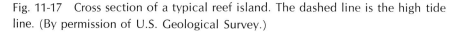

Fig. 11-17   Cross section of a typical reef island. The dashed line is the high tide line. (By permission of U.S. Geological Survey.)

Like most geologists I am more used to working on land than at sea and I do not by any means feel particularly comfortable when the sea is rough, but few hours of my life do I treasure as much as one late afternoon near the volcano in late November, 1963. The wind was high, the waves surged about us and the sea washed the small coastguard vessel I was on from stem to stern. The volcano was most vigorously active, the eruption column rushing continuously upwards, and when darkness fell it was a pillar of fire and the entire cone was aglow with bombs which rolled down the slopes into the white surf around the island. Flashes of lightning lit up the eruption cloud and peals of thunder cracked above our heads. The din from the thunderbolts, the rumble from the eruption cloud, and the bangs resulting from bombs crashing into the sea produced a most impressive symphony. High in the sky the crescent moon rushed headlong between racing clouds. As I write this passage I realize how hopelessly beyond my powers it is to do justice to such a grandiose performance of the elements. To do so one would need the romantic genius of a Byron or a Delacroix. What I can state is that no matter how green one's face has become, the seasickness is completely forgotten in the presence of such a performance.

The life of a new-born island is usually a pretty "iffy" affair. As long as the vent remains below sea level, the ejecta will be mostly ash and cinders, and the cone will be formed of this material which is easily eroded by wave action. If the island can survive and grow until the vent is safely above sea level, it may erupt fluid lava which is much more difficult to erode, and the island then stands a fair chance of becoming permanent land.

Surtsey is not an isolated volcano, but is one of a group which forms Iceland and its surrounding islands. It is especially interesting because this group is actually a part of the Mid-Atlantic Ridge which here rises above sea level where it can be easily studied, as most other parts of the ridge cannot. In effect, it is one of the most active parts of the ocean floor built up from the deep ocean so that its surface is exhibited, to the delight of geologists, as can be imagined.

Abyssal hills, whose existence no one even suspected until recently, are the most widespread physiographic feature on the ocean floor. They remained undetected for so long because they are quite small, not very high, and are thus difficult to find except through very precise electronic navigation. Even now, determining the shape of individual hills is difficult, but in general they are 50 to 1000 meters (164 to 3280 feet) high and 1 to 10 kilometers (.6 to 6.2 miles) wide. Typically, their sides are steep and their tops are broad and gently sloping. Most appear to be domes, nearly circular or elliptical in shape. Very little is known about the origin of abyssal hills, but Menard (1964) believes that they are small shield-type volcanoes or laccoliths. Drilling by the *Glomar Challenger* has shown the ones in the Gulf of Mexico to be salt domes.

Are there, then, no flat areas on the ocean floor? It might seem so from the above discussion, but there *are* regions of practically no relief, although they are not nearly so widespread as was once believed. Some continents are bordered, beyond the continental slope, by such *abyssal plains*. These are formed by very fine sediment derived from the continental areas and deposited by low-velocity currents. Another type of flat area is found around volcanic islands, and these *archipelagic aprons* are formed primarily by lava flows. Abyssal plains in tropical waters are probably formed by the burial of abyssal hills. These waters are especially favorable to the growth of small animals, and the deposition of their remains is fairly rapid under these conditions. It has been likened to a steady rain of fine particles. In some cases the hills have been completely buried to form an especially smooth plain; in others, only the hollows between the hills have been partially filled. Elsewhere abyssal hills have been buried by density currents carrying extremely fine-grained terrestrial sediments.

Composition of the Ocean Floor

Drilling of the deep-ocean floor has everywhere shown that thin layers of sediment filling the ocean basins are underlain by basalt. This is in great contrast to the continental rocks and points to completely different, though related, origins of the oceans and the continental areas. The ocean basins have, in fact, provided the keys with which scientists are beginning to unlock the myster-

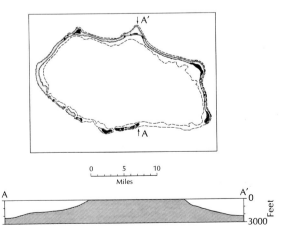

Fig. 11-18 A map and cross section of Bikini Atoll. (After K. O. Emery, J. I. Tracey, Jr., and H. S. Ladd, 1954. By permission of U.S. Geological Survey.)

ies of the origin and history of the earth's crust. These exciting and challenging theories will be discussed in Chapter 18.

The continental shelves and slopes are generally considered, geologically, as parts of the continents rather than as parts of the ocean basins. Where submarine valleys have been incised into them, dredges, scuba divers, and submersibles have brought to the surface samples of the bedrock from the valley walls. They are primarily sedimentary rocks and occasionally light-colored plutonic rocks such as granite. Offshore drilling has confirmed these findings. The sedimentary rocks are of various kinds and may be in any stage of consolidation.

Overlying the bedrock close to shore, i.e. on the shelves, slopes, and continental rises, are sands and muds, all of terrestrial origin. These sediments, however, seldom follow the ideal sequence with the coarsest particles closest to shore, grading into finer and finer materials seaward. Instead their distribution is extremely patchy and is determined largely by the migration of sea level across the shelf.

The slow steady deposition of very fine particles from the waters above over a very long time has blanketed most of the old features of the deep-ocean floors. These particles are called *pelagic sediments* and are composed of the remains of tiny marine animals, as well as dust blown from the continents, volcanic ash, and meteoritic dust. The deposits are clays and organic oozes which have considerable amounts of either calcium carbonate or silica, depending upon the composition of the animal skeletons. Minerals which have precipitated directly from the sea water are also present.

This steady accumulation of pelagic sediment is not allowed to proceed undisturbed. Because of the slowness with which deposition takes place, the average rate ranging from 1 to 10 millimeters (.04 to .4 inches) in 1000 years, there is ample time for changes to occur. Some sediments are altered and others are completely dissolved on their leisurely trip to the bottom. Those that do reach the bottom may be considerably reworked by bottom-dwelling organisms before they finally come to rest.

Turbidity Currents

This discussion of sedimentation on the ocean floor seems an appropriate place to investigate the especially controversial topic of turbidity currents. Much heated argument has been generated among geologists and oceanographers concerning these currents. Most would agree that a *turbidity current* is a moving stream of water that, because of its high content of suspended material, is much denser than the surrounding water.

Turbidity currents were invented by scientists in order to explain certain problems of present marine sedimentation. The principal troublesome fact was the discovery at depth of organisms and coarse sediments that normally are found only in shallow water. How did they get where they are? Possibly a current, triggered by an earthquake or by slumping, started near shore and billowed its way seaward, depositing sediments as it went and maintaining its existence as a separate entity because of its high density. This was an intriguing thought, and geologists set about trying to find an actual turbidity current. Unfortunately, such currents seem to be frustratingly elusive. In spite of the fact that nobody had ever seen one, they were blamed more and more for all kinds of otherwise inexplicable phenomena. Among these were the erosion of submarine canyons, the successive breaking of submarine cables after an earthquake, and the deposition in both the past and the present of graded bedding or certain repetitive bedding sequences. Some of these effects were even produced in the laboratory by man-made turbidity currents.

It has now been shown that most of these phenomena can be explained in other ways (van der Lingen, 1969). The existence of high-velocity currents, carrying very dense sediment loads and capable of extensive erosion and the transport of coarse materials, is considered doubtful at the present time. Low-velocity turbidity currents are still in good standing, and it is these that can transport very fine material great distances from the continents over almost level surfaces and deposit it in the deep-ocean floor to form abyssal plains.

The original difficult fact—the presence of shallow-water organisms and sediments in deep water—is now generally explained partly by slumping at the heads of submarine canyons and partly by slow creep down these natural chutes. Neither of these processes carries any material very far during any one movement, but they occur again and again and the over-all movement during geologic time can be great.

The turbidity current controversy admirably points up an ever-present danger in scientific investigation—that of inappropriate nomenclature. Because the hypothetical turbidity currents seemed a likely explanation of repeated sequences of graded bedding, such deposits were called "turbidites." Unfortunately, just what constitued a turbidite was never agreed upon, with the result that one could never be certain exactly what was being discussed. But even more damaging was the assumption that because they were called turbidites, then they must have been formed by turbidity currents. For many years this type of circular reasoning kept geologists from seeking other possible explanations of their origin. It now, at last, appears that they might be formed simply by ordinary bottom currents, with or without the aid of small earthquakes (van der Lingen, 1969). Thus the unwarranted use of a name that implies a method of formation can stifle imagination and investigation for a long time. It might even be profitable to take a skeptical look at other geological terms which imply genesis, not only those which are controversial, but even those which have gained acceptability.

Who Owns the Sea Floor?

Until technology made possible its exploration and exploitation, the management and ownership of the ocean floor was of little importance or interest. Now, however, it is obvious to anyone who reads newspapers and periodicals that the ocean and its resources will be of ever-increasing concern to man. This growing interest and its attendant problems were brought to the attention of the world by Arvid Pardo, Ambassador of Malta to the United Nations, in 1967, in a speech to the First Committee of the General Assembly. He made four proposals:

> 1. Ocean space, beyond the limits of national jurisdiction, is the common heritage of mankind.
> 2. Ocean space, beyond the limits of national jurisdiction, is not subject to the claims of national sovereignty.
> 3. Ocean space must be used for peaceful purposes only.
> 4. The resources of ocean space, beyond the limits of national jurisdiction, must be explored and exploited with a maximum of international cooperation, for the benefit of all mankind. (Borgese, 1968).

The United Nations is now making detailed studies in preparation for the drafting of an Ocean Treaty, similar to the Treaty on Outer Space which we already have. Other organizations as well are discussing the new ocean regime with the object of contributing pertinent information and ideas. Roger Revelle (1969) would add to Ambassador Pardo's four points two additional important ones. The first concerns an enemy we are just beginning to fight on land and in the air: pollution.

> Internationally coordinated action must be taken to prevent pollution of the ocean, including control of pollutants coming from the land or the air, such as pesticides, radioactive substances, poisonous chemicals, and sewage; from ships, submarines or other equipment used at sea, and from exploitation of marine resources, for example exploration, storage and transportation of oil and gas.

Proper regulation of these activities should be backed by sound research which is the topic of the second additional point.

> The freedom of scientific research in the ocean shall be kept inviolate. The exclusive rights granted to the coastal states shall not include the right to interfere with scientific research, provided that the coastal state is given prior notification of the plan to conduct the research, has full opportunity to participate in it, and has access to all the data obtained and samples collected, and provided that the research does not deleteriously affect marine resources or other uses of the sea.

To these might be added a plea for some areas of the ocean to be set aside as international wilderness parks. In this way samples of ocean ecology will be preserved for future generations to study and enjoy.

The ocean is indeed the earth's last frontier, and it is important to everyone that it not be exploited in the haphazard and often wasteful and damaging ways from which our other frontiers have suffered. While we are struggling to repair our mistakes on land, let us keep from repeating them in the sea. The global extent of the ocean requires planning and management on an international scale.

Selected References

Bascom, Willard, 1969, Technology and the ocean, Scientific American, vol. 221, no. 3, pp. 199–217.

Borgese, Elizabeth M., 1968, The ocean regime, Occasional Paper, vol. 1, no. 5, Fund for the Republic, Santa Barbara, Calif.

Emery, K. O., 1969, The continental shelves, Scientific American, vol. 221, no. 3, pp. 107–122.

Menard, H. W., 1964, Marine geology of the Pacific, McGraw-Hill Book Co., New York.

——, 1969, Anatomy of an expedition, McGraw-Hill Book Co., New York.

Revelle, Roger, 1969, The ocean, Scientific American, vol. 221, no. 3, pp. 55–65.

Shepard, F. P. and Dill, R. F., 1966, Submarine canyons and other sea valleys, Rand McNally, Chicago.

Thorarinsson, Sigurdur, 1967, Surtsey: the new island in the North Atlantic, The Viking Press, New York.

van der Lingen, G. J., 1969, The turbidite problem, New Zealand Journal of Geology and Geophysics, vol. 12, no. 1, pp. 7–50.

Fig. 12-1 Yosemite Valley, Sierra Nevada, California. (Photograph by Ansel Adams.)

12
Glaciation

Scenically, the world is more indebted to glaciation than to any other process of erosion. Without glaciation there would be few of the serrated peaks standing in isolated splendor along the crest of many of the world's lofty mountain ranges. Such a resplendent peak as the Matterhorn is an example, so familiar through endless repetition in calendars and travel posters as to verge on the trite until it is seen shining forth in reality.

Steepening of slopes in the summit regions of the Alps, the northern Rockies, Alaska, and the Sierra Nevada is but a single aspect of glaciation. The fact that a glacier can erode more deeply in some parts of its channel, less deeply in others, is responsible for the multitude of lakes that add such interest to the landscape of the north-central states and to Scandinavia. Deep valleys whose outlines have been sharpened by the glacial file—such as Yosemite Valley, the Lauterbrunnenthal of Switzerland, and the Norwegian fjords—are sufficiently spectacular to support large and flourishing tourist industries. These and many other testimonials to the effectiveness as well as the uniqueness of the glacial processes are widespread throughout the northern countries of the world.

The ice age, just ended, provided one of the more stirring chapters in earth history, and one whose effects to a greater or less degree have affected the lives of all of us. Soil was stripped from vast land areas, leaving them devoid of value for agriculture but enhanced as sources of minerals. The load of stripped soil and rocks was deposited elsewhere, notably in the Middle West in this country, thus making it a singularly productive land. Large lakes were created where none so extensive had existed before. Lake Agassiz is one of these, and its remnant, Lake Winnipeg, is a giant in its own right. The Great Lakes, also in large measure a product of glaciation, are still with us, and incidentally they flood thousands of acres of potential farm land.

Many of the areas blanketed by glacial ice were bowed down under its weight, and when this burden was lifted through the disappearance of the ice, the land rebounded. In Scandinavia, uplifted wave-cut features show that the central part of the peninsula has risen 244 meters (800 feet) or more. Historic records, and various shore-line structures such as ancient landing places, show that this rise is still continuing. It reaches the unusually high rate of about one meter (three feet) per century at the head of the Baltic and decreases to about zero in the vicinity of Copenhagen.

Sea level all over the world swung in rhythm with the waxing and waning of the ice sheets. When the ice sheets expanded, tens of millions of cubic miles of sea water were withdrawn to be locked up on land as ice. Sea level was lowered as a consequence, perhaps by 40 or 50 fathoms (240 to 300 feet). This may not seem like much, but it was enough to profoundly alter world geography. Land areas, now separated, were then connected, and with the climatic stringencies of the time not only were migrations of whole populations of animals stimulated but they were made possible by the appearance of *land bridges*. Among these natural causeways are such links as the ones that connected Tasmania and Australia, Ceylon and India, New Guinea and Australia, and others that united some of the islands of Indonesia. Most renowned of all was the land bridge joining Alaska and Siberia, now separated by the 55-meter-deep (180-foot-deep) waters of Bering Strait. In a way, this link must have been a veritable freeway, with all sorts of creatures, including human beings, pattering to and fro. Westbound from the New World, migrating into the Old, went the zebra, the camel, the tapir, and the horse. Eastbound immigrants to this hemisphere were elk, musk ox, bison, elephant, mountain sheep, and mountain goat. Not the least was man himself.

No wonder many of the evidences of past glaciation attracted the attention of observant men in Europe and New England in centuries past. The lavish supply of boulders in the farms of New England was a source not only of wonderment to the early settlers but also of wearisome, backbreaking toil. So much labor was involved in clearing fields strewn with glacially transported stones that more than one young man was readily convinced that a life at sea could be no harsher—even on a New Bedford whaler.

For many, the presence of these stranger-stones found far from their place of origin and very often completely out of harmony with their new environment —for example, granite blocks resting on a limestone terrain—was adequately explained as the work of that "vindictive affliction," the Great Flood of Noah.

In Great Britain, much of which was covered by only recently vanished glaciers, the unstratified, widely scattered glacial detritus was called the *drift*, a name betraying the belief that it originated as a deposit spread far and wide by an all-encompassing sea. In England there has been a tendency to assign a leading role to the sea as an agent shaping the surface of the earth. This is not surprising in view of England's insular position and the strength of the currents and waves that batter its coast. Some even accepted the idea that icebergs and ice floes may have transported these *erratics*, as such out-of-place rocks are called, for British whalers working off the coast of Greenland had seen such debris embedded in sea ice there, as had explorers elsewhere in the arctic.

Persuading English geologists that it was glaciers which had scoured the inland surface and transported rocks the size of small houses for scores of miles was a difficult task, for there were no existing glaciers to serve as models. Therefore, it is not surprising that the most eloquent advocates of the notion that ice could perform prodigies of work in shaping mountains and excavating valleys were Swiss. There are about 2000 glaciers in the Alps, and through the centuries their snouts have advanced or retreated, and alpine passes have been alternately ice-free or ice-blocked. Some villages, occupied in medieval times, are now buried by ice. The silver mine of Argentiere, active in the Middle Ages, is now covered by a glacier of Mont Blanc near Chamonix in alpine France.

Many alpine villagers must have been aware that when a glacier receded it left behind it a trail of barren, stony ground, interrupted by low, rocky ridges, diversified by lakes and ponds, and strewn freely with rock fragments, large and small.

In the early nineteenth century several European geologists became convinced that an ice sheet had covered much of northern Europe, and this aroused the curiosity of one of the leading Swiss naturalists of the day, Louis Agassiz. He persuaded one of these geologists to take him on an expedition in the Alps. Agassiz became a believer, and when in 1837 he came to America and to a professorship at Harvard, he spread the word far and wide of a "Great Ice Age" that had once refrigerated most of the Northern Hemisphere (Fig. 12-2).

A concept as novel as this aroused opposition, for some accepted the far more labored explanation that (1) the land sank, (2) the sea spread inland and boulders and other detritus were rafted by icebergs far and wide across its waters, and (3) the land rose, thus shedding its oceanic waters that left behind a residue of rocks and boulders scattered over the landscape.

Although Agassiz deserves full honor for carrying the word from Europe to America, as we have seen, he was not the first to believe that glaciers had once advanced across the European landscape. This problem of discovering who had the original idea is one of the manifold difficulties confronting the historian of science.

So it was with the glacial theory. A number of remarkably perceptive men had glimpsed the truth, and then were almost immediately forgotten. As early

Fig. 12-2 Extent of Pleistocene glaciation (white) in the Northern Hemisphere.

as 1802 erratic boulders in the Jura Mountains were recognized for what they were—glacially transported rocks. In Germany a professor of forestry, Bernhardi, employing the same reasoning, in 1832 wrote a paper stating his belief that the moraines and erratics that are significant terrain elements of the northern plain were evidence that glaciers had advanced southward from lands far to the north. He suffered the familiar fate of a prophet in his own country in that few people paid the slightest attention to him. Not until 1875 did the last German scientists accept the singular idea that their homeland had once been overridden by a sheet of ice.

Forward steps in science very often are the work of many men, rather than the brilliant inspiration of a single genius. Given a similar environment, or similar evidence, different people working quite independently of one another may come to the same general conclusion. The virtually simultaneous announcement of a mechanism of evolution by Darwin and Wallace is a classic example.

Little more than a century has elapsed since the glacial hypothesis won general acceptance in North America and Europe. An enormous amount of effort has been expended, chiefly in the last half-century, in the study of past and present glaciers. Today we know vastly more than our predecessors did of the extent of these vanished ice sheets, of the complexity of their major advances and withdrawals (there were at least four and possibly five), and something of the change of snow to ice and of the mechanics of glacier motion. The cause of glaciation, however, remains almost as big a mystery as it was a century ago. The riddle of glaciation is as challenging and inscrutable as many of the other more widely publicized problems of our day, such as the origin of the solar system or the mysteries of space.

Development of Glaciers

Extent and Distribution

Glaciers were vastly more important in the very recent geologic past than they are today, but even today their extent is not insignificant. Almost all of Antarctica is ice covered, 13 billion square kilometers (5,000,000 square miles) or so, an area about equal to the United States and Mexico. The Greenland ice cap is much smaller, approximately 1,820,000,000 square kilometers (700,000 square miles), but like the Antarctic ice cap its surface rises to a surprisingly high altitude—around 3000 meters (10,000 feet), and the ice is more than 3000 meters thick in places. This load of ice has succeeded in depressing the interior of the island below sea level. Thus, were the ice suddenly to be removed, Greenland would be a ring of islands enclosing a central sea.

A reasonable estimate of the frozen water piled up on land in glaciers and snow is that a land area of about 15 billion square kilometers (6,000,000 square miles) is covered; of this total, around 96 per cent is in Antarctica and Greenland. Were all this frozen water to melt, the level of the sea would rise all over the world between 30 and 60 meters (100 and 200 feet), with a devastating effect on property values in the world's seaports. In fact, the almost world-wide glacial recession that has occurred since the 1890's appears to be raising the sea level at a rate of approximately 12 centimeters (4.75 inches) per century.

Change of Snow to Ice

Without snow there would be no accumulation of extensive bodies of ice, and without ice there would be no glaciers. Hence glaciers are limited to those parts of the world where the temperature remains below freezing for a significant part of the year. This means that glaciers are active today in high mountains over much of the world, and in the farther reaches of the Northern and Southern Hemispheres. Thus, glaciers on the high upper slopes of such equatorial mountains as Kilimanjaro in Africa, the summits of the Andes in South America, and the Carstenz Toppenz Range in New Guinea are at altitudes of 4800 to 5400 meters (16,000 to 18,000 feet). In middle latitudes, as in the Sierra Nevada of California and the Swiss Alps, the regional snowline has descended to perhaps 2700 meters (9000 feet). Finally it reaches sea level in Antarctica (55° S), and stands at roughly 460 meters (1500 feet) in lands bordering the Arctic Ocean.

More than cold temperature is needed for snow to accumulate and remain from year to year. There needs to be heavy snowfall as well. This explains the absence of permanent snow on Mount Whitney (4418 meters—14,495 feet) whose summit rises above what should be the theoretical regional snowline. The mountain is in a dry region with a high evaporation rate and generally high summer temperatures. Many of the arctic lands lack a permanent snow cover because, although they are cold, they are also dry.

The local environment also plays a significant role. Since north-facing slopes in the Northern Hemisphere are shadier than south-facing ones, their snowline is lower. In many American mountain ranges the western side may be the windward slope, as it is for the Sierra Nevada and the Cascades. On that slope the precipitation is greater, cloud cover is more persistent, and snowline commonly is lower than on the eastern side.

The next problem to consider is the relationship of snow to ice. Snow, strictly speaking, is not frozen water, as ice is, but is frozen water vapor. In other words, it is the solid phase of water that has crystallized directly from water vapor in the atmosphere. Since it is a crystalline substance, snowflakes grow in regular geometric patterns, as everyone knows from seeing photographs in books and designs on Christmas cards. Although they seem to show an infinite variety, it probably is not strictly true that no two are ever alike. Actually, crystalline H_2O behaves like other mineral crystals in that there is an established crystal form for the compound. In snowflakes it is some variant of the hexagonal system—the same system of which quartz is a member.

When snowflakes settle to the ground, the individual crystals quickly interlock, and a dry, powdery, but surprisingly tough surface develops. The specific gravity of snow is much less than that of water, so that on the metric scale a centimeter of snow may be equal to a millimeter of rain water.

After snowflakes lie on the ground for a short while they ordinarily undergo a change. Individual flakes may sublimate (pass directly from a solid to a gaseous state), or they may melt. But since the water freed by melting is in the presence of a large mass of ice, it may refreeze into granules of ice. This gritty, granular snow, with a texture much like coarse sand, is a familiar phenomenon in snow banks that survive for a fair share of the winter behind a building, in the shade of a forest, or in the lee of a cliff. Such granular recrystallized snow is called *firn* (from a German adjective meaning of last year) in the German-speaking parts of Switzerland. *Névé* (a word going back to the Latin stem of *nix* for snow) is used in French-speaking areas. Both are used in articles in English.

Firn, which typically accumulates on the upper slopes of alpine mountains, goes through a gradual transition into glacier ice. Firn is usually white or grayish white, and the spaces between the grains of ice are filled with trapped air. At a depth equivalent perhaps to an accumulation of three to five years of firn, the pore spaces become smaller, or even may be lacking, and the transition into blue glacier ice is completed. This is accompanied by an increase in the specific gravity from perhaps 0.1 in newly fallen snow up to 0.9 in solid ice.

During the change from firn to glacier ice much of the air is forced out, sometimes under enough pressure to cause an audible hissing from the glacier. Some of the air remains trapped, however, and exists as bubbles in the glacier ice. When the pressure on the ice is released, as happens when the ice reaches the end of the glacier and is exposed to the air, the bubbles may burst, completely shattering large blocks of ice. Glaciologists working in Antarctica hope that trapped air bubbles can give them clues concerning atmospheric conditions in the past. Meanwhile they cool their drinks with the ancient ice; as the ice melts, the bubbles of air escape producing an effect similar to carbonation.

The change from firn to ice is aided by the increase in pressure with depth. This brings the individual ice grains into closer contact with one another, and thus reduces the amount of air space. At the same time the individual ice grains are undergoing a process analogous to recrystallization. Glacier ice, then, is a mosaic of interlocking ice crystals. In general, as the ice descends from the firn field at the head of a glacier down to the terminus, the ice becomes denser as more and more air is excluded and the ice crystals increase in size. Finally, they may reach such imposing diameters as 7.6 to 10 centimeters (3 to 4 inches).

Because water does not stay permanently locked into a glacier, it continues to play a part in the hydrologic cycle—ocean to atmosphere to land and back to the ocean again. Because of the slowness with which glaciers move, however, they can be considered as reservoirs. The length of time required for a water molecule to complete its trip through a glacier will depend, of course, on the speed with which the glacier moves.

Mechanism of Glacier Movement

Although the downslope motion of alpine glaciers has been known to scientists for nearly 150 years, we are still far from understanding its true nature. How is it that a solid substance, such as ice, can be said to flow, especially such a brittle solid as ice appears to be, judging from our own household experiences?

The problem is a perplexing one, for the surfaces of many glaciers are scored by cracks and fissures, known as *crevasses* (Fig. 12-3). As everyone is aware, fluids cannot fracture; whatever the mechanism of glacier advance may be, these bodies of ice do not move in the same way that the water does in a stream. When we speak of a river of ice, this is indeed using the language of metaphor. Yet to some degree there is an analogy between glacier and stream flow. We saw in Chapter 8 that in a normal stream channel the central part of a river has a higher velocity than its margins. This is true also of glaciers.

There are no hard and fast values for rates of glacier advance. Different ones move at different rates, and within a single glacier different parts may have a range of velocities. Glaciers move with ponderous majesty. The Great Aletsch (Switzerland's largest) advances around 50 centimeters (20 inches) a day near the middle section and about 25 centimeters (10 inches) per day about a mile upstream from the terminus. The Rhone glacier moves at the rate of approximately 91 meters (300 feet) per year. This is slow when compared with Greenland ice (Fig. 12-4) that makes its way from the central ice cap down through a narrow fjord to the sea at the rate of 24 meters (80 feet) per day, or 8 kilometers (5 miles) per year.

Before discussing the actual mechanics of glacier motion, we might consider

Fig. 12-3 Crevasses in glacier ice, forming where the surface of the glacier steepens abruptly over a bedrock irregularity. Blue Glacier, Mount Olympus, Washington. (Photograph by Tom Ross.)

how it is that the ice may be advancing at rates such as these, and yet have the lower end remain stationary or, as in many glaciers, be receding. The position of the glacier terminus represents an uneasy equilibrium between the rate of ice advance and the rate of melting. The glacier continually moves ice forward to the point where the rate of melting triumphs. Or we might compare a glacier to a side of bacon being fed into a slicer. Although the bacon is continuously being shoved forward it never advances beyond a fixed point because it is always being cut off by the oscillating blade.

To return to the way in which ice moves, it obviously does not flow as a liquid does. The crevasses, or surface cracks, rule that possibility out. However, in another respect much resembling a liquid, a glacier does essentially take the shape of its container—in this case the configuration given it by the walls of the enclosing valley. Also, in very nearly the same fashion as a stream,

Fig. 12-4   A continental ice sheet with glaciers reaching the sea. Inglefieldland, northwest Greenland. (Courtesy of Geodetic Institute, Copenhagen.)

Fig. 12-5   Glacial grooves cut in crystalline bedrock, Val Camonica, southern Alps. The polished rocks here lay along the ancient amber trade route, and are inscribed with prehistoric figures and symbols.

the center of a glacier advances more rapidly than the sides, and presumably the surface moves more speedily than layers near the bottom.

Although far from being demonstrated with certainty, some of the flow that occurs in glaciers appears to be the sort that is characterized as *plastic*. Ice crystals have a measure of built-in instability. They yield under pressure, and if the crystals are oriented properly this is easily accomplished. Ice is especially susceptible to slippage along planes parallel to the base of the typical hexagonal crystal. When enough of these are lined up, out of the millions present within a glacier, then a readjustment can occur through crystal gliding—a process akin to the multiple cleavage of minerals within a rock—and the whole solid substance makes a slight movement downslope. The cumulative effect of these individual slippages is enough partially to explain the downslope advance of the glacier as a series of minute readjustments under stress.

Another variety of glacial motion is actual slippage of the ice mass over the floor and margins of its trough. This mechanical sliding of the ice over its own bed is responsible for some of the familiar yet tremendously impressive end products of glaciation—grooves or scratches in the bedrock known as *glacial striations* (Fig. 12-5), and broad expanses of smoothed, barren rock called *glacier polish* (Fig. 12-6).

There is strong evidence that some motion of the ice, especially near the lower end of the glacier, is by shearing. This is displacement by fracture rather than by flow. When such shear planes are visible in the side of a glacier near its terminus, they can be seen curving upward in crescentic sheets. How large a percentage these shears contribute to the total glacial advance is hard to say. Probably it is rather slight, and very likely is less than the effective flowage achieved within the glacier by melting and recrystallization under stress and by crystal gliding.

These methods of movement are characteristic of ice in narrow valleys where the slope is sufficient for gravity to cause flowage. The ice in an ice cap, where the slope is very slight, also moves. This movement is generated by internal pressures which force the ice to move outward from a central point. It is called extrusion flow, as opposed to gravity flow, and takes place in the basal layers of the ice.

An interesting, and puzzling, phenomenon of some glaciers is their tendency to have short periods of greatly accelerated flow. These are known as *glacial surges* and are most characteristic of valley glaciers, although ice caps have been known to surge also. Surges generally take place after the glacier has been stagnant or even receding; suddenly it moves several miles in a few months. In 1953, for example, the Kutiah Glacier in the Himalayas surged 11 kilometers (7 miles) in three months. Various theories have been suggested to account for glacial surges. Some geologists have thought that earthquakes can shake great quantities of snow and ice down onto a glacier, or that increased snowfall for a few years at the head of a glacier could cause surging. Others have suggested that an increase in the amount of meltwater percolating down through the glacier would facilitate the sliding of the glacier along its bed. A very plausible

Fig. 12-6 Glacial polish and striations on igneous rocks in the Sierra Nevada, California. (Photograph by Philip Hyde.)

Fig. 12-7   An alpine-type glacier, showing the junction of tributaries. The glacial system occupies pre-existing stream valleys. Klinaklini Glacier, Coast Ranges, British Columbia, Canada. (Photograph by Austin S. Post, University of Washington.)

explanation which has been put forth recently is based upon the formation of a block of stagnant ice at the end of a glacier. Such a block would act as a dam until the pressure of the flowing ice behind it forced it to give way.

Alpine Glaciation

Although individual glaciers and ice fields are tremendously diverse, they can be placed in two broad categories: *alpine glaciers* and *continental glaciers*. Alpine glaciers typically are mountain glaciers that have their origin on mountain slopes and summits that rise above snowline. They advance downslope under the urging of gravity, and more often than not take the path of least resistance by following a pre-existing stream valley (Fig. 12-7). Unlike the continental glaciers that subdue many of the irregularities they encounter, alpine glaciers are more likely to accentuate irregularities in the landscape, making bold peaks even more jagged and steepening the walls of already deep canyons.

Although alpine glaciers are conspicuous elements of the world's snow-capped mountain ranges, they, too, are much smaller than they were in the Pleistocene. Then, they advanced northward far beyond the foothills of the Alps onto the lowlands near Munich and southward into the low hills marginal to the Po Valley of Italy. Reduced though they may be, some of the surviving alpine glaciers are impressively large. Several of the Himalayan ice streams are 40 kilometers (25 miles) long, and the Great Aletsch in Switzerland has a length of 22.5 kilometers (14 miles). The Seward glacier in Alaska with its tributaries has a total length approaching 80 kilometers (50 miles).

Glacial Erosion

There appear to be two leading ways in which alpine glaciers shape the land surface upon which they rest: *glacial quarrying* and *glacial abrasion*. In the first of these, rocks are sprung or pried out of place in much the same way they are in a commercial rock quarry, except, of course, at a far slower rate. In abrasion, the rock surface on which the ice rests is scoured or worn down in about the same way a wooden surface may be sandpapered. In general, quarrying seems to be most effective at the upper end of a glacier, essentially in the catchment area, while the effectiveness of abrasion is greater downstream.

The way in which this process operates is not completely understood. This is scarcely surprising because it is a process that takes place beneath the glacial ice in an environment inaccessible to us by any ordinary means.

The nearest access we have to this frigid, subglacial world is by crevasses that start at the surface and extend down through the ice to the bottom of the glacier. Such deeply penetrating fractures are rare, but are reasonably common in the catchment basin at the headward end of the glacier. A characteristically crescent-shaped crevasse at the head of a glacier is called a *bergschrund* (Fig. 12-8). As a rule it stands close to the boundary between the glacial ice which has started to move downslope and the stationary ice and firn frozen to the rocks at the upper end of the glacier.

In 1899 in the Sierra Nevada of California one of the leading topographers of that day, W. D. Johnson, had himself lowered on a line to a depth of 45 meters (150 feet) in such a bergschrund—certainly no venture for the faint-hearted—and what he saw there greatly impressed him. This was the ability of glacial ice to pry strongly jointed rocks loose from their foundations. Others have repeated such a descent, and many glaciologists have emerged equally convinced that the upper end of a glacier is a zone of active frost-riving.

The bergschrund and other crevasses provide an avenue by which meltwater supplied by melting of snow, névé, and surface layers of ice streams down into the inner recesses of the glacier. If such meltwater penetrates to the rocky

Fig. 12-8 The headwall of Palisade Glacier, in the Sierra Nevada, California, showing the well-developed bergschrund. (Photograph by Tom Ross.)

headwall, or to the glacier floor, it percolates into cracks and joints of the bedrock. Then, when the water freezes it expands and, upon expanding, it exerts a tremendous leverage against the enclosing rocks, which may then be pried loose.

Should these rocks be frozen into the glacier they are carried along with the glacial ice as it moves downslope, and a new surface is bared for the process to be repeated. The more that glacial meltwater freezes in these subglacial rock joints, the more effective the process of rock quarrying will be.

Differences of opinion center around whether or not any thawing takes place in the ice which is locked up in these crevasses in the rocks. Should freezing and thawing occur repeatedly, then a potent force is at hand to sunder the rocks as compared to what would occur were it to happen but once. According to many observers a considerable amount of melting does take place within and on the glacier surface. Anyone who has walked on a glacier on a windy day, or on a warm sunny day, will recall that a great amount of water is on the glacier surface, streams are everywhere, and water can be heard gurgling and roaring deep within the glacier's interior. Much of the surface meltwater makes its way down through crevasses, cracks, and other openings, large and small, all the way to the bottom. At night, when the air temperature drops below $0°$ C. ($32°$ F.), the streams dwindle away and the sound of running water is stilled.

Rocks ranging from blocks the size of boxcars down to fragments of flour-like dimensions, are embedded in the ice as the result of glacial quarrying and mass wasting from higher slopes. These are dragged along with the glacier as it moves downslope.

This entrained rocky debris acts as an abrasive, much like the sand grains on a piece of sandpaper. Where the embedded rock fragments are large, they gouge out long grooves and scratches in the bedrock. Where the included rock fragments are fine-grained, they may polish the surface of the overridden rocks, much like the fine emery powder a lapidarist uses. Visitors to the higher parts of Yosemite National Park are impressed by the broad expanses of smoothly polished, shining granite, as fresh looking as though it had been given its bright sheen only yesterday (Fig. 12-9).

The process of abrasion works both ways, for the bedrock and the abrasive are both affected. Many of the fragments embedded in the ice develop scratches on their surfaces, or one side of a boulder may be worn away until it is flat and smooth. Such a rock is said to be *faceted*, and several such flat surfaces may be planed off if the rock shifts its position within the ice as the glacier moves along. This selective abrasion results in the production of angular rock fragments (Fig. 12-10) with a markedly different shape from those rounded by running water in streams.

Fig. 12-9   Glacial polish formed on granitic rocks of the Sierra Nevada, California. (Photograph by Philip Hyde.)

Fig. 12-10 Glacial debris, showing typical faceted boulders and the large range in size of material. Sierra Nevada, California.

On a grander scale, as the debris-laden ice grinds away the surface over which it moves, it smooths most minor irregularities, so that a typically ice-abraded landscape consists of rounded-off rock knobs, and the lower ends of spurs and ridges are blunted or even worn away. Valleys are deepened, made more linear, and in the most striking examples their sides are steepened until they are almost vertical, as in Yosemite or the Lauterbrunnenthal, Switzerland (Fig. 12-11).

Landforms Produced by Glacial Erosion

The two major kinds of glacial erosion produce contrasting results. Quarrying makes for steepened higher slopes, spire-like peaks, and sharp ridge crests. Abrasion is more than likely to yield deep, trough-like, linear valleys, in which many of the exposed rock surfaces are rounded and smoothed off.

Of the features characteristically resulting from glacial quarrying, one of the more impressive is the *cirque* (Fig. 12-12). This is a horseshoe-shaped,

depositional irregularities in such glacially produced ground make for a distinctive, hummocky sort of landscape.

Low, nested, crescentically looped morainal ridges are a characteristic landscape element of the north-central states. They may make for rougher ground than the surrounding prairie, and thus may be the site of surviving wood lots, or be the location of farm houses placed to gain the advantage of the little height available in such terrain. The height of such ridges seldom exceeds 30 meters (100 feet), and in some places wide gaps may appear in the ridge, either because the ice at that point was not loaded with debris or perhaps because the moraine was eroded away.

The end moraine marks a line of comparative stability for the front of a continental glacier, where it may have stood its ground for many decades. When the ice wastes away so rapidly that such stability is lacking, the deposits blanketing the formerly ice-occupied area are the *ground moraine*. Broad expanses throughout the ground moraine may be barren rock, scraped nearly clean by the ice; the more resistant rock ledges stand higher than more deeply abraded weaker zones. These, if gouged out, may hold lakes. Over hundreds of square miles of Labrador, because of this rasping effect of the ice, the ground as seen from the air looks like an immense engraved geologic map.

On the other hand, much of the ground moraine, especially near the margins, may be an uninterrupted mantle of glacial deposits whose surface may be quite level if there was not too much rock debris in the melting ice. If there was an abundant supply then the debris is likely to be dropped indiscriminately and blocks of ice entrapped, which upon melting would produce a pock-marked landscape of kettle holes.

Some glacial forms on the ground moraine do show a regular geometry, and are not random heaps of glacial till. Among the shaped features are swarms of curious elliptical, rounded, low hills, resembling the bowl of a teaspoon turned upside down. These are *drumlins*, from an Irish Gaelic word *druim*, which means the ridge of a hill (Flint, 1957). Of these curious features, certainly the most renowned is Bunker Hill, although there are many others in the Boston region, including some of the islands in the harbor. There is a wide range in the size and shape of these whale-like ridges, but few are more than half a mile long or more than 100 feet high. In general, they appear to have a higher percentage of clay in their make-up than is characteristic of most glacial deposits. Although details of their origin are uncertain (no one ever saw them being made), there is little doubt that drumlins originated beneath the ice. They may represent accumulations made by sediment-laden and relatively slowly moving ice. As the excess load was being deposited, it also was being rounded off.

Eskers, from a Gaelic word used in Ireland, are elongate, narrow, sinuous ridges of glacial till which commonly show a measure of sorting, or rude stratification. They wander across the countryside, much like a canal levee or a railroad embankment laid out by a mildly inebriated surveyor. Their crests are rounded, their side slopes are moderately steep, and their longitudinal slope is gentle. Like conventional streams they may meander, occasionally they are joined by tributaries, but unlike ordinary streams they may climb up hill slopes, especially where they cross low ridges through passes. Seldom do their crests stand much more than 30 meters (100 feet) above their surroundings. Some may be as much as 483 kilometers (300 miles) long, although most are a great deal less. Once again we are confronted with an enigma because eskers are structures of uncertain origin. Small ones have been seen forming in the marginal zone of glaciers, but not the larger varieties. The consensus today is that eskers probably are deposits made by streams flowing in ice tunnels at the bottom of the glacier.

Lakes abound in areas scoured by continental ice. Anyone caught in a traffic jam behind a Minnesota car can read on its license plate that back home there are 10,000 lakes. A casual inspection of a moderately detailed map of the part of Canada bordering Hudson Bay, or of the Finnish and Russian portions of the Scandinavian Peninsula, shows a multitude of lakes, some large, some small, some round, and many elongate. Because of their great number, it is not surprising that they have a great variety of origins. The three leading categories are (1) lakes resulting from differential erosion by the ice, (2) lakes impounded behind dams such as accumulations of morainal debris, and (3) lakes in kettle holes and other undrained depressions.

The Great Lakes are a good example of a system involving two origins. In part they are ice-gouged, and their floors lie far below sea level. The bottom of Lake Superior is 213 meters (700 feet) below sea level and that of Lake Michigan 104.5 meters (343 feet). In part, they are blocked by moraines, especially around the south end of Lake Michigan, whose pendulous, lobate form is a close counterpart of the moraine-outlined glacial lobe whose place the lake has usurped. The Great Lakes had a different pattern during the last part of the ice age from what they do today; for one thing they were dammed to the north by the retreating wall of the receding ice sheet; for another, their levels were higher than those of the present-day lakes, and their outlets were quite unlike contemporary ones. One ice age outlet was via the Mohawk Valley and the Hudson, while another was down the course of the Illinois River to the Mississippi River, and thence to the Gulf of Mexico, rather than to the Atlantic by the Gulf of St. Lawrence, the present route.

Glacial lakes have a great deal to tell us of the chronology of the latter part of the ice age. This is especially true of lakes whose surface was frozen in winter and open water in summer. In such strongly seasonal lakes, two distinctive sedimentary layers presumably accumulate each year; a thicker, light-colored band laid down in summer when streams are active, and a darker, thinner layer with a higher organic content that accumulated under the ice in winter when fine material, such as clay, has an opportunity to settle out through the still water. These rhythmic laminae are *varves*, from a Swedish word *varv*, meaning a periodic repetition. Although they are by no means restricted to glacial lakes (such seasonally controlled sediments also accumulate in the sea), they most commonly occur in cold, fresh-water lakes. A half-century of patient comparison of the varves in the glacial lake deposits of Sweden enabled geologists to establish a chronology extending back 17,000 years from A.D. 1900 (Flint, 1957).

Multiple Nature of Glaciation

When the glacial hypothesis was proposed over a century ago, people spoke of a Great Ice Age. The idea was generally held that in some mysterious way the ice advanced across the northern lands, lingered a while, and then withdrew. Men of that distant frigid day, if indeed there were men at that time, were called upon to endure this affliction but once. Today we know that there were men on earth during that actually not-so-distant day, and that the ice age was a vastly complex event, involving multiple advances and withdrawals of the continental ice sheets.

The map (Fig. 12-22) shows the maximum extent of the North American ice sheet. How do we really know, though, that there was more than one advance, and how do we determine how many there actually were?

Most geologists interpret the available evidence to indicate that there were at least four major advances of the ice, separated by three interglacial phases when the ice withdrew, perhaps completely. The evidence for this complex succession is based in part on (1) the way in which moraines and other deposits of a later glacial stage may overlap those of an earlier advance, and (2) the degree of weathering of these glacial and interglacial deposits. Older deposits may have a soil profile developed on them, younger ones almost surely will not. The original constructional pattern of older moraines and other glacial landforms will be blurred, or perhaps even obliterated, compared to younger ones. Weathering may have progressed to depths of 2.5 or 3 meters (8 or 9 feet) in older glacial deposits, and some boulders, even though they appear solid and intact, are so deeply decayed that they readily can be sliced through by a bulldozer blade. Such weathered glacial material is known by the eminently descriptive but etymologically unpleasant word, _gumbotil_. This was originally defined as a gray, thoroughly leached clayey soil that characteristically is sticky when wet, but hard and firm when dry.

The accompanying graph (Fig. 12-26) makes evident the unequal distribution in time of the four glacial stages. There were two earlier ones, Nebraskan and

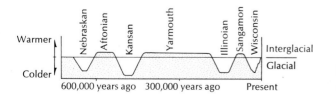

Fig. 12-26 Glacial and interglacial stages of the Pleistocene.

Kansan, separated by a long interglacial interval—the Yarmouth—from two later advances, the Illinoian and Wisconsin. The years assigned to the various episodes are estimates. No one really knows whether or not the Pleistocene was 1,000,000 years long, but probably the order of magnitude of this estimate is about right, although a prevalent opinion, based on radioactive dating of volcanic rocks, is that it may have been as much as 3,000,000 years.

The names of the glacial stages also indicate the extent of the various glaciations—the earlier, Nebraskan and Kansan ice sheets spread farther south than the last, the Wisconsin. As mentioned before, the deposits and landforms produced by the Wisconsin glaciation appear so fresh that they look as though they were created only yesterday. It was thought that the ice age ended about 25,000 years ago, but since the announcement in 1947 of the time-keeping value of carbon-14, the estimated time-lapse since the withdrawal of the continental ice from the vicinity of the Great Lakes has decreased to around 11,000 years.

One of the more significant reference points in North America in dating the last phase of the Pleistocene is the Two Creeks forest bed near Manitowoc, Wisconsin. There, in a deposit of glacial till are the splintered logs of spruce, pine, and birch trees, all aligned in the direction of ice advance. Beneath the glacial debris, stumps of trees still stand in the positions they held before being overridden by this last short-lived advance of the ice. Since these logs have a radiocarbon age of 11,400 years, this gives us a date in this isotopic chronology for the close of the Pleistocene in the Great Lakes region.

The effect of this telescoping of time since the last ice retreat on the thinking of archeologists, climatologists, geographers, geologists, and historians has been profound. The time available since the deglaciation of northern Europe in which to deploy many of the peoples of that wonderfully diverse population and to develop the various cultural patterns of both the Old and the New World seems much too short. This is especially true when one considers how stubbornly most primitive societies resist change, and to progress from a simple hunting and gathering existence to the development of agriculture, the domestication of animals and plants, and the erection of complex hierarchal social structures in the surprisingly short number of generations available is difficult to believe.

The interglacial intervals, whose names are taken from Afton Junction, Iowa, Yarmouth, Iowa, and Sangamon County, Illinois, were times when the glaciers may have disappeared completely. In addition, there is strong evidence to suggest that during some of them the climate was warmer than it is now.

In unglaciated regions which today are arid, such as the Mojave Desert in California, the interglacial intervals were generally warm and dry, but when the glaciers advanced they were cooler and wetter. These wetter periods are called *pluvials* in reference to the increased rainfall.

The diagram (Fig. 12-26) shows that the Wisconsin stage was itself composed of a series of warmer and colder periods. In the western part of the United

States these glacial substages are called the Tioga, the Tenaya, and the Tahoe, and they may or may not be the same as substages of the Wisconsin in other parts of the country. Undoubtedly the other stages had their substages also, but evidence of them has been blurred or destroyed by erosion and subsequent glaciation, while remains of the younger Wisconsin substages are still fresh.

Postglacial Climatic Changes

Determining the causes of the climatic fluctuations responsible for the waxing and waning of the ice sheets during the Pleistocene is a major scientific problem. We shall discuss some of the possible explanations, but before that we should consider the question of whether or not we are living in an interglacial time, and whether or not the whole chapter in earth history built around the successive encroachments of the ice sheets ended with the Wisconsin glaciation.

The answer to this last question is simple—we do not know. For one thing, the length of time during which climatic observations have been made is far too short. Reliable instrumental records date only from the middle of the nineteenth century.

Enough is known from a mass of information of all kinds to tell us that the climate since the end of the ice age has not always been the same. Some of this scattered information seems authentic, some of it is interpretive, and some of it is frankly speculative. We are working in the realm of human history, though, and historic records have survived of many of life's bitter experiences which are reflections of climatic stringency, such as droughts, floods, crop failures, blocking of alpine passes by long-enduring accumulations of ice and snow, and successions of unusually savage winters. Many of these things were written down, others have to be inferred from such occurrences as finding village sites once built on a lake shoreline, now far up the slope laid bare through the evaporation of the lake waters. In an opposite sense, the original site of lake dwellings may now be concealed by a rising lake level.

The doomed Norse colony in Greenland is an example of the impact of a climatic change. The colony was founded in A.D. 984 and perished around 1410. In its early history the Arctic seas were unvexed by ice and Viking ships could make a passage where today ice floes and stormy seas bar the way. The colonists raised cattle and hay, built permanent habitations, and the colony flourished to such a degree that it had its own bishop. With a climatic change that brought the Greenland ice southward again, with the pressure of the Eskimos at their gates, with a succession of crop failures, with the rise of permafrost in the ground—so that even such shallow excavations as graves were no longer possible—and with the perils of the ocean crossing too great for the frail vessels of that day, the colony and all its inhabitants perished.

Piecing much of the fragmentary evidence together is a task yet unfinished, although a vast store of information has been accumulated, chiefly from Europe where a long and often turbulent record can be sifted. However, we know

little of what postglacial climatic changes were like in lands that are more remote, or with radically different climates—such as the monsoonal tropics.

In western Europe and the Mediterranean region postglacial climates appear
to have oscillated somewhat as follows: From about 8000 to 5000 B.C. the glaciers of Europe receded, until a warmer climate than today's prevailed between
5000 and 1000 B.C. This happier time is the so-called *climatic optimum*, and
during it many of the alpine glaciers disappeared and the higher passes were
open. Temperatures were lower between 1000 B.C. and A.D. 1, then warmed
until A.D. 400 (about the fall of Rome), cooled gradually, but were generally
still warmer than they are today, until A.D. 1300. You may recall that North
Europeans were prowling over parts of the northern world then that are completely inhospitable today. For example, it was roughly in this period that the
Greenland colony flourished, and that Leif Ericson may have reached America
by a northerly route (about A.D. 1000).

From A.D. 1300 on, glaciers readvanced, alpine passes were blocked, and living conditions in northern Europe were harsh. To some, the adversity of the
northern winters may go a long way to explain the hardships that were accepted as part of ordinary living in that more brutal age.

Alpine glaciers were at their maximum during the seventeenth, eighteenth,
and nineteenth centuries, and about 1850 reached their greatest peak since the
end of the Wisconsin. This time of advancing glaciers has been called the
"little ice age." Evidence suggests that some of our modern glaciers are products
entirely of this little ice age, i.e. that they disappeared completely during the
climatic optimum and are not remnants of Pleistocene glaciation, but newly
formed ice streams.

Since 1900 the rate of recession for most, although not all, Northern Hemisphere glaciers has been rapid. The last half-century has seen a significant
warming of the atmosphere, accompanied by shifts of marine currents, fish
populations, migration of land animals, and changes in annual temperatures
and precipitation. As an example, according to Flint (1957), measurements of
Swiss glaciers show a reduction in area from 1853 square kilometers in 1877
to 1384 square kilometers in 1932, a loss of 25 per cent in fifty-five years.

Where this trend is leading, no man now living can say, but it will be interesting to watch the pattern unfold. Will the Earth's atmosphere continue to
warm, will temperatures level off, or will the air chill once more and the glaciers start their march again?

Other Ice Ages

A single catastrophic deep-freeze of the earth would not fit very well with our
ideas of uniformitarianism, and, when we look back through the geologic record, we do indeed find evidence for at least two periods of extensive glaciation
preceding the Pleistocene. One occurred about 280 million years ago, or in the
Permian period. The evidence for it consists largely of the glacial deposits

known as *tillites*, which are unsorted mixtures of sand, gravel, boulders, and clay that have been lithified. Some of the boulders show striations, and many of the tillites rest on striated rock surfaces. Other, and extensive, tillite deposits are marine in origin. In these, boulders and cobbles from melting sea ice appear to have dropped into much finer bottom sediments, deforming the horizontal layers of those sediments. The presence of a number of distinct tillite layers indicates that the Permian glaciation, like the Pleistocene, had stages and substages.

There are several puzzling things about this glaciation. One is that it seems to have been limited to the Southern Hemisphere. Another is that the directions of the striations indicate a number of centers of glaciation, unless the continental masses of that hemisphere were not separate at that time but together formed one large land mass, and unless the earth's poles were not where they are today. Far-fetched as this may seem, there is some evidence for it in the paleomagnetic record which was discussed in Chapter 2. This puzzle will concern us again in Chapter 18.

The next oldest ice age occurred about 600 million years ago and is called the Infra-Cambrian because it came at the very end of the Precambrian but before the extensive fossil record of the Cambrian. Evidence for it, too, consists of tillites, many of which were marine, indicating a widespread distribution of sea ice. In contrast to the Permian ice age, the Infra-Cambrian was not limited to one hemisphere, but was amazingly world-wide.

The boundary between the Cambrian and the Precambrian has been defined by a "sudden" change, geologically speaking, in the fossil record. Precambrian fossils are practically non-existent, consisting almost entirely of simple plants called algae. In the Cambrian, however, there was a tremendous development of animal life, complex animal life in a great variety of forms. Some geologists have suggested that the Infra-Cambrian ice age may have prevented the normally slow evolution of life, but at the end of the ice age, with the improved climatic conditions and the rising sea level, evolution proceeded rapidly.

Causes of Glaciation

The purpose of the short discussion concerning earlier glaciations in other times and other places is to point up the apparent circumstance that in the decipherable span of geologic time glaciation is a rare event. We shall attempt to unravel the tangled skeins of fact and speculation surrounding our near-contemporary, the Pleistocene glaciations.

Among the things to be kept in mind about the Pleistocene glacial record before attempting an explanation are the following:

1. Pleistocene glaciation was a multiple rather than a single event, and four major advances commonly are recognized in North America and Europe.
2. These advances were not of equal size, nor were the interglacial times of

equal length. The record is bifurcated, with two earlier and more extensive glacial stages separated from two later and smaller episodes by the longest interglacial.

3. Glaciation appears to have been synchronous on both sides of the Atlantic. North American and European glaciers advanced or receded in harmony. The evidence is less clear for the Southern Hemisphere; but apparently when Northern Hemisphere glaciers advanced or receded, Southern Hemisphere glaciers did the same. In other words, the entire earth seems to have responded to the same climatic pulses.

4. Although the evidence is not completely certain, nonetheless it strongly suggests that Pleistocene glaciation is a relatively unusual event, at least in the latter part of the earth's history.

5. Times of glacial advance appear also to have been times of lowered temperature, as demonstrated by the nature and fossil content of cores recovered from sediments at the bottom of the sea, and by evidence of lowering of the regional snowline on higher mountains of the world.

Naturally, a phenomenon as challenging as an ice age has brought forth a host of attempts at explanation. No single explanation has won unanimous acclaim; each has its adherents and its detractors. At least fifty hypotheses must have been advanced over the years, but most of them contain some inconsistency that turns out to be a fatal flaw.

Astronomical explanations make much of the fact that the eccentricity of the earth's orbit changes slowly with respect to the sun; also the inclination of the earth's axis with respect to the sun undergoes a slow change—part of the year the earth is tipped toward the sun, part of the year away from it. The Northern Hemisphere is inclined away from the sun in the winter when the earth and sun are a little closer to each other than they are in the summer. The effect is to give the Northern Hemisphere slightly warmer winter temperatures than the Southern. A third variable is that the amount of the earth's inclination to the plane of its orbit may change slightly over millennia. These various eccentricities in the relation of the earth to the sun do not combine in any systematic fashion, with the result that there will be variations over the years in the distribution of solar energy at any given place on the earth's surface, but not in the total amount received from the sun.

Without carving our way through a bristling thicket of technicalities, one of the objections to an astronomical cause for the ice ages is the strong possibility that glaciation in the Northern and Southern Hemispheres would be in opposite phase. While this would fit well the Permian ice age, all the accumulated evidence concerning the Pleistocene glaciation seems to indicate that when northern glaciers advance or recede, their southern counterparts do the same thing at the same time. Another critical point is that were such a repetitive and essentially geometric process to be effective in inducing climatic oscil-

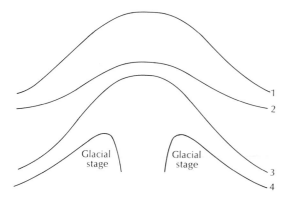

Glacial stage

Glacial stage

Fig. 12-27 The effects of fluctuation of solar energy on temperature, precipitation, and snow accumulation as shown by curves: (1) solar energy; (2) atmospheric temperature; (3) precipitation; (4) snow accumulation. Two such increases in solar radiation would be required to account for the four Pleistocene glacial stages.

lations, then it would very likely have produced ice ages on a periodic schedule in the geologic past. We have already mentioned the extreme rarity of glaciations through geologic time, which makes the rhythmic repetition of astronomically induced large-scale glacial and nonglacial cycles seem implausible. There is mounting evidence, however, that astronomical variations could be responsible for climatic fluctuations within a glacial epoch, producing glacial and interglacial stages and substages.

Variations in the amount of solar radiation received on earth appear to hold a leading position as a cause in the thinking of many geologists. What the nature of these variations may be is unknown at the moment, or even for that matter that they actually occurred. Whether the total energy output from the sun remained constant, or whether changes may have been selective, is debatable; for example, did the amount of ultraviolet light change while other parts of the spectrum remained constant?

Assuming that solar radiation did vary, then an ingeniously argued hypothesis is one advanced by Sir George Simpson, a British meteorologist. The accompanying diagram (Fig. 12-27) illustrates the steps in his argument, if we assume, first, that two marked increases in solar energy occurred in the Pleistocene, as shown by Curve 1, and that this would cause an increase in the Earth's atmospheric temperatures, as shown by Curve 2.

We know from our own experience that an increase in temperature will bring about an increase in evaporation, and this is reflected commonly by an increase in precipitation, as shown by Curve 3. Along with an increase in precipitation, very likely there would be a world-wide increase in cloud cover, and this would tend to keep the temperature at the ground surface lower than it might otherwise be.

Precipitation occurs as both rain and snow, and Curve 4 is concerned with snow accumulation since this is the crucial factor controlling the growth of glaciers. Snow must accumulate from year to year, and be converted to neve and into ice. Ice will accumulate, as shown by the curve, up to a point as the temperature increases, but there will come a time when the temperature rise is great enough that this no longer is possible. Then the snow accumulation curve drops off suddenly, as shown, although the precipitation curve continues on up, closely paralleling the temperature (2) and radiation (1) curves. As the amount of solar energy begins to decline, atmospheric temperature and precipitation also decrease, but the amount of snow accumulation will increase when the temperature drops far enough—the second hump in Curve 4.

Translating snow accumulation into glacial terms, this means that two increases in the amount of solar energy received on the surface of the earth would yield four glacial stages, which would be divided into doublets separated by a long interglacial interval. Incidentally, should this hypothesis prove to be valid, then the long interglacial should be cool and dry relative to the two shorter, warmer, and more humid interglacials. Such a relationship has not been established, but this may be no more than an indication of our ignorance at this moment in the history of science.

Simpson's theory points up one of the great difficulties confronting scientists attempting to explain an occurrence as complex as the ice age. An understanding of its cause will come essentially through a knowledge of climatology and meteorology, but evidence for the existence of former widespread glaciations is geologic. Any valid theory of causes will have to fit the patiently worked out pattern of distribution of moraines and other glacial deposits.

Geologists are prone to criticize Simpson's theory on the grounds that glacial advances seem to be accompanied by lower temperatures, rather than increased ones, as is shown by the chilling of ocean waters revealed by cores of bottom sediments, by regional lowering of snowlines on mountains, by the broad southward displacement of taiga forest and tundra into more temperate lands, and by the southward migration of hordes of arctic animals, such as the musk ox and the extinct woolly mammoth.

A logical explanation appears to be one proposed by R. F. Flint of Yale University (1957). He points out in his theory, which is admittedly tentative, that two factors may have been operative in the Pleistocene which may not have coincided so effectively in earlier geologic times. These were (1) an unusual prevalence of high mountain ranges and therefore of high, snow-accumulating uplands, and (2) fluctuations in the amount of solar energy great enough to produce an average annual lowering of temperature in the middle latitudes of 6° to 8°C.—the amount estimated as necessary to produce an ice age. Since there are two elements in this hypothesis that are essential for its operation— high, snowy uplands and fluctuations in solar energy—Flint gives it the dual name of *solar-topographic concept.*

Much needs to be done before the validity of the concept is firmly established, but these things seem reasonably certain. Fluctuations in solar energy appear to correlate with advances and retreats of glaciers today. The historical record also indicates that postglacial climate has not been constant; at times it has been warmer, and at others—notably the thirteenth century—it has been colder, almost cold enough to herald the onset of a glacial advance had it endured long enough.

However, pulsations of solar radiation very possibly may have occurred with about the same frequency in the remoteness of the geologic past as they have in modern times. The factor that sets the Pleistocene apart is its seeming prevalence of mountains and uplands. Most of the world's better known mountain ranges were elevated to their present heights then, including such lofty summits as those of the Himalayas, Andes, Caucasus, and Alps. Not only was the Pleistocene a time of unusual crustal activity, but the north-south pattern of such significant mountain ranges as those of North and South America became firmly established, and since these lie athwart the general planetary circulation of the atmosphere, they may have had a significant effect upon the growth and dispersal of glaciers.

An additional theory for the ice age was introduced recently by Maurice Ewing of Columbia University and William L. Donn of New York University (1956). Their theory involves two major elements: (1) the North Pole migrated from a position in the North Pacific to its present location at the beginning of the ice age, and (2) the Arctic Ocean alternated between ice-free and ice-covered phases during the Pleistocene.

Polar wandering, which we met in connection with the Permian ice age, is involved in this hypothesis to explain the sharp contrast between generally equable climates prevailing through most of geologic time, and the refrigerated episode, the ice age, immediately preceding the present. A North Pole located in a large body of water such as the Pacific, with free circulation, would not cause the severe climatic stringencies of one located in a closed basin such as the Arctic—hence the abrupt break with climatic patterns of the past.

The hypothesis states that, were the Arctic ice cover to melt, sea level would rise, and the Atlantic water could more readily penetrate the Arctic basin. This would result in a great supply of moisture for arctic-generated storms and consequently increased precipitation over the Canadian barrens and northern Europe. Were this precipitation to be in the form of snow, ice sheets would grow and continental glaciation would be the consequence.

With the expansion of glaciers, sea level would be lowered, the interchange of water between the oceans would be impaired, the Arctic Ocean would freeze once more, and the source of moisture needed for continental glaciers would disappear. Thus, "temperature changes in the surface waters of the Arctic and Atlantic Oceans are the cause of, rather than the consequences of, the waxing and waning of continental glaciers (Ewing and Donn, 1956)."

Another theory which has had its ups and downs has recently been revived by Gilbert N. Plass. It suggests that climatic variations extensive enough to trigger glacial epochs may be caused by variations on the amount of carbon dioxide in the atmosphere. A marked increase of CO_2 would produce a so-called hothouse effect or general rise in temperature. Energy from the sun can reach the earth's surface because it comes as visible energy—light—to which our atmosphere is transparent. Solar energy radiates from the earth back into space, however, as heat, i.e. in the infra-red portion of the spectrum. CO_2, as well as water vapor and ozone, is partially opaque to infra-red radiation, thus keeping the solar heat close to the earth. The more CO_2, the warmer the atmosphere. In fact, the warming up that is generally recognized as having characterized the last fifty years is considered by some to result from the enormous quantities of CO_2 added to the atmosphere through the burning of coal and oil once the industrial revolution hit its stride.

In order to initiate a glacial epoch, the CO_2 must be decreased; how is this accomplished? One way, theorizes Plass, requires an extensive period of mountain building. This in turn would lead to the increased exposure and weathering of igneous rocks. This weathering would increase the formation of carbonate rocks which would remove significant amounts of CO_2 from the air. It is true that both the Permian and Pleistocene ice ages were preceded by periods of extensive mountain building, and, in addition, large amounts of CO_2 would have been removed from the atmosphere during the Carboniferous when much of the earth's coal and oil were being formed in marshes and shallow seas.

Most recently pollutants from our industrialized civilization have been suspected of being more important than CO_2 in changing climatic conditions. It has been discovered, for example, that the town of La Porte, Indiana, has 31 per cent more rain, 38 per cent more thunderstorms, and 245 per cent more days with hail than towns nearby. La Porte is directly downwind from the steel plants of South Chicago and Gary, and, in fact, there is a definite correlation between the amount of precipitation in La Porte and steel production in Gary. Atmospheric dustiness tends to lower the earth's surface temperature, and the last decade has seen a slight lowering of this temperature. In addition, the 1968 ice coverage in the North Atlantic was the most extensive in sixty years.

Much additional work needs to be done to establish the validity of any glacial hypothesis. Nonetheless, such theories are important in provoking debate, in forcing scientists to marshall arguments pro and con, and in encouraging them to seek new evidence. Such work might even prevent us from initiating an irreversible climatic trend and starting a man-made ice age.

Selected References

Broecker, Wallace S., 1966, Absolute dating and the astronomical theory of glaciation, Science, vol. 151, no. 3708, pp. 299–304.

Ewing, Maurice and Donn, William L., 1956, A theory of ice ages, Science, 15 June, pp. 1061–66, reprinted *in* Study of the earth, J. F. White, ed., Prentice-Hall, Englewood, N.J., 1962, p. 203.

————, 1958, A theory of ice ages, Science, 16 May, pp. 1159–62, reprinted *in* Study of the earth, see above, p. 217.

Flint, R. F., 1957, Glacial and Pleistocene geology, John Wiley, New York.

Lewis, Richard S., 1965, The frozen frontier, *in* 1965 Science Year, Field Enterprises Educational Corporation, Chicago, pp. 210–25.

Nielson, Lawrence E., 1968, Some hypotheses on surging glaciers, Geological Society of America Bulletin, vol. 79, no. 9, pp. 1195–1201.

Plass, Gilbert N., 1956, Carbon dioxide and the climate, American Scientist, July, pp. 302–16, reprinted *in* Study of the earth, see above, p. 224.

Wright, H. E., and Frey, D. G., eds., 1965, Quaternary of the United States, Princeton University Press, Princeton, N.J.

Fig. 13-1   Arroyo de la Parra, between San Xavier and Loreto, Baja California. (Photograph by William Aplin.)

13

Deserts

Least familiar of land areas, aside from the extreme arctic, are the world's deserts. Perhaps their seeming mystery lies in their relative remoteness from lands such as western Europe and the Atlantic coast of North America, where the modern pattern of western civilization developed. Had western life remained centered on the Mediterranean, deserts would be much closer to our daily lives because the limitations imposed by aridity bear heavily on such bordering countries as Spain, Morocco, Algeria, Libya, Egypt, and Israel.

In earlier days much of the southern shore of the Mediterranean was the granary of Rome, and once-flourishing cities, such as Leptus Magnus in Libya, are now stark ruins half buried in the sand. One of the problems in studying deserts is that the boundaries are not fixed inexorably, but may expand or contract through the centuries. In fact, there is much evidence from paleobotany—the study of fossil plants—that deserts are relatively late arrivals among the earth's landscapes. Deserts require a rather specialized set of circumstances for their existence, and in a moment we shall inquire into what some of these are.

First of all we need a working agreement as to what constitutes a desert. Temperature is not the only factor; some are hot almost all the time, others may have hot summers and cold winters, and some are cold throughout much of the year. Since drought is their common factor, in a general way we might

Fig. 13-2   The desert crowds the edge of the irrigated flood plain of the Nile, near Giza, Egypt. (Courtesy of Trans World Airlines.)

call those regions deserts where more water would evaporate than actually falls as rain. In other words, the criterion we are using is relative rather than absolute, such as saying all regions are deserts that have less than 25.4 centimeters (10 inches) of rainfall in a year.

Drought is their prevailing characteristic, and deserts notably are regions of sparse vegetation. Few are completely devoid of plants, but some come very close to this ultimate limit. Typically, desert plants are widely spaced. Their colors tend to be subdued and drab, blending with their surroundings. Their leaves may be small and leathery in order to reduce evaporation. In fact, some, as the *saguaro* of Arizona or the barrel cactus of the Sonoran Desert, may have no leaves at all. Other desert plants, such as the ubiquitous sage, may develop an extraordinarily deep root system in proportion to the part of the plant that shows above ground. Plants with these adaptations of extensive roots, leathery leaves, and large water-holding capacity are called *xerophytes*, from a combination of Greek words meaning dry + plant.

Every gradation exists in deserts, from those that are completely arid and that are essentially barren expanses of rock and sand, devoid of almost all visible plants, to deserts which support a nearly continuous cover of such plants as sagebrush and short grass. Dry regions with such a characteristic seasonal cover are best referred to as *steppe*, and commonly are marginal to the more desolate wastes.

The accompanying map of the world (Fig. 13-3) shows that the dry regions are concentrated in subtropical and in middle latitude parts of the earth's surface. For example, there is nearly continuous desert from Cape Verde on the west coast of Africa, across the Sahara, the barren interior of Arabia, the desolate mountains of southern Iran, and on to the banks of the Indus in Pakistan. All told, 46 or 48 billion square kilometers (18 or 19 million square miles), or 36 per cent of the land surface of the earth, might be classed as arid.

One region of deficient precipitation that is difficult to classify is the barren land of the Arctic. The precipitation may be 25.4 centimeters (10 inches) a year or less, yet with a low evaporation rate, the tundra appears to be far better watered than it is. This illusion is aided by permafrost which keeps surface water from sinking very deep into the ground.

These cold deserts are so unlike the more typical dry lands of middle and low latitudes that we shall leave them out of this chapter in order to concentrate on the familiar sort of desert. The map (Fig. 13-3) shows that the preponderance of the dry areas of the earth—exclusive of the Arctic—are on either side of the equator, chiefly around latitude 30°, and that they tend to favor the western side of continents.

Fig. 13-3 Dry regions are found chiefly in the subtropics and the middle latitudes. They occur (1) below the high pressure cells in the atmosphere where descending air is being heated; (2) along cold-water coasts where moisture-laden air (fog) is transported to the warmer land; and (3) to the lee of high mountain ranges where descending air is being warmed. (After P. Meigs, "Future of Arid Lands," 1956. By permission of the American Association for the Advancement of Science.)

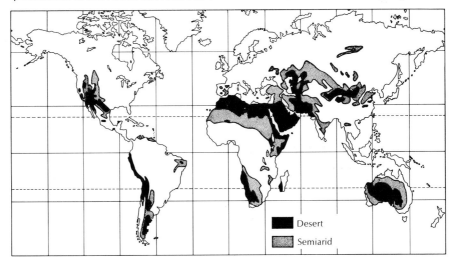

Contrary to the popular image, most deserts are not vast shimmering seas of sand across which such picturesque characters as Foreign Legionnaires slog along while sheiks on spirited stallions sweep by. Although many deserts are sand covered, the majority are not. Most desert regions are more likely to be broad expanses of barren rock, or of stony ground with only a rudimentary soil profile developed. Ground colors are largely those of the original bedrock. They lack the red colors of tropical soils, especially those that are alternately wetted and dried, or the blacks and dark grays of humid temperate regions where the organic content in soil may be high. The bright red color that we associate with such places as Grand Canyon and Monument Valley comes in large part from coloring matter within the rocks themselves, rather than from red-soil forming processes active there today.

It is typical of many desert regions, especially those in continental interiors, that streams originating within the desert often falter and die within the desert's boundaries. This pattern of streams that do not reach the sea is called *interior drainage*, and is an unusual feature to a visitor from a well-watered region with through-flowing streams. Some desert streams simply wither away and sink into the sand. Others may carry enough water to maintain a lake at the end of their course. Since this will be a lake without an outlet, it is almost universally salty or brackish—the Dead Sea, about 396 meters (1300 feet) below sea level at the end of the River Jordan, is a renowned example. A larger water body without an outlet is the Caspian Sea, covering about 426 million square kilometers (164,000 square miles), and even though it is supplied by such a mighty river as the Volga not enough water reaches it to overcome the inexorable losses of evaporation and to allow the lake to spill over the low divide separating it from the Don River and the Black Sea.

At many arid parts of the world, where the water brought in by streams cannot hold its own against evaporation, desert lakes may be only short-lived seasonal affairs, or may be completely dry for decades. Such ephemeral lakes, so characteristic of drought-burdened lands, are called *playa lakes*, in the southwestern United States, and they evaporate quickly leaving a *playa*. Some playas may be glaring expanses of shimmering salt, such as the Bonneville Salt Flats near Great Salt Lake in Utah, or they may be broad, dead flat, clay-floored dry lakes—seemingly created for landing fields, they have such an ideally level surface.

Causes of Deserts

Before we launch into a discussion of the landforms in deserts and the processes that operate there, it might be well to consider briefly what special circumstances are responsible for causing some parts of the earth's surface to be deprived of normal rainfall. Omitting the polar regions of deficient precipitation, there are three major types of arid regions. These are: (1) Horse Latitude deserts, (2) rainshadow deserts, and (3) deserts produced by cold coastal currents in tropical and subtropical regions.

Many of the world's deserts are centered about 30° north and south of the equator, in the so-called Horse Latitudes (Fig. 13-3). The origin of these deserts is intimately connected with the circulation of the earth's atmosphere. Near the equator, the sun's rays strike the earth's surface more directly than at other places, resulting in greater solar radiation which heats the air. The hot moist air rises, cools, and loses its ability to hold water. The excess water falls as rain in the equatorial regions. The cooled high air spreads northward and southward from the equator into the areas of the Horse Latitudes, where it descends and is warmed, enabling it to hold much more water. The Horse Latitude deserts, then, are deserts because they are continually parched by warm, dry winds which suck up any moisture that they can find.

A second cause of aridity is the interposition of a mountain barrier in the path of a moisture-bearing air current. We have a striking example in western North America with the desert stretching eastward in the so-called rain shadow of the Sierra Nevada of California. The profile (Fig. 13-4) across this part of central California at the latitude of San Francisco shows the great disparity between precipitation on the western slope of the Sierra Nevada and Owens Valley. As moisture-laden air from the ocean is cooled going up the front of the Sierra Nevada, it is forced to give up the water that it can no longer hold. It passes down the eastern slope, is warmed again, and evaporates any water it comes across.

The third type of deserts, those formed because of cool coastal currents, are perhaps the least familiar to North Americans. Such deserts flourish along the middle latitude coasts of continents whose shores are bathed by cold coastal currents, such as the Humboldt Current off the coast of Chile and Peru, or the Benguela Current off the Kalahari Desert of southwest Africa.

These deserts are exceptionally impressive for the very dramatic climatic contrasts encountered within extremely short distances. The desert of southern Peru and northern Chile, for example, is one of the very driest lands on earth, yet its seaward margin is concealed in a virtually unbroken gray wall of fog. Winds blowing across the cold waters of the coastal current are chilled, their moisture condenses, and thus a seemingly eternal blanket of fog stands over

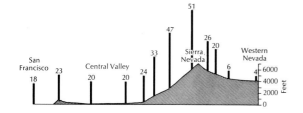

Fig. 13-4 Total rainfall at stations from coastal California across the Sierra Nevada, showing the effect of the mountain barrier in reducing precipitation to leeward. (After Bowman, Forest Geography, 1911. By permission of John Wiley & Sons, Inc.

the sea. Once this fog drifts landward to where the air temperature is higher, the fog burns off almost immediately, and the water-holding capacity of the air current increases rather than decreases as it moves across the heated land.

Few deserts, however, are the product of a single cause operating to the exclusion of all others. For example, the Atacama Desert on the west slope of the Chilean Andes is not only affected by the cold Humboldt Current, but it is also in the Horse Latitudes. Furthermore, cold air sweeps down from the Andes, so that all three desert causes operate at one place.

In summary, we can say that deserts do not have a simple reason for their existence. Their origins are complex, but they are well worth trying to understand, not only for their own sake as an intellectual challenge but also because so much of the future of mankind is dependent upon the utilization of arid lands. An equally intriguing question is whether the boundaries of deserts are stationary, or whether they are contracting or expanding. As we have seen, the climate of the world has slowly changed in historic time. This is strikingly true of deserts, and much of the most compelling evidence comes from the Sahara and the Middle East (Fig. 13-5). Artifacts, stone implements, and rock paintings of extraordinary subtlety and sophistication testify to the presence of early man in what are desolate expanses of the Sahara today. At a later date much of this barren region was the granary of Rome, and colonial cities of that day as well as roads along which the legions marched are now overrun by drifting sand. The expansion of the desert broke the slender thread of communication lines linking the cultures of the Mediterranean and African worlds. Apparently thereafter the two had a vague and uncertain awareness of each other over the centuries, but it was not until the introduction of the camel caravan that the land connection was re-established. By this time each culture had evolved along a different path.

Stream Erosion in an Arid Region

Paradoxical as it may seem the leading agency responsible for sculpturing the landforms of a desert is running water, just as it is in humid regions. Puzzling indeed is how all the work was done of removing the enormous volumes of rock that once filled the canyons of desert mountains, or of shaping the mountains themselves, or even of wearing them away completely to bare rock plains stretching endlessly to the empty horizon, when there appear to be no streams at all in this seemingly timeless land.

Part of the answer may lie in the brief discussion in the immediately preceding section. Not all landforms in all deserts were necessarily formed under the climatic regime we see today; they may be fossil landscapes in a sense, survivals of erosional patterns carved during a time more humid than ours. We shall see one result of a recent climatic shift a little farther on when we talk about the Pleistocene (ice-age) lakes of the desert. Their abandoned shorelines

Fig. 13-5a The present desert regions of North Africa and the Near East were the sites of flourishing civilizations only a few thousand years ago, as Leptus Magnus in Libya, which is now isolated in a sand sea. (Photograph by A. E. L. Morris.)

Fig. 13-5b Another example of desert encroachment is Mirgissa, Sudan, near the second cataract of the Nile, which dated from 1500 to 1300 B.C. (Photograph by C. W. Meighan.)

contour the slopes of many of the mountains and their saline deposits whiten the floors of many of the western desert basins.

Not all the erosion of desert landscapes, however, was done in the remote geologic past and under climatic controls alien to the contemporary world. Almost all deserts have some rainfall, even though ten to fifteen years may elapse between showers.

Nature of the Run-off

Contrary to popular belief, cloudbursts are relatively rare in arid lands—they are much more common where rainfall is greatest, as for example, the rainy tropics, or the southern Atlantic coastal states. When a moderate rain does fall on the desert, it is likely to assume the apparent proportions of a cloudburst in a more humid region and also to do a very effective job of erosion, because there is no vegetation to protect slopes from the spattering effect of rain drops, from rill wash, and from the rapid cutting of the ravines and arroyos that are so typical an aftermath of desert cloudbursts.

Anyone caught in one of these sudden downpours is likely to have an unforgettable experience should he fail to make his way to higher ground in a hurry. In a matter of moments, a dry sandy arroyo, bordered by low but steep cliffs, is filled with a surging, mud-laden flood. Such a stream swirls and churns forward violently, sweeping a great mass of debris along with it. Such flash floods make deserts impassable, until the arroyos drain. This they do almost as rapidly as they filled, because there is no continuing source of water supply for them as there is in regions of plentiful rain and perennial springs. In a matter of a few hours beneath the desert sun, an arroyo floor covered by 3 meters (10 feet) of water may be dry sand again, interrupted by only occasional pools of muddy water.

Such torrential floods as these short-lived desert ones may on occasion overtop the low banks of desert dry washes and spread out as a sheet of muddy, turbulent water over the desert floor. Such a _sheet flood_ is vastly effective in picking up loose sediment and shifting it around the landscape.

Thus, deserts strike a curious balance when their rates of erosion are contrasted with those of humid regions. Much less rain falls in arid regions, but slopes are correspondingly more vulnerable because they lack the stabilizing effect of vegetation. Badlands, or slopes scored with great numbers of gullies, large and small, are characteristic desert landscape elements.

Depositional Forms

Wherever erosion occurs, there is deposition close by. This is especially true in arid regions because here ordinary streams cannot escape beyond the confines of the desert. Their water sinks underground into sandy stretches of their own normally dry stream beds, or it evaporates, or it may be withdrawn by the fiercely competitive water-seeking plants, such as mesquite and tamarisks that line the banks of many desert watercourses. Much of the desert landscape is dominated by stream deposits, in large part because of the inadequacy of running water to move great quantities of debris out of a desert basin and on to the sea.

A most characteristic desert landform is the _alluvial fan_ (Fig. 13-6). These are accumulations of boulders, gravel, and sand deposited by dry-climate streams which begin at the point where such a stream leaves the rock-walled defile it has eroded in the mountains and starts out across the basin at their feet. The stream velocity is further reduced because much of the water sinks into the porous, sandy subsurface layers of the fan, thereby decreasing the volume which is one of the factors controlling the velocity. The stream quickly becomes overloaded where it starts across the fan. Its load does not diminish, but its volume fades rapidly. The main channel very soon separates into a score of distributaries, and it acquires a braided pattern.

Streams may be able to cross the fan in time of flood, but under normal conditions they sink into the sandy ground almost as soon as they leave the

Fig. 13-6   The surface of an alluvial fan in Death Valley, California, is marked by a braided pattern of small channels. Drying patches of mud have begun to crack and curl. (Photograph by Philip Hyde.)

bedrock of the mountains. This is almost always a surpirse to visitors to a dry country—to see a stream waste away, growing thinner and thinner downstream and then vanishing completely—quite the reverse of humid climate streams, whose volume commonly increases downstream.

Actually the water has not vanished but is percolating slowly through pore spaces in the fan. Far down the fan where the material is finer—chiefly sand and silt—the permeability is less. The gradient is diminished, too, and the thickness of the fan is correspondingly reduced. The result of all these factors is that water which was deep in the ground in the mid-section of the fan is forced to the surface at the toe, or lower end. Here it may seep out in springs, or *cienegas* (from a Spanish word for swampy ground), commonly marked in the desert by clumps of mesquite trees—one of the best guides to water in the arid Southwest.

If there is a single alluvial fan at the base of a desert range, probably there will be others—in fact, there probably will be one at the mouth of each principal canyon. These fans overlap like palm fronds, to make a nearly continuous apron sloping away from the mountain front to the basin floor (Fig. 13-7). Such a constructional surface built up by overlapping, or coalescing alluvial fans

Fig. 13-7 The gently sloping constructional surface along the east front of the Sierra Nevada, Owens Valley, California, is composed of a series of overlapping alluvial fans which form a bajada. Note the size of the boulders in the foreground, about two miles from the mountain front; to the east (left) they give way to finer sediments. (From the Cedric Wright Collection, courtesy of the Sierra Club.)

is called a *bajada* (pronounced as though it were spelled bahada), from a Spanish word meaning a gradual descent. We use the same root when we speak of Baja California for the peninsula in Mexico on the west side of the Gulf of California.

Driving across a bajada on a road paralleling a mountain front and near the base is a singular experience. The road continually rises and falls, much as though it were laid across an immense ground swell at sea. The high points are opposite canyon mouths where the road is on the axis of the fan, and the low points are about midway between canyon mouths where the road is in the swale separating two fans.

A desert basin bordered by mountains whose lower slopes are partially buried in their own debris is called a *bolson*. This is a Spanish word meaning purse, but used locally in the southwestern United States for a mountain-girt desert basin or valley without an outlet, in other words a closed basin with interior drainage. In a humid region—where precipitation is greater than evaporation—such a basin would be occupied by a lake whose waters would rise until they spilled over the lowest point of the basin rim.

This will not happen in a desert. There, since more water evaporates than falls as rain, a large permanent and continually expanding lake cannot form. There are lakes in deserts, though, and some, such as Great Salt Lake, are well known. Characteristically, their shorelines fluctuate with climatic variations, and their waters are saline. Sometimes, as in Great Salt Lake, when the surrounding region is underlain chiefly by sedimentary rocks, the leading dissolved constituent in the water is NaCl. In others, where the principal source of soluble material has been the erosion of volcanic rocks, the water may hold a much higher content of Na_2CO_3 (sodium carbonate). Since everything soluble in the rocks of a desert region may be carried into such a lake, no wonder some desert lakes are natural chemical factories. Saline lakes in volcanic areas are likely to be especially prolific sources of unusual substances, including potash, potassium salts, and in the eastern desert of California, boron compounds.

During a dry climatic cycle such a saline lake may evaporate completely, leaving an achingly white residue of salts, as complex chlorides, sulphates, and carbonates. The surface of such a salt layer may be smooth, or if it is undergoing solution, it may be very rough. The incredibly jagged terrain of salt pinnacles on the floor of Death Valley (Fig. 13-8), nearly 90 meters (300 feet) below sea level, was a fearsome stretch for the first party of emigrants to cross, with their battered wagons and patient, long-suffering oxen.

The playas we have just been discussing are saline, and these chemically complex accumulations of soluble salts commonly are (1) the residue left behind through the evaporation of a once larger lake, or (2) the salt accumulation that gradually builds up at the end of an essentially continuously flowing desert stream system. There are some desert rivers that do have a fairly continuous flow of water throughout the year, even though they are underground through the porous sand and gravel of the stream bed, and thus not visible from the surface at all. Actually, most saline playas are combinations of the two types. A large number of them are the floors of dessicated lakes and are also at the end of a salty, perennial stream. This is true, for example, in Death Valley, which once held a lake about 193 kilometers (120 miles) long and whose surviving saline playa is also at the end of the Amargosa River, which name appropriately enough means bitter.

Clay playas are more likely to be found in smaller isolated bolsons, rather than in larger basins at the end of a long and integrated drainage system. Most of the year they are dry and their surface is baked as hard as a brick; in fact, they make ideal emergency landing fields. Or the whole lake floor may be con-

Fig. 13-8   The Devil's Golf Course, Death Valley, California, is composed of salts—complex chlorides, sulphates, and carbonates—which were concentrated in Lake Manley at the end of a long drainage system. The salts were then precipitated as the lake disappeared. (Photograph by Ansel Adams.)

verted into a gigantic multidirectional landing ground, as at 24-kilometer-long (15-mile-long) Rogers Dry Lake at Edwards Air Force Base in eastern California.

When a typical short-lived downpour is over, desert streams waste away as suddenly as they sprang into being. Then the arroyo bottoms quickly return to their seemingly unchanging state of dry, shifting sand enclosed between steep arroyo or canyon walls. The playa may endure a bit longer, but ultimately its murky, dark-hued water evaporates or sinks underground. Before this happens, the suspended load of finely divided silt and clay particles has been disseminated throughout the entire body of the ephemeral lake. Thus, when the lake is gone its newly revealed floor is a dead flat expanse of uniformly distributed, tan-colored, fine-grained sediment. Very often, too, the surface layer of playa clay shrinks as it dries, cracking into thousands of small polygonally bounded blocks. These have much the same sort of pattern one sees on the tops of basalt columns, such as those in the Devil's Postpile. Both patterns are the result of shrinkage. In the case of basalt, it is the loss of volume on cooling; and in playa clays it is the loss of water.

Erosional Landforms

In addition to gullies, valleys, and canyons, there is in the desert another erosional landform, the _pediment_. Pediments are bedrock surfaces, stripped bare, gently sloping away from low desert mountains toward the lower part of the _bolson_. From a distance these broad encircling surfaces look like a uniformly sloping bajada or a constructional apron of overlapping alluvial fans, rather than a product of long-continued degradation and removal of many thousands of cubic meters of bedrock.

The process by which pediments are formed has long been a puzzle to geologists who have debated it at great length ever since pediments were first described in 1897. Very briefly, there are two principal schools of thought on the subject. Some geologists think that a pediment is formed by lateral planation of streams, i.e. by the erosion of a stream as it swings back and forth over the rock surface. Other geologists believe that the pediment surfaces weather and the products of weathering are removed by rillwash or sheetwash which is the unconcentrated flow of water during a rainstorm. Most recently it has been suggested that both processes operate in varying degrees.

Erosional Cycle in an Arid Region

We have a broad desert of our own in western America, one well worth studying in its own right. Geologically it provides us with a superb sample of a desert environment and, although it might be more exciting to go to Timbuktu in the Sahara to study deserts, we can see almost every desert landform displayed in our own country. Only within recent years has the desert

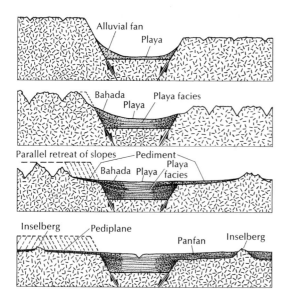

Fig. 13-9 The erosional cycle in an arid region.

become a place to be visited instead of avoided. Now, for example, motels and swimming pools flourish where once dust-plumed wagon trains were targets for the Apache.

The mountainous landscapes of our own Southwest perhaps are not fully representative of the great deserts of the world—many of these are vast empty plains. Our desert possibly has more relief than most, and the mountain-girt bolson with a playa in its lowest part is a typical landform.

The diagrams (Fig. 13-9) showing the erosional cycle in an arid region are modeled on the typical terrain of the American Southwest. We start with a recently uplifted mountain range, which has an upland surface cut by narrow gullies separated by broad flat divides. As weathering and erosion proceed, the mountain slopes recede nearly parallel to their original slopes; there is not the flattening and rounding caused in humid regions by soil creep. Also the waste material is not removed from the area but accumulates at the base of the mountains in desert basins. Because streams are infrequent and flow only for a short while, an integrated drainage never has a chance to develop, and, as a consequence, base level is the valley floor which is occupied by a playa. This basin gradually fills with sediment, thus raising base level so that the streams lose gradient very rapidly and dump their loads at the edge of the playa. As these deposits build up, the streams are forced to leave their alluvium farther and farther upstream, until a thick alluvial plain or bajada is formed at the mountain front.

With the retreat of the mountain slopes, a pediment is formed which is extended back until only a remnant of the former mountain still exists—an *inselberg*. Eventually the inselberg also will be eroded away. This extensive pediment is called a *pediplane* and corresponds to the peneplain of a humid region.

Wind

Wind Erosion

Over many of the dry lands of the world the wind blows seemingly without restraint, adding a note of harassing melancholy to an already unbearably lugubrious scene. Such a desolate region, with the wind keening ceaselessly across it, is the drier part of Patagonia of the far southeastern reaches of Argentina. Other deserts are perhaps as windy on occasion, but in most of them times of extreme windiness alternate with times of calm.

During the intervals the wind blows, its erosional effectiveness is likely to be greater than in humid regions. In the latter, the ground surface is protected by vegetation and by a more tenacious mantle of weathered soil which also may be damp throughout most of the year. Vegetation plays an important role indeed because the potency of the wind drops off very rapidly at the ground, the plants acting as an extremely effective baffle.

The wind is a most effective agent of transportation for certain sizes of rock particles, and scarcely at all for others. It is this high degree of selectivity that makes the wind such an unusual agent of erosion.

Obviously there is an upper limit to the size of particles that the wind reasonably can be expected to move. Very few boulders of the size that are handily transported by streams, waves, or glaciers will be moved by the wind— even by a tornado. This immobility is true for loose material down to the size of large pebbles, although the smallest pebbles, about 4 millimeters (.15 inch) in size, are commonly moved. Since the wind winnows out and whisks away all the finer material, the larger pieces settle down in place to build a mosaic veneer of pebbles called *desert pavement* (Fig. 13-10). Many desert surfaces are sheathed with this protective armor of rock fragments, an armor that can be remarkably smooth and flat. If for some reason a chink in the armor develops,

Fig. 13-10 The surface of the stony desert in Jordan is composed of a blanket of pebbles and gravel which were left behind after the finer sand had been blown away by the wind. (Photograph by A. E. L. Morris.)

Fig. 13-11 Post eroded by wind heavily laden with sand. Near Palm Springs, California. (Photograph by D. C. Strong.)

perhaps by having individual gravel fragments shoved aside by a jeep breaking a trail across the desert, then the newly exposed sand will be blown away by the wind. The new rock fragments eventually settle down, and the desert pavement is stabilized once more.

Odd as it may seem, very fine-grained particles, such as silt or clay flakes, are difficult for the wind to start in motion. In general, the same principles are operating here that also operate in streams of water. Silt and clay, with their small dimensions and strong cohesiveness, are difficult for a moving current of air to pick up. Once lifted out of place, however, they remain in suspension many times longer than sand. Their tenacity when in place is demonstrated when the wind blows full strength across the surface of a clay playa. Very little dust is stirred up as a rule, the hard-packed clay particles hold firm, although some of the looser sand and silt around the margin may be picked up—especially if the playa surface is sun-cracked.

This difficulty of the wind in picking up fine material is no longer true once the natural tenacity of these small particles is overcome. The way in which such particles may be loosened is illustrated by such familiar examples as the running of a band of sheep across clayey ground, or the driving of a car along the dusty tracks of a desert road, or above all else, the plowing of dry and dusty soil in advance of the windy season. To anyone flying over an arid or semi-arid region on a windy day, it is most impressive to see the dust blowing in long streamers from some fields while others are dust free.

Once silt- or clay-sized particles are picked up by the wind, there are no restraints imposed on their travel comparable to the lot of their waterborne contemporaries, which are limited to the drainage pattern of some stream or to some glacier's course. Windborne dust swept from the fields of Colorado in the *Grapes of Wrath* days of the 1930's was carried as far east as the New England states. In fact, some dust was blown far beyond, out over the North Atlantic.

There are scores of other examples of the efficacy of air currents in moving fine particles over vast distances. Volcanic eruptions of the explosive sort are the most telling kind because they constitute a point source for the volcanic dust, and also, because of the unique nature of this material, it can be traced over vast distances all the way back to the origin. Krakatoa, in Java, hurled dust into the upper atmosphere to circle the earth several times and to produce an appreciable fall in regions as remote from the source as western Europe. Icelandic eruptions have made appreciable deposits in Europe, too, as well as in eastern North America.

More typical, however, of the aspect of dust transport with which we are concerned here is the dust whirled high into the air by turbulent winds of the Sahara and broadcast far and wide over the Mediterranean, southern Europe, and on occasion as far as England, 3200 kilometers (2000 miles) away.

The sheer quantity of material that can be in the air at any given time is sur-

Fig. 13-12 Telephone poles sheathed against wind erosion. Near Palm Springs, California. (Photograph by D. C. Strong.)

Fig. 13-13   The base of Mushroom Rock may have been carved by sand blasting. Death Valley National Monument, California. (Photograph by A. M. Bassett.)

prising. In a dust storm of average violence a cube of air 10 feet on a side might well have 1 ounce of dust suspended in it. This seems trifling, but if we increase the size of our cube of air until it is a mile on each side and maintain the same saturation of dust, then each cubic mile of air is supporting 4000 tons of solid material. Thus, a storm 300 or 400 miles across might well be sweeping 100,-000,000 tons of solids along with it.

All this is strong evidence that the wind can be an effective erosional agent and one of especially great significance in arid regions. It is the only agent that can transport material beyond the confines of the typical desert. Not only can the wind carry sediment beyond the rim of a desert basin, but it is also the one agent that can overcome the rising base level imposed on desert streams as desert bolsons gradually fill up with their own debris.

Some of the closed basins in the desert supposedly excavated, or, as we say, *deflated*, by the wind reach very large dimensions. A famous example is the great oasis of Kharga in the Sahara west of the Nile. This depression is about 193 kilometers (120 miles) long by 19 to 80 kilometers (12 to 50 miles) wide and 183 to 305 meters (600 to 1000 feet) deep. The walls are flat-lying layers of readily eroded sandstone, apparently unbroken by any sizable faults. A series of longitudinal sand dunes trails off downwind from this great depression whose sandy floor has almost certainly been the source of supply. This is not to say that the wind hollowed out the whole basin; its origin unquestionably is more complex. Rain wash and streams may erode the friable sandstone and sweep the loose sand down to the floor of the trough. There the sand rests until it is picked up by the wind to accumulate in dunes to the leeward of the basin.

Ground water sets a lower limit to wind erosion. Once the desert surface is deflated to where the water table is reached and the ground is kept damp, then the wind can no longer pick up loose material with the same ease, and deflation slows to a virtual halt. In Kharga the beneficial effect of this has been the appearance of springs around the margins of the great depression. The springs appear at the level that the water table intersects the ground surface at the base of the escarpment. Such springs are the source of water for the true oases, for with only a little ground water to draw on, date palms flourish and make a startlingly green contrast to their stark surroundings.

Another impressive closed basin in the Sahara is the Qattara, 298 kilometers (185 miles) long with a floor that is a searingly forbidding quagmire of salt and shifting sand covering about 19.5 million square kilometers (7500 square miles). Whatever the origin of the Qattara Depression may be, the wind almost certainly played a prominent role in enlarging it and in deepening it to the water table. Evaporation of this water through centuries is responsible for the accumulation of the salt deposits. This immense, impassable saline trough acted as a barrier in World War II that made it impossible for Rommel's Afrika Korps to turn Montgomery's flank and compelled the Germans to attack where the British forces were concentrated at El Alamein on the narrow neck of land separating the Qattara from the Mediterranean.

At some future date this sink may have surprising utility if an unorthodox proposal should be developed of leading Mediterranean water to it through canals and a tunnel, and then using the 457-meter (1500-foot) drop available from the rim of the depression to generate power from the unlimited supply of water available in the sea. Gradually the depression would fill with water to form a concentrated saline lake whose surface it is estimated would stabilize around 45.7 meters (150 feet) below the level of the Mediterranean when the input of sea water balanced the loss of lake water through evaporation beneath the Saharan sun.

It must not be forgotten, however, that most of the modeling of a desert landscape is accomplished by water, not wind. Wind, even wind charged with a great deal of sediment, erodes hard bedrock very little. The most it can do is to remove the finest material from rock that is already weathered. The contrast between landforms found in humid regions and those found in arid regions is caused by the difference in climates; the agent of erosion remains essentially the same—water.

Wind Deposition

Dust lofted out of the desert by winds may be swept for scores of miles before it sifts down to accumulate in a tawny blanket spread far and wide over an area which may be enormously distant from the source. Such a deposit of airborne silt is called *loess*, from the German *löss* or *lös*, a fine, yellowish-gray loam characteristic of the Rhine and other river valleys.

The most renowned of such deposits are those of northeast China, and it was to these that a German geologist gave the name of loess while on an exploring expedition to the outermost parts of the Russian and the Chinese empires. The loess of China has been exported by the wind out of the Gobi and across the Kalgan Range upon whose barren ridges the Great Wall was built. Deposited on the North China plain by the dust storms of centuries, loess lies deep in the valleys of this ancient land, often to thicknesses of hundreds of meters, although it lies much thinner on the crests of divides. This tan-colored silt when picked up by the Yellow River gives the stream its name, as well as the Yellow Sea whose waters are stained for hundreds of kilometers from shore by suspended dust particles.

There are many other parts of the world where loess is found; among them are the Mississippi Valley and Central Europe, neither of which possibly can be construed as an arid region. This widespread distribution is simply an indication that like other things in nature, loess has more than one origin. Presumably these widely distributed blankets of dust were spread far afield by strong winds sweeping across the barren ground that was marginal to the glaciers once covering northern North America and Europe.

A characteristic result of the work of the wind in arid lands is the sand dune. More misconception than understanding surrounds these intriguing features. For one thing they are by no means limited to deserts. Many of the larger and more renowned of the world's dunes are along shorelines, such as those on the shore of Lake Michigan, the length of Cape Cod, and the coast of Somalia. Dunes also border the sandy plains of some large rivers—the Volga is an outstanding example.

Few deserts are completely sand-covered. Most, as we learned earlier, have surfaces of stripped, planed-off bedrock, or else are protected with a carapace of stones, each touching the other, the so-called desert pavement (Fig. 13-10).

Nevertheless, there are sandy areas in almost all deserts; the south-central part of Arabia and the western part of the Sahara are among the better known. Broad areas in the Sahara which are covered with sand dunes are called *ergs*, and this curious name is now used rather generally for similar sand seas elsewhere (Fig. 13-15).

Sand dunes, whether coastal or desert, share many characteristics. Since they normally consist dominantly of sand-sized grains they are a testimony to the extraordinary sorting ability of the wind. Finer material such as silt may be blown far away, while coarse rock fragments, such as pebbles and gravel,

Fig. 13-14 Small ripple marks make intricate patterns on the surface of sand dunes in Death Valley. The wind that formed these ripples blew from the right, but the wind that shaped the dunes in the background, at the right, blew from the left. (Photograph by William Aplin.)

Fig. 13-15 Great sand seas called ergs, shown here in eastern Rub' al-Khali, are typical of the desert of Arabia, as well as parts of the Sahara and other deserts of the world. In the foreground is the camp of a self-sufficient geophysical party exploring for petroleum. (Courtesy of Aramco.)

may lag behind the sand which has been moved some intermediate distance.

Dunes are neither stable nor permanent features of the landscape, but may be continuously on the march. Wind-drifted sand blows up the gentler slope of a sand dune. When it reaches the crest it may be carried a short distance over it—the tops of dunes when the sand is driving across them sometimes seem to be smoking. Behind the crest the sand drops out of the wind stream to accumulate on a steeper slope, the *slip face*. With dry, well-sorted sand, the inclination of the slip face may be as much as 34°. Should the slope become steeper, the sand becomes unstable and shears along a slightly gentler plane with the result that a small avalanche of dry sand glides to the base of the dune. When new sand falls on the slip face, the slope steepens once more, and so the process repeats itself again and again. The net result is a transfer of sand from the upwind to the downwind side of the dune with the consequence that the dune slowly migrates, in a sense rolling along over itself.

Dunes show a fascinating variety of shapes and patterns (Fig. 13-16). Where the wind is variable in strength and inconstant in direction sand may be heaped up in ever-changing forms, although in most dunes the general rule persists that the downwind slope is steeper than the upwind one.

Where the wind holds more constantly from a single direction, as it may along a sea coast, dunes are likely to have a more persistent geometry. For example, they may be aligned at right angles to the wind in which case they are called *transverse dunes*. These are likely to be quite short and to flourish where an abundant supply of sand is available and where the winds are strong. Typically, many coastal dunes are in this category. Or if dunes are lined up parallel to the prevailing wind they are known as *longitudinal dunes*. These are most likely to form when the wind blows strongly out of a single quarter, the supply of sand is sparse, and vegetation is virtually absent. These conditions are fulfilled admirably in the remote Great Sandy Desert of Australia. Some of the individual longitudinal dunes there are said to be more than 96.5 kilometers (60 miles) long.

A curious and esthetically appealing sand dune is the *barchan*. These are beautifully symmetrical, crescent dunes—sometimes as perfectly proportioned as the crescent moon, symbol of Islam. In a barchan the horns of the crescent point downwind and the steeper slip face lies between them. Barchans, too, need a nearly constant wind direction, and also a wind with not too high a

Fig. 13-16 Long lines of dunes stretch across the Arabian Desert. The dunes are shaped by winds blowing diagonally across the lines. (Courtesy of Arabian American Oil Company.)

Fig. 13-17 Old shoreline at Mono Lake, at the eastern base of the central Sierra Nevada, California. The level of the lake was highest during the last glacial age, for in nearby areas the old lake and glacial deposits interfinger. (Photograph by John Haddaway.)

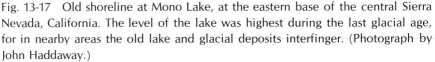

velocity. For these reasons barchans flourish in the trade wind deserts, or in coastal deserts such as the Atacama. Barchans are migratory dunes, as a rule, and may march across the desert landscape at a rate of as much as 15 meters (50 feet) per year. This is done by sand streaming up the gentle upwind slope of the crescent and sliding down the steeper slip face—a procedure typical of most dunes. The chief variation here is that sand swept around the ends of the dune tails off downwind to make the tips of the crescent. As the dune migrates these points continue to pace its progress.

The desert surface between the individual barchans is more likely than not to be barren bedrock, almost completely devoid of sand. The wind is a remarkably tidy housekeeper, whisking up the loose sand from the bedrock surface of the desert between the dunes. Part of the reason seems to be that grains bounce along much more readily over a bare rock surface than they do over sand. Sand has a retarding effect on bouncing grains, and once they strike an accumulation, such as in a dune, their independent free-roving days temporarily are ended.

Desert Lakes

Among the many distinctive features of such a dry and furrowed landscape as our own Southwest are the desert lakes. They owe their existence in large part to the inability of desert streams to develop through-flowing courses. Where such streams are blocked, even though by no more than the advancing toe of an alluvial fan, their water is ponded and a lake results. Some of these, such as Great Salt Lake with a surface area of about 5.2 million square kilometers (2000 square miles), are quite large. Others are little more than saline ponds.

Almost all desert lakes have the common attribute that their water is brackish or saline to a greater or less degree. The concentration of salt in Great Salt Lake ranges from as little as 13 per cent to as much as 27 per cent of the weight of the water, as compared to 3 per cent for the oceans.

Desert lakes are extremely sensitive climatic indicators. In a dry cycle their water wastes away through evaporation and the drought-diminished streams are not able to hold their own against the loss. The lake level drops and the shores are bordered by an ever-widening band of salt. Most famous of such expanses is the Bonneville Salt Flat adjacent to Great Salt Lake, the scene of many a determined assault on land speed records.

Another interesting feature of deserts is the evidence that they were once the site of far larger lakes than the shrunken remnants that survive today. The most redoubtable of these now vanished inland seas in the United States was Lake Bonneville. This was the precursor of present-day Great Salt Lake, and shorelines of this one-time inland sea now scar the higher slopes of the Wasatch Mountains up to 304 meters (1000 feet) above the modern lake. The area flooded by Lake Bonneville was close to 52 million square kilometers (20,000 square miles), compared to the 5.2 million or so of Great Salt Lake. During part of its history, Lake Bonneville had an outlet north to the Snake River and thence to the Pacific by way of the Columbia River.

An interesting contemporary of Lake Bonneville was Lake Lahontan, located mostly in western Nevada not far from Reno. In this ruggedly mountainous area, all the intervening valleys were filled with long narrow arms of the lake. Pyramid, Walker, and Winnemucca lakes are the chief surviving remnants of Lake Lahontan, but both Lahontan and Bonneville left their imprint on the landscape in an impressive array of wave-cut and wave-built landforms. Among these are wonderfully well-preserved beaches, gravel bars, sea cliffs, deltas, and limy tower-like deposits known as _tufa_. These last are built up underwater by lake-dwelling calcareous algae. In the arid climate of the western states these relics of a more humid time in the immediate geologic past are almost as perfectly preserved as though the lakes were in existence only yesterday. Radio-carbon dates indicate that indeed they were, since their lower levels were at the sills of caves occupied by human beings who lived there approximately 11,000 years ago.

Fig. 13-18 Panamint Mountains and Panamint Valley, eastern California. This valley, like others nearby, held a large lake during the last ice age when precipitation was higher than at present. Lake shorelines can be seen at the base of the mountains and below them are large fan-deltas built out into the lake. The white materials of the playa are salts concentrated as the lake desiccated. At nearby Searles Lake, a large chemical plant is based on mining such salts. (Photograph by Roland von Huene.)

A remarkable set of ice age lakes briefly was a part of the California landscape in the desert east of the Sierra Nevada. Individually these lakes, far smaller than such giants as Lahontan and Bonneville, were part of a whole system of connected lakes and streams. One series extended north from the site of modern Lake Arrowhead in the San Bernardino Mountains out across one of the driest parts of North America, the Mojave Desert, to Death Valley. This now dessicated depression then held a lake perhaps 193 kilometers (120 miles) long and nearly 122 meters (400 feet) deep, to which the name

Lake Manley is given. This honors the memory of Lewis Manley, a mountain man of tremendous strength and resolution, who saved the first party of pioneers to reach Death Valley. In a period of six weeks he hiked all the way to the coast and back again, and then back to the coast, in order to bring supplies and to lead the survivors out of their trap.

To the west of Death Valley a similar set of lakes and streams led from Mono Lake (Fig. 13-17), at the base of the Sierra Nevada, down the length of Owens Valley to the basin of Searles Lake (Fig. 13-18) and on to Death Valley. Searles Lake acted as a gigantic chemical processing plant, concentrating an enormous tonnage of dissolved material which is now being recovered from the dazzlingly white expanse of the saline playa.

The American desert has no monopoly on lakes, past and present, as they are equally characteristic of other arid regions throughout the world. Among well-known examples are Lake Chad in Africa, Lake Eyre in Australia, and Lop Nor, Lake Balkhash, and the Aral Sea in Central Asia. Not only were the Aral and Caspian seas larger in the recent geologic past than they are today, but they were connected with one another as well as with the Black Sea. Many others, such as the Dead Sea, are rimmed by abandoned shorelines that scar the barren slopes of the bordering desert hills much like gigantic flights of steps.

The obvious recency of these expanded lakes, coupled with the fact that their shorelines in a few favored locations, such as the flanks of the Wasatch and the Sierra Nevada, actually cut glacial moraines, leads many geologists to conclude that the last high stand of the lake coincided with the time of ice advance. Thus, these ancient lakes are sometimes called pluvial lakes. The meltwater increase appears also to have been the cause of lake expansion, and as the glaciers of the world receded to their present diminished extent, the level of the desert lakes of North America fell. Many of them vanished almost entirely, leaving a barren expanse of salt or shrunken alkaline ponds as relics of what once was an inland sea.

Selected References

Blackwelder, Eliot, 1954, Geomorphic processes in the desert, California Division of Mines Bulletin No. 170, Chapter V, pp. 19–90.

Hadley, Richard F., 1967, Pediments and pediment-forming processes: short review, Journal of Geological Education, vol. 15, no. 2, pp. 83–89.

Leopold, A. S., and the Editors of Life, 1962, The desert, Time Inc., New York.

Sharp, Robert P., 1966, Kelso dunes, Mojave Desert, California, Geological Society of America Bulletin, vol. 77, no. 10, pp. 1045–73.

Fig. 14-1 Metamorphosed sedimentary rocks at Convict Lake, California. Small contortions and dislocations of the bedding surfaces along faults are visible at the contacts in the center of the picture. (Photograph by John Haddaway.)

14

Metamorphic Rocks

A geologist studying igneous and sedimentary rocks has an advantage over one investigating the metamorphic rocks, in that some of the two other kinds form on the earth's surface in environments where they can be observed.

A difficulty with understanding the origin of the metamorphic rocks is that no one has ever seen a metamorphic rock being formed, and for this reason much of our thinking about them is pure conjecture. This is not to say, however, that it is as fanciful as the speculations of science fiction; there are strongly limiting physical and chemical boundaries within which any theory of metamorphism must operate.

What, then, are some of the distinctive characteristics of these difficult rocks that serve to make them a group apart? In the first place they are derived rocks, as the name (taken from the Greek, meaning to change in form) indicates. In a more familiar usage, the same basic word is employed when we speak of the metamorphosis of a caterpillar into a butterfly.

Secondly, they are crystalline rocks, in roughly the same sense that the nonglassy igneous rocks are, but metamorphic rocks have not crystallized from a molten phase, such as lava. Most of them seem to have changed over from whatever they were originally to their new crystalline condition without becoming fluid, although many have been deformed plastically during recrystallization. Thus, they show many of the patterns, such as contorted parallel bands resembling the layers in marble cake, that are associated with flowage but not necessarily with liquidity.

Since recrystallization is the dominant element in the formation of the metamorphic rocks, accompanied in some cases by readjustment or rearrangement of many of the new minerals so that they occupy less volume, many hold the opinion that the process of metamorphism takes place within the crust—often at considerable depth—and that heat, pressure, and chemical activity operating through long periods of time are essential elements in the formation of metamorphic rocks, developing new minerals, crystal alignments, and structures that are in equilibrium with their new environment.

Let us illustrate with slate, one of the more familiar metamorphic rocks. It has been used for centuries as a roofing material, and, despite the advent of many synthetic substitutes, it remains the nearly ideal material for blackboards. Two properties are responsible for its desirable attributes: (1) good quality slates are dense and uniformly fine-textured rocks, and (2) they split, or cleave, along nearly perfect plane-parallel, closely spaced surfaces. This property is called *rock cleavage* to distinguish it from the *mineral cleavage* of such things as mica crystals.

The two attributes, mineral cleavage and rock cleavage, are related, however. Rock cleavage develops best when a number of minerals with highly developed cleavage are so lined up that if one crystal fails under stress, the next gives way, then the next, and so on until through a kind of chain reaction the rock splits along a plane resulting from the cumulative effect of myriads of failures along the cleavage surfaces of parallel and aligned crystals.

It is in the near-perfect parallel alignment of mica flakes that slate differs from any of the rocks that we have studied heretofore. In shale the clay flakes and minute sedimentary particles are aligned, it is true, but not with the perfection found in slate.

Is there a significant difference between shale and slate in their chemical composition? The answer is no, with the exception, perhaps, that mica contains somewhat less water of crystallization than do the clay particles of shale. Clearly, then, something has happened to shale to convert it to slate. This has been a change involving a recrystallization of the minerals from clay to mica and a realignment of the individual mica flakes so that their cleavage planes are oriented parallel to one another instead of at random as in an igneous rock where they crystallized directly from solution.

Another important distinction between a shale and a slate is that the cleavage, which is indeed the hallmark of a slate, does not necessarily coincide with the original stratification. A close look at a slate sometimes will show sandy streaks, or flattened-out fossils, or some relic pattern inherited from the original fabric of the sedimentary rock, all cutting across the cleavage planes in the slate.

Geologists interpret all these lines of evidence to mean that an original rock, such as a fine-grained, laminated shale composed of clay particles, when subjected to high pressure was recrystallized to form a slate. In the process of recrystallization the clay flakes were altered to mica, and the new minerals were rearranged so that they were aligned parallel to one another and at right angles to the principal direction of the pressure which was applied to them.

This example of the conversion of shale to slate by recrystallization under directed pressure and heat within the earth's crust is an illustration of only one type of metamorphism. Unfortunately, the products of metamorphism are even more difficult to classify than the igneous or sedimentary rocks. Reasons include the inaccessibility of the environment where these changes occur, their complex nature, and, most importantly, metamorphism generally involves at least three major elements—heat, pressure, and chemical activity. These may vary enormously; in some circumstances pressure clearly has been ascendant, in others chemical processes appear to have a leading role, yet in others the changes appear to be primarily the result of local heating.

The processes, their consequences, and the resultant types of some representative metamorphic rocks are listed in the table below so that their relationships relative to one another are made more apparent.

Type of Process	Rock
a) Heat dominant	
Contact metamorphism	*hornfels*
b) Chemical-fluids dominant	
Hydrothermal metamorphism	*serpentine*
c) Directed-pressure dominant	
Dynamic metamorphism	*mylonite*
d) Directed pressure and heat	
Regional metamorphism	
foliated	*slate* *schist* *gneiss*
nonfoliated	*marble* *quartzite*
e) Uniform pressure and heat	
Plutonic metamorphism	*migmatite*

Contact Metamorphism

This variety of metamorphism is one in which heat played the leading role with chemical activity and pressure cast in secondary parts, the heat being supplied by an igneous body which has invaded the earth's crust. The name is derived from the fact that this sort of metamorphism is most likely to be found in the shell, or halo, or *aureole* as it is called, in contact with and surrounding an igneous intrusion.

The simplest illustration of what is meant by contact metamorphism commonly will be found in the zone immediately adjacent to a dike or sill. If these hypabyssal igneous rocks have been injected into shale, for example, very commonly there will be a baked or hardened band a few centimeters or a few meters wide immediately next to the now-solidified igneous rock. The clay minerals in the original shale have been changed in much the same way that clay is fired in a kiln to make bricks or pottery.

Larger intrusions, such as batholiths and stocks, make their influence felt over a wider range. The wall rock next to the intrusion may be converted into a dense, hard, nonlayered rock called *hornfels.* Normally, these metamorphic rocks are not layered or banded, although they may be cut by such great numbers of closely spaced fractures that they separate readily into small angular fragments. Their mineral composition is highly variable, because they can be derived from such a host of original rock types—depending in part upon what kinds of rock were brought in contact with an intrusive body. Very often their texture is so fine-grained that new minerals formed through recrystallization are too small to be recognized by the eye but have to be identified under the microscope.

Other changes induced by heating a rock may be brought about by reactions between minerals already present. An example would be the conversion of dolomite into an olivine-bearing marble:

$$2CaMg(CO_3)_2 + SiO_2 = 2CaCO_3 + Mg_2SiO_4 + 2CO_2$$

dolomite + silica = calcite + olivine + carbon dioxide

This means that a dolomitic rock with some quartz sand included in it when heated will have carbon dioxide liberated and driven off as a gas, leaving behind a rock composed of calcite and olivine crystals.

Hydrothermal Metamorphism

Not only is an immense quantity of heat liberated around an igneous intrusion, but great amounts of high-temperature gas and fluids also are freed. Very often these volatile elements of a magma travel for long distances through the enclosing host rocks. These fluids and gases are chemically potent and react readily with many of the minerals they encounter. This means that new material may be introduced into a rock as part of the processes of metamorphism, rather than simply a chemical rearrangement and recrystallization of the minerals already there. Or, conversely, material may be subtracted.

An example is the change of olivine to serpentine. Olivine is an unstable mineral chemically, and in rocks where it is abundant, as in dunite, the rock is altered readily from gabbro to serpentine when it is attacked by chemically active hot waters coming from an igneous intrusion. The fact that this type of metamorphism is one chiefly involving the addition of water is brought out by comparing an average formula of serpentine, $Mg_3Si_2O_5(OH)_4$, with olivine, $(Mg,Fe)_2SiO_4$, for here the principal change has been the loss of iron and the addition of the hydroxyl (OH) ion from water.

Since the hydrothermal metamorphism of dunite and related rocks which are high in iron and magnesium and low in silica is achieved largely through the addition of water, this means serpentine usually takes up a larger volume than the parent rock from which it is derived. As a result of this swelling, serpentine commonly is brecciated, and may be cut by a multitude of cracks and fractures. Sometimes these fractures will be filled in later with white veins of dolomite, and these make a bold and striking contrast to the prevailingly dark green of the serpentine. This unusual green rock, cut by white veins which in turn cross-cut one another in a most complex pattern, bears the picturesque name of *verde antique* and in a bygone age was greatly favored for the walls of florist shops, funeral parlors, and the lobbies of small-town hotels.

Serpentine itself is more likely to be a somber, dark green, or even black or red rock that will take a high polish. For this reason it is favored for bank lobbies, store fronts, and the foyers of buildings. It can be seen in the United States perhaps most strikingly in the imposing columns in the rotunda of the National Gallery of Art in Washington, D.C., as well as in the United Nations building in New York.

Dynamic Metamorphism

In localized parts of the earth, crustal stresses may have built up to the point where rocks are sheared and crushed along large fracture planes which are known as faults. We shall see some of the effects of the operation of these forces when we discuss faulting. Here we are concerned primarily with the small-scale changes brought about by the crushing and granulation of rocks at moderate depth under the application of severe local stress, but without the high temperatures that are characteristic of deeper burial.

The principal change in rocks subjected to the lateral stress developed along the slippage planes of large faults is the shearing, rolling out, and grinding up of mineral and rock fragments to produce ultimately metamorphic rocks with markedly parallel, lens-shaped, and banded patterns. This kind of metamorphism does not cause chemical changes to any pronounced degree—it is primarily responsible for a mechanical rearrangement, principally a realignment or even crushing of the more susceptible minerals.

Not all the minerals in rocks are equally resistant to such duress; they are much like people in this regard—some are stubborn and unyielding, others are pliant and readily molded. In a rock reshaped by dynamic or stress metamorphism, susceptible, flake-like minerals such as mica may be drawn out in parallel streaky bands or layers, while the more obdurate ones, such as feldspar, stand out as rounded or even lenticular, eye-like clots—in fact, they are called by the German word for eyes, *augen*.

A representative example of dynamic metamorphism is the rock *mylonite*. The name comes from the Greek world for mill, and these are rocks that in a figurative sense were caught in the geologic mill and ground until they were reduced to powder. The minerals in the original rock are crushed into minute fragments and may be completely pulverized. These are not loosely consolidated or friable rocks, however, but are completely recrystallized and may be as hard and durable as flint. They actually may be converted into glass-like, streaky material resembling obsidian.

A microscope usually is needed to determine the character of a mylonite, for it will appear as a hard dense rock with few, if any, visible minerals. It is only under magnification that they can be seen to consist of angular, minutely brecciated mineral fragments which are recrystallized to form a metamorphic rock.

Regional Metamorphism

This somewhat ambiguous term is used for metamorphic rocks that characteristically are exposed on the earth's surface over broad areas, sometimes many thousands of square miles. Such wide expanses of crystalline rock—both igneous and metamorphic—were given the name of "shields" many years ago, and they were regarded as the foundation of the continents. Two well-known examples are the Canadian shield, the broad expanse of igneous and metamorphic rocks marginal to Hudson Bay and extending southward into Minnesota and Wisconsin and eastward across Labrador, and the Fenno-Scandian Peninsula, which includes most of Finland, Sweden, and Norway. In both of these shields, as well as in similar areas elsewhere, metamorphic rocks of great variety crop out, to the virtual exclusion of most other types. This is interpreted to mean that these rocks have been laid bare through erosion on a subcontinental scale and we know of them only through the stripping away of their original cover. Such rocks, with so wide a distribution, must have a more general cause than the heat generated by a single igneous intrusion, or the reactions produced by chemically activated water, or the grinding effect of movement along a fault plane.

Because of its widespread occurrence, this type of metamorphism is sometimes given the name of regional, but a more meaningful term is *dynamothermal*, which emphasizes that recrystallization is brought about by heat and directed pressure working in unison.

The outline near the beginning of this chapter lists two major categories of rocks in this class: nonfoliated and foliated. Foliated rocks—the word comes from the Latin, *foliatus*, meaning leaved—are those having well-developed rock cleavage in some varieties, and characterized by the parallel orientation of their tabular minerals and varying degrees of banding, or color layering.

FOLIATED ROCKS   *Slate*, from the old French word, *esclate*, or slat, demonstrates the nature of dynamo-thermal metamorphism. As already mentioned, slates are usually derived from finely laminated rocks such as shale, when the original clay particles were recrystallized to mica flakes under directed pressure and heat. Slates can be made from other fine-textured rocks, as volcanic tuffs, and very often these produce the more colorful varieties, such as red, green, and dark brown slates.

Schist, from the Greek word, *schistos,* cleft, or *schizein,* to split (we see the same word in schizophrenia, meaning a split personality, or in schism, meaning a division of opinion within a group), is probably the most widely occurring of metamorphic rocks. Slates grade into schists with increasing grain size. All schists include tabular, flaky, or even fibrous minerals in their composition, and the extent to which these are developed in parallel orientation determines to a considerable degree whether or not schistosity, the characteristically wavy or undulatory rock cleavage, develops. Many schists split readily into tabular blocks. These are the familiar flagstones (Fig. 14-2) so widely used throughout Europe in courtyards and for castle walls, and in this country for fireplaces, patios, and barbecues. Some schist makes a good building stone because of this tabular habit; its edges are easily trimmed and it can then be used in much the same way as bricks.

Schists can be made from a wide variety of rocks by recrystallization under directed pressure and moderately high temperature. In the main, schists are rocks whose original grains and minerals were small, many of which on recrystallization were converted to plate-like minerals such as mica. Among the many rocks that are likely candidates for metamorphosis to schist are mudstone, muddy sandstone, basalt and other dark igneous rocks, and clayey limestone (Fig. 14-3).

Gneiss, an old Saxon miners' term for a rock which is rotted or decomposed (pronounced as though it were spelled nice), is a banded rock, usually with layers of light-colored minerals alternating with dark-colored bands (Figs. 14-4, 14-5). This is a coarser-grained rock than the other two foliates described above. In fact, the size of the quartz and feldspar crystals it contains is about the same as in granite—the rock from which very often it is made. Gneiss can be derived from other coarse-grained igneous rocks, too, such as diorite and gabbro. Recrystallization of the rock has rearranged the minerals so that most of the light-colored ones are concentrated in one layer, while the dark ferromagnesian ones are grouped in another; these light- and dark-colored bands alternate rhythmically through the rock. These bands, unlike the cleavage planes of slate, do not make uniformly parallel planes that continue for long distances, but more commonly are strongly distorted. They probably were deformed plastically—that is, although the rock was still in the solid phase it was able to flow, in about the same way that butter or sheet lead can be made to do without becoming liquid at all.

Gneisses do not have the highly developed rock cleavage of slate or schist. In spite of the roughly uniform spacing of their bands, they break in about the same unpredictable fashion as a piece of granite does when struck a hammer blow.

Fig. 14-2 Foliated rocks (schist) used as roofing material in the Alps near Zermatt, Switzerland. The mountain in the background is the famous Matterhorn, which is composed of highly foliated, regionally metamorphosed rocks.

Fig. 14-3 Folded calcareous schist derived from a thinly bedded clayey limestone. (Photograph by William Garnett.)

Fig. 14-4   Gneiss from Julian, California, clearly showing the dark ferromagnesian bands and the light quartz-feldspar bands. (Photograph by William Estavillo.)

Fig. 14-5   Biotite gneiss from Uxbridge, Massachusetts. (Courtesy of Ward's Natural Science Establishment, Inc.)

Fig. 14-6   Gneiss from the Wind River Mountains, Wyoming, intruded by several small dikes of granitic material. (Photograph by A. M. Bassett.)

NONFOLIATED ROCKS   These are rocks that were shaped by the same processes as those responsible for the foliated rocks, but they are not banded nor do they have well-developed rock cleavage. Two leading examples are *marble*, which consists of the mineral calcite whose composition is $CaCO_3$, and *quartzite*, which has quartz as its leading constituent.

Marble is the coarsely crystalline equivalent of limestone, and thus in the majority of cases very likely had an organic origin. In limestone, the organic material often is still visible as shells or structures built by lime-secreting plants, such as algae, or animals, such as coral. In marble, these traces of past life very largely are erased. Marble is a good example of a metamorphic rock which has undergone a physical transformation without necessarily having undergone any drastic change in chemical composition.

Under sufficiently high temperatures and pressures the lime of the original rock was recrystallized—in many cases by accretion—around earlier formed seed crystals of calcite. If this process continues long a coarse-grained rock of nearly equidimensional crystals of calcite results, and these will be visible to us in a rock which has a rather sugary texture.

Pure marble is snow white, and one of the most highly prized varieties from ancient times down to our day comes from the quarries of Carrara on the west coast of Italy. This is a remarkably uniformly textured rock which is ideal for sculpture because of its freedom from impurities, as well as the fact that its hardness is no more than 3 since it consists of the single mineral, calcite. Granite, by contrast, is nearly as hard as steel.

All marble is not pure white and this is immediately clear to anyone who has observed it in banks, building façades, lobbies, public lavatories, and old-fashioned table tops and dressers. In general, the black and gray areas in marble probably are colored by carbonaceous matter, brown and red zones are made by iron oxide, and green reflects the presence of various iron- or magnesium-bearing silicate minerals.

Marble is the result of the metamorphism of limestone, and since many of these limy rocks once contained muddy or sandy layers, the silica present in such beds will be recrystallized, too. A fairly typical reaction under suffi- ciently high temperature is the following:

$$CaCO_3 + SiO_2 = CaSiO_3 + CO_2$$
lime + quartz = wollastonite + carbon dioxide

Wollastonite is a colorless, bladed mineral, commonly arranged in fan- shaped radiating needles that penetrate the host rock, marble. It is repre- sentative of a wide variety of related minerals that develop with increasing clay and sand content in impure limestone to yield the so-called *calc-silicate rocks* when they are metamorphosed. Commonly these rocks have higher specific gravity than marble, and of course become harder with increased silicification.

Quartzite is metamorphosed quartzose sandstone, and in it the pore spaces once separating the individual grains are filled with newly formed quartz. In some rocks, the so-called sedimentary quartzites, this quartz has been added to the already existing sand grains. The ghost-like boundary separating the original quartz grain from the silica added to it may be barely discernible. The silica cement which fills in the pore spaces of the original sandstone may prove to be stronger than the sand grains themselves, and when the rock is struck a hammer blow it breaks through the grains, rather than around them.

Quartzites are nearly always light colored; light pinks or reds are very characteristic. Many are white or light gray, and with increasing amounts of impurities their colors darken until some may be black. Very often quartzites are interbedded with marble, calc-silicate rocks, and other rocks derived from sedimentary sources. Relic sedimentary structures, such as cross-bedding, are sometimes preserved and these are emphasized by slight color differences which superficially may resemble the banding in a gneiss.

Plutonic Metamorphism

This is the type of metamorphism believed to occur deep within the crust under conditions of very high pressure and elevated temperature. The pressure is more likely to be hydrostatic than directed as was the case in the making of the foliated rocks. By hydrostatic pressure we mean the pressure, for example, that bears down on a submarine deep in the sea. This sort of pressure, trans- mitted through liquids, is equal in all directions, not greater in one and less in another as is true of directed pressure. Such a pressure—equal in all directions —is sometimes called the confining pressure, in order to eliminate the idea that it is necessarily restricted to objects immersed in a fluid.

The great pressures in these deeper realms of the crust are responsible for the formation of more compact, or denser, varieties of minerals. Minerals crystallizing in this metamorphic zone are likely to be stubbier or more equidimensional than elongate.

A striking characteristic of plutonic metamorphic rocks is their intimate association with intrusive igneous rocks. Often these two types of rocks will alternate with one another in a single outcrop; for example, there may be a layer of granitic material, then of schist, then of granite, then of schist, and so on.

An extreme example of plutonic metamorphism is the rock *migmatite*, taken from the Greek word *migma*, a mixture, and this, indeed, is what the rock looks like. In part, these rocks have the banded or layered appearance of gneiss, and yet in other parts of the outcrop, the constituent minerals will have the nonoriented, random, scattered pattern so typical of granite.

With the problem of the origin of migmatites we have traveled full circle, and once again are confronted with the vexing question of the origin of granite. In a migmatite in which igneous material interfingers with metamorphic in such an intimate fashion, how was the granite formed? Was the granite injected as a magmatic fluid into a pre-existing rock, such as gneiss; or was the gneiss transformed into granite by a replacement process? Strong support for the latter theory is given when the shadowy parallel alignment of minerals, say in a bordering schist, can be traced across an igneous contact and out into the granitic body itself. This is especially likely to be the case where the igneous body is a concordant one and the contact conforms to structural patterns within the body of metamorphic rocks.

Contrariwise, if a granitic body has sharp, clear contacts cutting abruptly across the grain of the enclosing wall rocks, and if in addition the granite holds unmodified angular inclusions or *xenoliths* (a Greek word meaning stranger stones) within it, then very likely the granite had a truly magmatic origin and shouldered its way into the metamorphic shell encasing it.

To summarize, then, there are at least three very plausible ways in which granite may originate (there may be others, too, but these three are the most likely): (1) it may be truly magmatic and have invaded the upper levels of the earth's crust as an igneous intrusion; (2) it may result from the melting or fusion of pre-existing rocks through the application of enough heat to melt them; and (3) it may be the result of recrystallization of some other wholly different kind of rock by the movement of ionic solutions through it and its conversion into granite by the essentially metamorphic process of *granitization*. To this metamorphic process the Greek-derived name of *metasomatism*, or change in body, is given, and in this application the word means the transformation of one rock into another with a different chemical composition.

Thus, rocks may progress through an evolutionary cycle of their own. They may be created in Vulcan's forge, speaking metaphorically, by crystallizing from a molten solution, magma. Exposed at the earth's surface through long stretches of geologic time, their constituent minerals when weathered out are redeposited as sedimentary rocks. Should these later be buried deeply enough, they may be recrystallized as metamorphic rocks, such as hornfels, slate, schist, marble, or any one of a wide variety of types. Then if circumstances are favorable, they may be remelted, or they may be recrystallized through granitization to be reborn again, indistinguishable in every respect from their original igneous incarnation.

Selected References

Barth, T. F. W., 1962, Theoretical petrology, John Wiley, New York.

Jackson, Kern C., 1970, Textbook of lithology, McGraw-Hill Book Co., New York.

Williams, H., Turner, F. J., and Gilbert, C. M., 1954, Petrography, W. H. Freeman and Co., San Francisco.

Fig. 15-1   Deformed sedimentary strata. Murdafil, Iran. (Photograph by Serofilms and Aero Pictorial, Ltd., London.)

15

Structural Geology

We saw in the preceding chapter how some rocks are drastically altered by heat and direct pressure. The rocks in themselves are one piece of evidence that tremendous forces are at work in the earth. Regionally metamorphosed rocks indicate also that these forces act over large areas of the crust. Additional evidence of their strength and extent is supplied by the attitude of sedimentary beds which were horizontal at the time of deposition. Many rocks are not at all horizontal now, but have been tilted, even to the point of standing vertically (Fig. 15-2), and some have been actually turned upside down. In addition, rocks that were deposited in oceans and seas now are many feet above sea level. Mount Everest, for example, is composed of marine sedimentary rocks. Finally, of course, there are faults, or fractures in the earth's crust, along which movement has occurred.

The results of these forces are the concern of the structural geologist. He wants to know first: what is this structure I'm looking at? Then: how did it form?

Describing Structural Features

The structural geologist in the field is faced with several problems. While there are some small features that he can hold in his hand, major geologic structures involve large areas of the crust, and certainly cannot be viewed in their entirety from one vantage point. Thus, the geologist has a problem arising from sheer size. In addition, much of the structure he wants to study is underground and therefore invisible. A third difficulty stems from the fact that

Fig. 15-2   Steeply tilted strata adjacent to a fault, Dinosaur National Monument, Utah. (Photograph by Philip Hyde.)

Fig. 15-3   Block diagrams illustrating strike and dip of beds.

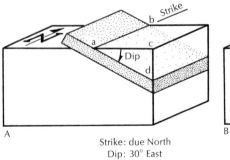

A

Strike: due North
Dip: 30° East

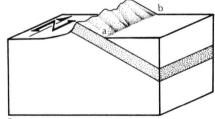

B

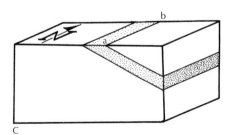

C

Fig. 15-4   Strike and dip of tilted beds. (Photograph by William Estavillo.)

part of the structure has probably been removed by erosion, and is invisible on that account.

By making a map which shows where the different rocks occur on the surface and what their attitudes are, he has the basis to reconstruct the structure in his mind and express it in the form of cross sections. Only when this has been done can he proceed to interpret the conditions which may have interacted to produce this form.

Geologists use two measurements, strike and dip, to describe how sedimentary beds and other planar structures are arranged. These terms are best explained by means of block diagrams.

The *strike* of a bed is the compass direction of any line made by the intersection of the inclined bed with an imaginary horizontal plane (Fig. 15-3a). In the photograph (Fig. 15-4), the surface of the water is such a horizontal plane, and the intersection of the tilted bed with the water is the strike of that bed, line a-b. The strike remains the same even if the ridge crest is irregularly eroded (Fig. 15-3b) or even reduced to a flat surface (Fig. 15-3c). The strike is expressed as the number of degrees the horizontal strike-line deviates east or west from due north, i.e., forty degrees west of north = N 40° W.

The *dip* is the angle that a bed is inclined below an imaginary horizontal plane (Fig. 15-3a). It is shown on the photograph also (Fig. 15-4). As you can see, the dip is measured at right angles to the strike; this is most easily determined by rolling a marble down the inclined surface of the bed. If the dip is 30° E, 30° is the amount of inclination, while E is the direction of the inclination.

Fig. 15-5   Geologist measuring strike. The strike here is N 10° W. (Photograph by J. R. Stacy, U.S. Geological Survey.)

Folds

Rocks within the earth's crust can react to the forces of deformation in several ways. When the deformation is only temporary and the rock returns to its original shape and size, it is called *elastic deformation*. This occurs when an earthquake wave ripples outward from its point of origin, without breaking rocks or distorting them permanently. Since the rocks are only temporarily changed, the geologist never sees the results of this type of deformation, except as it damages man-made structures.

Under other conditions, however, rocks may be deformed into permanent folds, and this is called *plastic deformation*. Then, in layered rocks especially, the geologist has something to study. Because the beds were originally horizontal, he can, through careful examination and classification of the folds, determine what kinds of forces have acted on them.

Fig. 15-6   Close-up of geologist measuring dip. The dip here is 55°. (Photograph by J. R. Stacy, U.S. Geological Survey.)

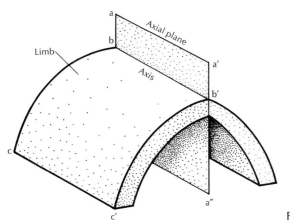

Fig. 15-7   Parts of a fold.

STRUCTURAL GEOLOGY

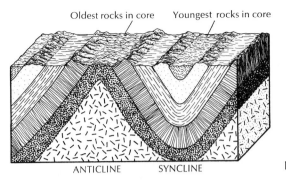

Oldest rocks in core   Youngest rocks in core

ANTICLINE   SYNCLINE

Fig. 15-8   Anticline and syncline.

Fig. 15-9   Anticline on the banks of the Potomac River, Washington County, Maryland. (Photograph by I. C. Russell, U.S. Geological Survey.)

Fig. 15-10   Anticlinal and synclinal folds of the "Grande Chartreuse," north of Grenoble, French Alps. (Photograph by Swissair-Photo, AG Zurich.)

Folds may vary in size from a fraction of an inch to several miles wide, but, whatever their dimensions, they have the following parts (Fig. 15-7). The *axial plane* divides the fold in half as symmetrically as possible. In Figure 15-7 a-a'-a" is the axial plane. The *axis* is a line formed where the axial plane cuts any layer (Fig. 15-7). The segments of the fold on each side of the axial plane are the *limbs*.

There are several ways of naming and classifying folds. One way is best seen in cross section (Fig. 15-8). An *anticline* (Fig. 15-9) generally has its limbs dipping away from each other (anti = opposite), and the oldest beds after erosion are nearest the central core. A syncline generally has its limbs dipping toward each other (syn = together), and the youngest beds after erosion are nearest the central core.

Then a fold can be modified by looking at and describing its axial plane and its limbs (Fig. 15-11). These terms—symmetrical, overturned, etc.—can be applied to both anticlines and synclines. A second glance at the recumbent fold will show how difficult it can be to tell an anticline from a syncline unless the age of the beds is known.

So far we have discussed folds almost as if they existed only in two dimen-

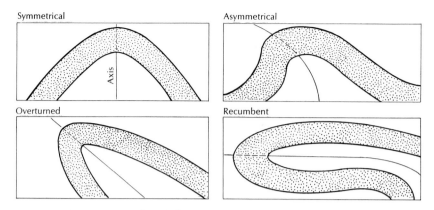

Fig. 15-11   Some types of anticlines.

sions, but, of course, this is not true. One of the major skills of a good geologist is the ability to visualize geologic structures in three dimensions. Here is a favorite puzzle to test your skill in this respect. Imagine a cube of cheese three inches on a side which has been covered with aluminum foil. It is now cut into one-inch cubes. Using your 3-D mental visualization, determine how many little cubes have three foil-covered surfaces; how many have two foil-covered surfaces? one? none?

Folds, of course, do have three dimensions, and in order to describe them in this aspect, we use block diagrams (Fig. 15-12). The axis, a-a', is a line, and so we may speak of its *bearing*, or compass direction. The axis in this case bears about due north. In addition, the axis need not be, and often is not a horizontal line, but can be a line which plunges downward (Fig. 15-12). The *plunge* of the axis is measured in degrees below the horizontal. Axis b-b' bears due north and has a plunge of about 20°.

Many folds plunge in two directions and are known as *doubly-plunging* (Fig. 15-13). A good way to visualize a group of doubly-plunging folds is to imagine a tablecloth which has been mussed up into gentle wrinkles. The central Appalachian Mountains are good examples of doubly-plunging anticlines and synclines (Fig. 15-14). This diagram shows another interesting point. Erosion has been so extensive that anticlines do not form ridges, and synclines do not form valleys. Instead, the relief of the area is determined by the resistance

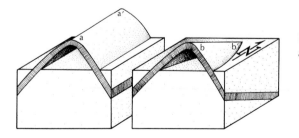

Fig. 15-12   Plunging and non-plunging anticlines.

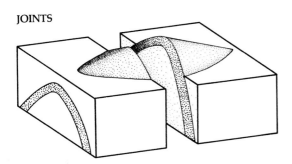

Fig. 15-13 Doubly plunging anticline.

of some beds to erosion, and the susceptibility of others. Only by mapping dips and strikes and by knowing the age of the beds can the geologist tell which are indeed anticlines and which are synclines.

Joints

Nearly all the rocks visible on the earth's surface are cut by cracks or fractures. They are so commonplace that few people grant them more than the most casual notice. If noticed at all, they probably are regarded as something that has always been and that requires no explanation.

Fig. 15-14 Block diagram of Appalachian-type fold mountain system. (Modified from Nevin M. Fenneman, *Physiography of the Eastern United States,* 1938.)

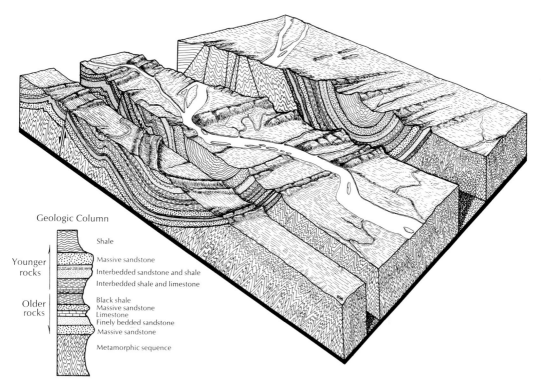

Geologic Column

Younger rocks

Older rocks

Shale

Massive sandstone
Interbedded sandstone and shale
Interbedded shale and limestone

Black shale
Massive sandstone
Limestone
Finely bedded sandstone
Massive sandstone

Metamorphic sequence

Fig. 15-15 Widely spaced joints mark out the chimney-like pinnacles of thick-bedded sandstone in Monument Valley, Arizona. (Photograph by William Aplin.)

Actually, the nature of such cracks, or *joints*, does require an explanation. The spacing between individual cracks may be less than an inch, or it may be scores of feet. Large intervals between cracks are rare, however. Joints very often are spaced rather uniformly, and the spacing interval may be quite constant for a given rock type. In general, the interval is less in fine-grained rocks, more in coarse-grained ones. Hornfels, shale, and mudstone, as well as igneous rocks such as basalt and obsidian, are likely to be closely jointed. Granite, gneiss, and thick-bedded sandstone, such as the sandstone that makes the sheer cliffs in Monument Valley, Arizona, may be widely jointed (Fig. 15-15). In some cases joints control the landscape. This is especially true in Arches National Monument in Utah, and the regularly spaced joints intersecting at nearly right angles look from the air like the ruin of an ancient city. This is a striking feature on the air route from Los Angeles to Denver.

Joints typically occur in *sets*, and in many cases these intersect each other

Fig. 15-16   Sets of joints split the granitic rocks on the east face of Mt. McAdie, Sierra Nevada, California. (Photograph by Tom Ross.)

Fig. 15-17 Rectangular joints in the Moenkopi Formation. Capitol Reef National Monument, Utah. (Photograph by J. R. Stacy, U.S. Geological Survey.)

Fig. 15-18   Jointing in the Philmont Scout Ranch area, Colfax County, New Mexico. (Photograph by J. R. Stacy, U.S. Geological Survey.)

Fig. 15-19 A fault scarp in alluvium, produced during the Hebgen Lake earthquake, Montana. (Courtesy of Montana Highway Dept.)

in three directions approximately at right angles to one another (Figs. 15-16, 15-17). Should these fractures be uniformly spaced, a rock outcrop may resemble masonry. In fact, the word joint is said to have been used long ago by British coal miners because such rhythmically fractured rocks resembled the way in which bricks are joined in a wall.

Some joints are associated with folding and other types of deformation of the earth's crust. Other joints, and these are among the more perfectly formed, result from contraction on drying such as mudcracks, or cooling on crystallization such as lava. The columns that are typical of basalt flows, and sills such as the Devil's Postpile (Fig. 4-27) and the Giant's Causeway, are outstanding examples of this *columnar jointing*.

Sheeting is a kind of jointing often seen in quarries. The joints are parallel to the topographic surface and resemble stratification. They are fairly close together near the surface, but become farther and farther apart with depth until they disappear. A widely held theory is that these slab-layered joints are the result of the release of pressure. The rocks crystallized at great depth within the earth under very high confining pressure. Now that they are exposed at the surface, through uplift and the stripping away by erosion of the covering rocks, these fractures have developed as the result of rebound. It is quite logical to expect that the rocks would spring outward since the atmospheric pressure is so much easier to overcome than the confining effects of the adjacent rocks.

Faults

Faults, like joints, are breaks in the earth's crust, but along faults movement has taken place. Sometimes faults, or evidences of them, show on the earth's surface. A cliff, called a *scarp* (Figs. 15-19, 15-20), may in rare instances actually be a fault, but it is almost always only the surface expression of a fault at depth. Such fault-line scarps often cut across other features of the landscape in a straight line and in many instances show up beautifully on aerial photographs as a line which does not fit with the rest of the surface. Sometimes

Fig. 15-20   The eastern fault-line scarp of the Sierra Nevada, California, showing the accordant summit of the range and the steep eastern face. Mt. Whitney is in the left background. (Photograph by Roland von Huene.)

Fig. 15-21   Streams have been offset right laterally where they cross the trace of the San Andreas fault. Carrizo Plain, California. (Photograph by A. M. Bassett.)

streams are offset along a fault—they will change direction and run along a fault trace a short distance before they resume their former direction (Fig. 15-21). On a smaller scale the actual fault surface can sometimes be found. It is often slippery and polished-looking, an effect of frictional sliding called *slickensides*, and may have grooves indicating the direction of its most recent movement (Fig. 15-22).

In some areas the region of the fault is is a jumbled mass of broken rock blocks called a *fault breccia*. Usually the rock fragments in a breccia range from a few centimeters to several meters, but in Death Valley, California, the breccia fragments may be as much as half a kilometer long. This particular formation bears the official name, Chaos, which gives some idea of its appearance. The brecciated zone may be tens of meters thick. On a much smaller scale the fault may be marked by a zone of mylonite, or by a zone of finely ground rock resembling clay, called *gouge* (Fig. 15-24). Sometimes the only evidence for a fault is a break in the continuity of rock units, so that they end abruptly against quite different rocks.

Faults are generally classified according to the type of movement that has occurred—up, down, or sideways. Often the movement is *oblique*, or a combination of any of the others, in which case it is classified according to the direction of principal movement. The classification is summarized in Figure 15-23.

Unfortunately, in the field one cannot always determine the actual movement that has taken place. If the geologist can see the fault only on the surface of the ground, he sees only the horizontal part of the movement. If he can see the fault only on a mine wall, for example, he sees only the vertical component of the movement. Even slickensides may be misleading, for they show only the direction of the most recent movement. Where only the *apparent* movement can be determined (which is most of the time), the word *slip* is omitted from the fault name.

Dip-slip Faults

Normal Faults may be part of very large structures or they may be a smaller local system (Fig. 15-23). Sometimes they are associated with other structural features such as folds (Fig. 15-26), and can occur any place, in fact, where the surface is being locally stretched. A series of normal faults breaks a region into blocks of which the lower are called *graben*, and the higher *horsts* (Fig. 15-

Fig. 15-22 Slickensides; grooves indicate vertical movement. Mojave Desert, California. (Photograph by A. M. Bassett.)

Before Faulting   After Faulting

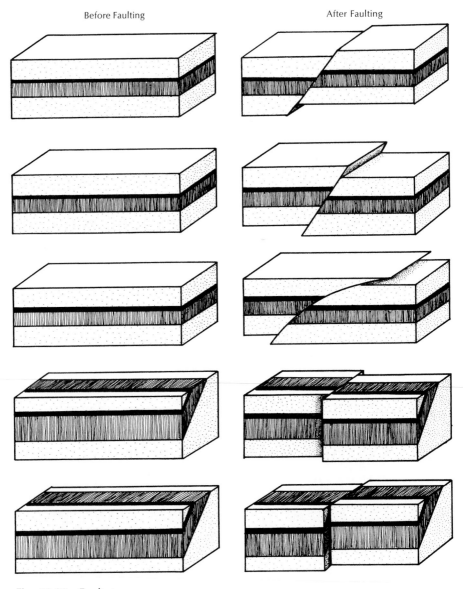

Fig. 15-23   Fault types.

27). The blocks often are tilted so that one edge is higher than the other. Such structure is responsible for the Basin and Range area of the western United States, where the short, rugged mountain ranges are tilted fault blocks (Figs. 15-28, 15-29).

Great zones of normal faulting are found in several places in the world, where repeated faulting has left a jumble of tilted blocks. One such area is in East Africa, where the rift zone is 6000 kilometers (3700 miles) long, has two branches, and appears to connect with oceanic ridges in the Red Sea and the Gulf of Aden.

Reverse Faults (Fig. 15-23) are often, but not always, found in the hard,

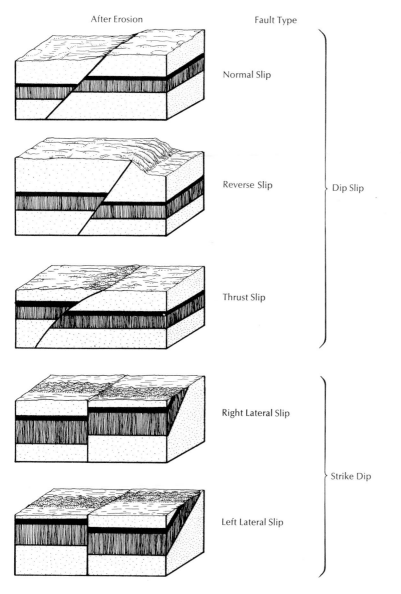

After Erosion

Fault Type

Normal Slip

Reverse Slip

Dip Slip

Thrust Slip

Right Lateral Slip

Strike Dip

Left Lateral Slip

highly metamorphosed, crystalline rocks that compose the central portions of all the continents, and are often associated with regions which show extensive folding. Sometimes an uplifted block of the crust appears to be bounded on all sides by reverse faults. Whether these faults retain their reverse character at depth, or whether they change to vertical faults, can be debated (Fig. 15-30). Such a raised block comprises the bulk of the Beartooth Mountains in Montana and Wyoming. This block of Precambrian rocks is about 64 by 128 kilometers (40 by 80 miles) in size, and in places has been raised as much as 1000 meters (3280 feet).

Thrust Faults (Fig. 15-23) are a type of reverse fault in which the angle of

Fig. 15-24   Fault, with gouge zone producing gully and talus cone in center. Beds on right dip steeply right, parallel to the fault. Beds on the left dip left, arching up at fault; this drag shows that the rocks on the right moved relatively upward. Johnson County, Tennessee. (Photograph by W. B. Hamilton, U.S. Geological Survey.)

dip of the fault plane is less than 45°. They occasionally appear as a rupture in overturned fold (Fig. 15-31), but they may take other forms also. Some merely follow a bedding plane, while others appear not to have been associated with folding at all, as the fault cuts directly across horizontal beds or previous structures. In many cases older beds come to rest upon younger ones.

When thrusting takes place on a vast scale, it can be spectacular indeed, for the dip of the fault is usually less than 10° and the amount of movement is measured in kilometers. This is called an *overthrust*. The image of a great sheet of rock moving almost horizontally for kilometer after kilometer is truly intriguing. Where erosion has worn away a small part of the overthrust sheet so that we can see the younger rocks below, we have what is known as a *fenster*, or *window* (Fig. 15-32). When erosion has removed most of an overthrust sheet, a remnant of it is called a *klippe*.

Although much of the spectacular scenery of Glacier National Park is the result of glacial erosion, the rocks comprising the park were moved into their present position, before the glaciation, along the great Lewis Overthrust. Precambrian sedimentary rocks were uplifted and then thrust over shales and sandstones of Mesozoic age. This huge sheet moved at least 19 kilometers (12 miles) and may have moved as much as 48. Chief Mountain is a klippe left from the erosion of the forward part of the overthrust sheet.

Strike-slip Faults

If you place two bricks side by side on a table and move them backward and forward, you will have reconstructed the movement of a typical strike-slip fault. Whereas normal, reverse, and thrust faults have most of their movement up or down the dip surface, strike-slip faults move parallel to the strike of the fault plane (Fig. 15-23). Like the fault plane between the two bricks mentioned above, the fault planes of strike-slip faults are usually vertical or nearly so.

Fig. 15-25 Small faults displace the rock layers near Zuma Beach, California. (Photograph by William Aplin.)

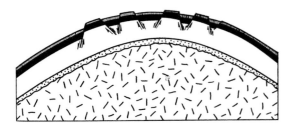

Fig. 15-26   Normal faulting on the crest of an anticline.

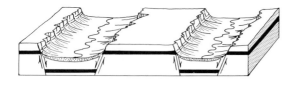

Fig. 15-27   The block diagram illustrates the structural and physiographic relations of a horst (center) and adjacent graben (valleys).

These faults are further classified as right-lateral or left-lateral, depending on the apparent relative motion of the blocks. An easy way to determine this is to stand facing across the fault trace; if the further block apparently moved to the left, it is a left-lateral fault; if the further block apparently moved to the right, it is a right-lateral fault. Use your imagination to confirm the fact that it does matter which side of the fault trace you stand on.

Strike-slip faults come in all sizes, from hand-specimen size to fault zones that extend for many kilometers. They are not limited to crystalline basement rocks, but can involve all types, and they are commonly not associated with major folding.

Examples of strike-slip faults are world-wide, the Great Glen fault in Scotland (Figs. 15-35, 15-36), the Dead Sea rift, and the Alpine fault in New Zealand, to mention a few. Probably the most intensively studied is the San Andreas fault in California. The San Andreas extends for nearly 966 kilometers

Fig. 15-28   The structural and physiographic relations of the Sierra Nevada and adjacent regions, shown diagrammatically.

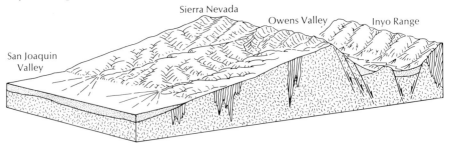

Fig. 15-29 Fault block mountains. Lake County, Oregon. (Photograph by W. C. Mendenhall, U.S. Geological Survey.)

(600 miles) in a generally northwest direction from the Salton Sea near the Mexican border to Point Arena on the coast of northern California where it appears to enter the sea (Fig. 15-37). As the map shows, it is a remarkably persistent, narrow fault zone, rather than a single fault, and it changes compass directions several times.

In spite of the intensive study the San Andreas has received, geologists are still in disagreement about it in many ways. The late Professor N. L. Taliaferro of the University of California believed that most of the movement on the fault was very ancient and vertical in sense, and that the horizontal movements in late Pliocene and Quaternary time are very small. In contrast are the conclusions of Mason Hill and T. W. Dibblee (1953) who feel that movement on the fault has been primarily horizontal, and that the east block has moved

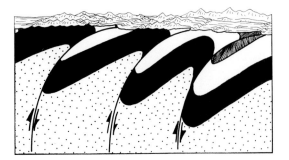

Fig. 15-30 Reverse faults changing to vertical faults at depth.

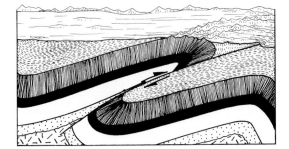

Fig. 15-31 Overturned fold rupturing to form a thrust fault.

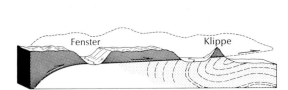

Fig. 15-32 A cross-sectional diagram showing a low-angle thrust fault, above which old rocks (heavy dots) have moved from left to right. An outlier (klippe) and an inlier (fenster) have been produced by erosion of the thrust sheet. This diagram illustrates conditions prevailing along the Lewis overthrust, Glacier National Park, Montana.

south (right-lateral fault) 16 kilometers (10 miles) since the Pleistocene, 105 kilometers (65 miles) since the Miocene, 362 kilometers (225 miles) since the Eocene, and 536 kilometers (350 miles) since the Jurassic. They came to these conclusions by matching rock units of similar type and age across the fault and measuring the distance they had been displaced from each other. The main trouble is that the fault has been such a prominent feature for so long that it may control the type of deposition in its vicinity; these deposits may originally not have matched across the fault.

Fig. 15-33 The Keystone thrust; dark Cambrian limestone has been thrust over light-colored Triassic sandstone. Spring Mountains, Nevada. (Photograph by A. M. Bassett.)

The Meaning of Structures

After a geologist determines what it is he is looking at, he then comes to the second big question: how did it form? There are several ways of going about answering this. He may start by examining under the microscope the fabric, or pattern, within the rocks involved. The physical condition and the distinctive arrangements of the components of a rock give clues as to the forces which have acted upon it. Here laboratory techniques can be revealing. By subjecting a cylinder of rock to artificially created stresses we can find out what conditions produce what effects.

Fig. 15-34   The Playground thrust; Paleozoic limestone has been thrust over varied Mesozoic rocks. Old Dad Mountain, San Bernardino County, California. (Photograph by Gary W. Vogt.)

Laboratory Experiments

Sometimes single crystals are used in these experiments, and the ways in which these undergo deformation can be discovered. Calcite, quartz, and biotite are among the minerals tested most extensively. Rocks, however, are made up of many crystals, usually of several different kinds, so that in addition to changes within single crystals, there are important reactions at the interfaces of crystals. Several factors determine the way a rock specimen will react. The ones usually considered in testing are (1) rock type, (2) temperature, (3) confining pressure, (4) directed stress, (5) the presence of fluids, and (6) time. It is customary to keep all but one of these variables constant while varying that one in a systematic manner.

It is obvious from looking at structural features in the field that rocks do not behave at depth as they do at the surface. The "silly putty" effect is especially noticeable. Silly putty will bounce off the floor, or it will smash into bits

Fig. 15-35 Loch Ness, in the Northwest Highlands of Scotland, is a lake in the basin along the trace of the Great Glen fault. (Photograph by Aerofilms and Aero Pictorial, Ltd., London.)

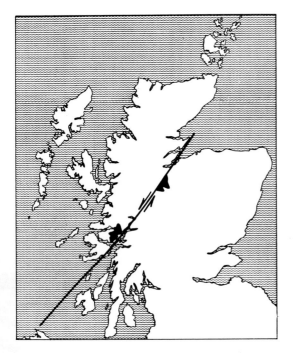

Fig. 15-36 Left-lateral offset on the Great Glen fault, Scotland, as determined by the separation of two granite bodies (black) interpreted to have been once contiguous. (After W. Q. Kennedy, "The Great Glen Fault," *Quarterly Journal of the Geological Society of London,* 1946. By permission.)

if hit with a hammer; yet, given a few hours, it will behave like an extremely viscous liquid and form a puddle or assume the form of its container. Rocks at depth seem to act in much the same way. One of the important results of laboratory studies is the determination of when a given rock will cease to act like a fluid and break. The conditions producing flow and rupture have been reproduced for many rock types, and by far the most important condition, as with silly putty, is time. The geologist naturally wants to see the results of his experiments within a few days, while actually rocks are subjected to the forces of deformation for millions of years. This difference in time naturally has a great effect on the importance of some of the conditions, for some are time-dependent and others are not. So that, although these laboratory experiments are beginning to be helpful in interpreting field observations, the great number of variables involved, including the time factor, makes their application of limited usefulness.

Some of the significant conclusions, however, can be stated. (1) Most rocks are brittle and fracture easily when the confining pressure and the temperature are low and the rate at which the stress is applied is high. An analogy would be the behavior of silly putty when struck with a hammer. (2) Most rocks flow plastically when the confining pressures and temperatures are high, even though they are not actually liquid. Silly putty, too, will flow when it is warmed. (3) Even when confining pressures are low, rocks will exhibit some deformation if the stress is maintained over a long period of time, just as silly putty will form a puddle if left alone for a while.

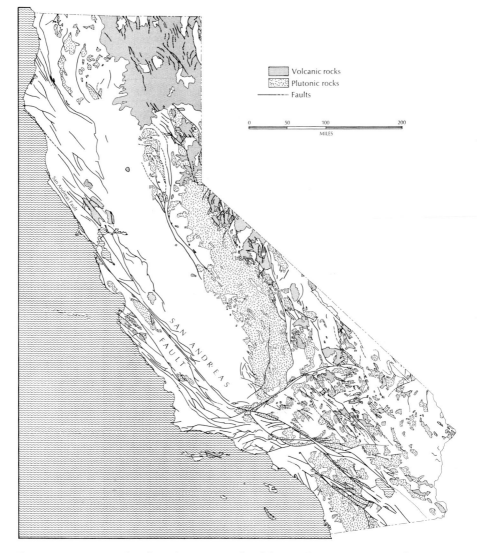

Fig. 15-37 A generalized geologic map of California shows the extent of the San Andreas and other faults and the distribution of plutonic and volcanic igneous rocks. (After "Tectonic Map of the United States," U.S. Geological Survey and American Association of Petroleum Geologists, 1961. By permission.)

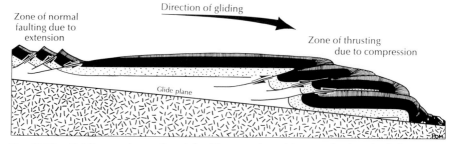

Fig. 15-38   Folding and overthrust faulting produced by gravitational gliding.

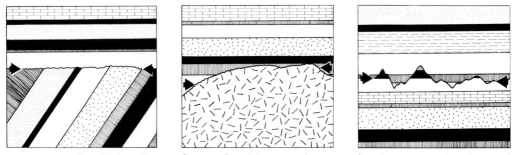

Fig. 15-39   Types of unconformities: angular unconformity, nonconformity, and disconformity.

Mechanisms

By observing larger structures it may be possible to determine the movement that occurred and perhaps the direction of the forces involved. These latter are usually classed as tensional or compressional. *Tensional forces* tend to pull the rocks apart, such as happens when any unit is stretched. For example, when a portion of the earth's crust is bowed upward into a fold, the surface at the top of the fold is stretched and normal faults may occur (Fig. 15-26). Normal faults in general are the result of tensional forces, although they are not always associated with folds.

Compressional forces tend to push rock together, and may even attempt to move one portion of the rock past another portion. Under different conditions and in different rocks, these forces may produce folding, jointing, or faulting, or a combination of these. Most folds, reverse faults, thrust faults, strike-slip faults, and some overthrusts are the results of compression.

Overthrusts, however, are a special and very interesting case. For a long time geologists assumed that there were horizontal compressional forces pushing each overthrust block from behind. But suppose a block of rocks were to be uplifted quite a bit and then tilted. If there were an especially incompetent bed, such as shale or salt, in the sequence of layers, the upper strata might slide down the slope under the influence of gravity alone, forming an overthrust fault (Fig. 15-38). The origin, called *gravitational gliding*, is generally accepted among geologists today for *some* overthrust structures. Others are

Fig. 15-40   Angular unconformity; a conglomerate lying unconformably on tilted sandstone beds. Park County, Wyoming. (Photograph by C. A. Fisher, U.S. Geological Survey.)

obviously the result of horizontal compression. Gravitational gliding probably does not operate at depth, but is limited to rocks at or near the surface. This is supported by the fact that such thrusting rarely involves the crystalline basement rocks. As Figure 15-38 shows, some types of folds may be produced by this process.

The Scale Model

Attempts have been made to duplicate these larger structures in the laboratory by the use of scale models. Some of these have been fairly successful, especially those involving faults. A German geologist, Hans Cloos, produced good normal fault systems and graben. He placed wet clay on a table which could be lengthened, and the resulting cracks in the clay closely resembled the pattern found in the field. Several investigators have produced both tensional and compressional faults in sand as a result of vertical uplift.

Efforts to produce folding have been somewhat less rewarding. It is necessary to scale down mathematically all the factors involved, such as stress, strength, and time. When this is done, the exact material that ought to be used in the model can be determined. Unfortunately, it often happens that there *is* no substance with those exact properties. To compound the difficulty, not just one but several such substances must be found, for crustal deformation

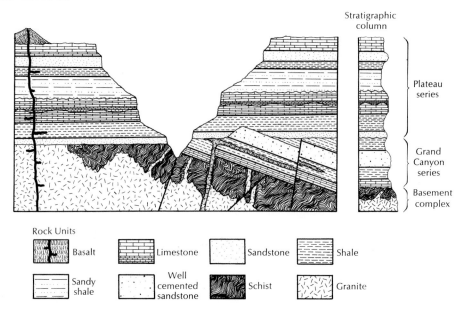

Fig. 15-41   Cross section of the Grand Canyon.

usually involves more than one rock type. In these circumstances, substances which only approximate the correct ones are often used. Another difficulty arises because not all of the many variables can be considered, so that the resulting model is necessarily oversimplified. The principal factor which is difficult to scale down is, as always, time. So that, even if natural-looking folds can be produced, it can be legitimately argued whether they were produced by the same conditions in the laboratory as in the crust.

Unconformities

Orogeny vs. Epeirogeny

The deformations we have been discussing—faults and folds—are all a part of the processes of mountain-building, or orogeny. Periods of mountain-building are called orogenies and are characterized by strong forces operating over large areas to produce extensive linear belts of distortion, such as the Appalachian Mountains. Even the stable interiors of the continents, however, where mountain-building is not going on, have their ups and downs, known as *epeirogenies*. While the amount of localized distortion is slight, the amount of downwarping or uplifting may amount to over a kilometer, and the areas involved may be vast. Just as orogenies leave their records in the rocks as folds and faults, so epeirogenies leave their records, too, as buried erosion surfaces known as *unconformities*. It should be noted that unconformities are not restricted to areas of epeirogeny, but occur in orogenic belts as well.

Unconformities can be of several types, as shown in Figure 15-39, and all represent gaps in the sequence of sediments. When an area is low, it may be invaded by a shallow sea, where sediments are deposited over a long time. When the area is uplifted, the sea retreats, and the newly emerged surface is exposed to the agents of erosion. Erosion may remove a little or a lot of the sedimentary cover and may create a surface of great relief or one that is almost flat and featureless. Then, when the land again subsides and the sea again invades the dry areas, new sediments are deposited upon the erosion surface, which becomes an unconformity.

If the rock layers above and below the surface are essentially parallel, the feature is known as a *disconformity*. If the erosion surface was developed on a series of tilted sedimentary beds, we have an *angular unconformity* (Fig. 15-40). A *nonconformity* occurs where the younger beds overlie an erosion surface developed on crystalline, non-stratified rocks such as plutonic and metamorphic rocks.

The cross section of the Grand Canyon (Fig. 15-41) shows the principal rock units exposed in its walls. There are three great divisions, and these are: (1) the basement complex of various kinds of metamorphic and plutonic rocks which are cut by small intrusions and dikes; (2) the Grand Canyon series of tilted sedimentary rocks; and (3) the Plateau series of flat-lying sedimentary rocks. Notice that the Grand Canyon series is separated from the rocks above and below by profound unconformities, indicated by the letters A and B. A is

Fig. 15-42 Grand Canyon, showing the basement complex of crystalline rocks exposed in the inner gorge, separated by a nonconformity from the flat-lying beds above. (Photograph by L. F. Noble, U.S. Geological Survey.)

a nonconformity, and B is an angular unconformity. There are two major disconformities within this pile of strata, and these lost intervals are shown by the letters C and D.

The Grand Canyon provides a superb illustration of the kind of reasoning followed by a geologist in unraveling the past story of a region from the kind of rocks, the inferred nature of the environment of their origin, and their structural pattern. The Grand Canyon is also an outstanding place to demonstrate how a single segment of the earth's crust alternately may be elevated and depressed.

To emphasize this last point, the major events in its geologic history may be recapitulated briefly:

1. *Deposition* of the sedimentary and volcanic rocks which today make up the schists of the Inner Gorge.

2. *Metamorphism* of these sedimentary and volcanic rocks and the *intrusion* of small granite bodies, both occurring at great depth.

3. *Uplift* and profound *erosion* and reduction of the entire region to a nearly level lowland.

4. *Deposition* of the sediments of the Grand Canyon series, chiefly on the floor of the sea that spread inland over the nearly level surface of erosion (nonconformity A) produced in Step 3.

5. *Deformation* of these rocks, chiefly by large normal faults with the resulting production of tilted fault block mountains.

6. *Erosion* of these mountains until they, too, were reduced to a nearly flat surface.

7. *Deposition* of at least 365 meters (1200 feet) of sandstone, shale, and limestone in the sea that spread inland across the erosion surface (angular unconformity B) of Step 6.

8. *Uplift* and *erosion* with the removal of an indeterminate thickness of sedimentary rocks and the cutting of an erosion surface.

9. *Deposition* of about 183 meters (600 feet) or more of limestone on the sea floor, creating disconformity C. This rock, incidentally, makes the most conspicuous cliff within the canyon.

10. *Uplift* and *erosion* with almost no accompanying deformation, since the strata above and below this surface, discontinuity D, are virtually parallel.

11. *Subsidence* and *deposition* of about 60 meters (2000 feet) of limestone, sandstone, and shale, partly on land and partly in the sea.

12. Here the local record that can be read in the rocks ends. The limestone that makes the rim on which stand the various park hotels and campgrounds is at least 185 million years old. Obviously, a great deal has happened in the chapters that are missing here, but from the record that can be pieced together elsewhere in nearby regions, there was deposition throughout much of the remainder of geologic time, then uplift and the stripping off of several thousands of feet of rocks, followed by volcanic activity (represented by the neighboring San Francisco Mountains), and finally downcutting by the Colorado

Fig. 15-43   Grand Canyon. The upper horizontal Plateau series is separated by an angular unconformity from the older, tilted Grand Canyon series. (Photograph by L. F. Noble, U.S. Geological Survey.)

River to excavate its present course across the Kaibab and Coconino Plateaus. These plateaus have been uplifted vertically for more than a mile with almost no internal disturbance of the underlying strata, an excellent example of epeirogeny.

In reconstructing the history of an area from the rocks and structures, the geologist must use every available piece of evidence, but, even so, many pieces of the puzzle are missing, and the actual record may be slim indeed.

As Herbert L. Hawkins has remarked, "Just as a net has been described as a set of holes held together with string, so a series of strata must often represent a succession of non-sequences separated by films of sediment."

Selected References

Eardley, A. J., 1951, Structural geology of North America, Harper and Brothers, New York.

Hill, M. L., 1947, Classification of faults, American Association of Petroleum Geologists Bulletin, vol. 31, pp. 1669–73.

——, and Dibblee, T. W., Jr., 1953, San Andreas, Garlock, and Big Pine faults, California, Geological Society of America Bulletin, vol. 64, pp. 443–58.

Oakeshott, Gordon B., 1966, San Andreas fault: geologic and earthquake history, 1966, Mineral Information Service, October, pp. 159–65, California Division of Mines and Geology, Sacramento, Calif.

Spencer, Edgar, W., 1969, Introduction to the structure of the earth, McGraw-Hill Book Co., New York.

Fig. 16-1 Earthquake damage in San Francisco, California, April 18, 1906. (Courtesy of the California Historical Society.)

16

Earthquakes and the Earth's Interior

Early in the morning of April 18, 1906, the schooner *John A. Campbell* was running on a southeast course before a fresh NNW breeze, with Point Reyes in northern California due east 233 kilometers (145 miles). With no warning the vessel shuddered suddenly with almost the same sensation as if she had run aground. The startled crew could scarcely believe their senses because the chart showed a depth of 2400 fathoms (14,400 feet) at their position.

Although the crew had no way of knowing it, they were only a few among the many whose daily routine was disturbed, or even ended forever, by the events set in motion at 5:12 a.m. throughout a region covering about 975 million square kilometers (375,000 square miles) surrounding San Francisco. This disaster of more than a half-century ago is still a most instructive example today because it devastated an essentially modern city. The population of the United States is becoming increasingly urbanized, and many of the problems faced by our predecessors in San Francisco are exactly the same ones that might confront Civil Defense agencies today, with the additional burden of immense traffic jams, unimaginable a generation ago.

San Francisco in 1906 in some ways resembled the city of today, but in others it was very different. As today, its rows of closely crowded houses swarmed over the many hills. The cobblestoned streets along the waterfront were clangorous from the iron-tired wheels of horse-drawn drays. The pungent

Fig. 16-2   Damage in the city of San Jose resulting from the San Francisco earthquake. (Courtesy of the California Historical Society.)

atmosphere of the waterfront was compounded in large part of the aromatic by-products of this multitide of horse-powered vehicles, as well as the fragrance of roasting coffee and the beery blast that emanated from the dark, block-long sawdust-floored saloons. The San Francisco waterfront was an infinitely more picturesque sight then than now, with the delicate tracery of the upper yards and rigging of the wind driven Cape Horners rising above the pier sheds, and the waters of the bay whitened by the paddle wheels of ferries and Sacramento River boats.

Much of this colorful world, including the "Barbary Coast," was obliterated in a series of violent shocks in the early morning hours when most people were asleep. Had the earthquake struck later when people were up and about, the casualty roll would have been far greater. How many people died will never be known, but it may have been as great as seven hundred. Many transient residents in such structures as sailor's boarding houses simply vanished, and since even the sketchy pre-1906 records disappeared in flames, many of the former permanent inhabitants could not be traced. In several of the investigations made after the event, almost all the destruction was attributed to the fire. It was indeed the leading destroyer, but earthquake damage was not negligible, perhaps averaging as much as 25 per cent.

Newer buildings in the San Francisco of 1906 looked much like those in the older downtown sections of American cities today. By 1906 riveted steel frames were coming into wide use for taller structures. Exterior walls were more commonly faced with masonry than they are today, and reinforced brick was widely used for smaller commercial buildings. Windows were narrower than we are used to now, ceilings were higher, and since this was a time when the Victorian influence prevailed, most buildings were crusted with ornamentation and gingerbread—real earthquake hazards. San Francisco was unusual in one regard, and that to its sorrow, in that it was one of the larger wooden cities of the world. Typically, the residential section consisted of block-long rows of multistoried houses or apartments crowded next to one another on 7.6-meter (25-foot) lots.

A significant lesson emphasized by the San Francisco earthquake was the importance of the kind of ground in determining the extent and nature of damage to buildings. Those founded on solid rock showed slight damage when compared with virtually identical structures built on waterlogged or unconsolidated ground. This environmental control was especially impressive in downtown San Francisco where an area about twenty blocks square is built on ground reclaimed from the sea by filling in this part of San Francisco Bay after the Gold Rush of 1849. Here on this sludgy foundation, made up of sunken ships, water-soaked refuse, bottles and bodies, all buried under poorly consolidated mud and silt, the most severe damage was concentrated.

Fig. 16-3 Damage in the city of San Jose resulting from the San Francisco earthquake. (Courtesy of the California Historical Society.)

Fig. 16-4   Sacramento Street in San Francisco, April 1906, following the great earthquake. Notice how the brick fronts of the buildings spilled across the street. The people are watching one of the great fires resulting from the earthquake. (Photograph by Arnold Genthe, courtesy of California Palace of the Legion of Honor.)

Fires broke out at many points almost immediately following the strongest shocks, which occurred within three minutes. The fires started from a variety of causes: overturned stoves, ruptured gas lines, and short-circuits from the primitive wiring systems of the time. At first there was little awareness that these fires were the true enemy. People either gawked at them or attacked them on a piecemeal basis—and not too successfully, because the alarming discovery was made almost at once that there was no water for the fire hoses. For one thing, most of the pressure lines below the streets were ruptured at innumerable points. Of even more fundamental importance, the main reservoirs of the city were Crystal Springs Lake and San Andreas Lake, which gives its name to the San Andreas fault which runs beneath the two lakes for their full length. Since it was slippage along this fault that caused the earthquake, a less favorable location for a city's water supply scarcely could be imagined. Fortunately, the dams held so that a flood was not added to the other afflictions of that unhappy day; the failures were in the pipe lines.

Fire started near the waterfront and swept inland across the broken city. Should you ever visit San Francisco, try to visualize the swath, eighteen blocks deep, swept by fire from the Embarcadero at the waterfront inland to Van Ness Avenue, the first wide street where a fire line could be held. Elsewhere vain attempts were made to check the advancing flames by dynamiting whole rows of buildings in their path, to keep the fire from leaping from roof to roof in the same fashion that a crown fire does in the forest.

From technologic and scientific points of view the San Francisco earthquake provided a big forward step. The California State Earthquake Commission, appointed by the governor but supported financially by the Carnegie Institution, conducted an exhaustive investigation. Hundreds of people were interviewed and evidence was collected from every damaged area, as well as records from virtually every seismograph station in the world.

Sudden slippage along a fault was the primary cause for this earthquake, and this was demonstrated beyond any reasonable doubt. Not only was faulting established as the mechanism, it was shown unequivocally that the displacement was strike-slip. Before 1906 the importance of this type of fault movement was only dimly appreciated. But here for all to see was a nearly continuous trail of furrowed ground, fractured barns, and various other mishaps strewn across the northern California countryside, with all the evidence showing that the western side of the San Andreas fault moved horizontally northward with respect to the eastern side which moved southward. The fault has a length of perhaps 966 kilometers (600 miles) on land, and in 1906 around 400 kilometers (250 miles) were in motion—from Point Arena north of San Francisco to San Juan Bautista to the southeast.

The maximum offset of 6.4 meters (21 feet) was near Tomales Bay; elsewhere in the neighborhood of San Francisco it held rather consistently at around 4.5 meters (15 feet). This meant that once the dust subsided along the active part of the fault, roads, fences, lines of trees, and buildings intersected by the

Fig. 16-5 The 1906 San Francisco earthquake toppled some of the giants of geology. Stanford University, California. (Photograph by W. C. Mendenhall, U.S. Geological Survey.)

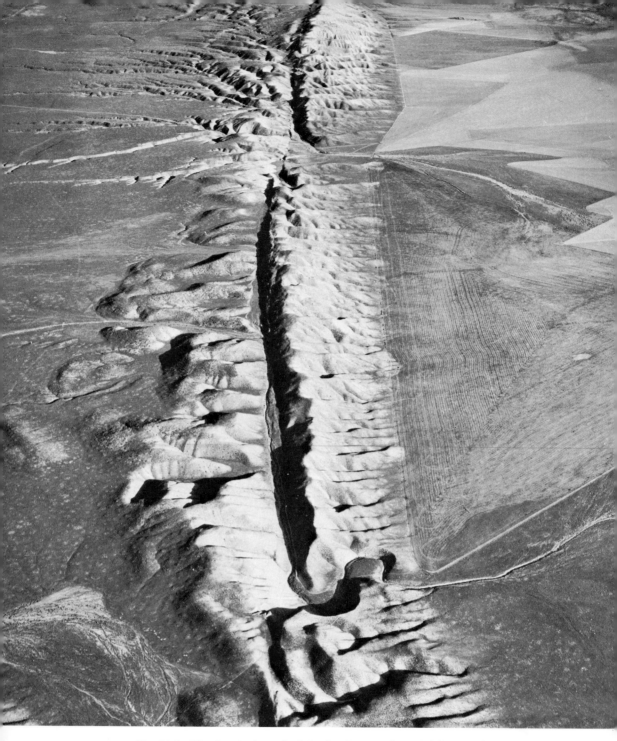

Fig. 16-6 The San Andreas fault in the Carrizo Plain, California. The stream in the left center of the picture turns and flows along the fault toward the observer. In the center foreground it turns to the right. Such streams are said to be "offset" and may result in part from movements along the fault. This part of the fault did not move in 1906. (Photograph by William Garnett.)

fault line were sliced through as neatly as though by a saber and everything west of the fault no longer matched up with anything east of it.

This brief description of the most famous of American earthquakes serves to introduce the subject of earthquakes in general. Its effects on man and his works was tremendous, and its effects on the earth itself have also been mentioned. These latter, however, were neither so varied nor so widespread as those that resulted from the earthquakes that struck southern Chile in 1960.

Effects of Earthquakes on the Earth

At 6 a.m. on Saturday, May 21, 1960, the first shock hit. The city of Concepcion suffered most of the damage from this first tremor, and as the citizens started cleaning-up operations on Saturday and Sunday, they were repeatedly jolted by aftershocks. At 2:45 p.m. Sunday an unusually strong shock was felt, and many persons throughout southern Chile left their homes and stood about in the streets. Fortunately, they were still standing there at 3:15 p.m., when the main shock rocked the entire region.

"The motion of the ground during the main shock was as if one were at sea in a small boat in a heavy swell. The ground rose and fell slowly with a smooth, rolling motion, smaller oscillations being superimposed on larger ones. In Concepcion, cars and trucks parked by the side of the road rolled to and fro over a distance of half a meter while they bobbed up and down in response to the movement of the ground. The tops of the trees waved and tossed as in a tempest. Some already damaged buildings fell. The earthquake itself was silent; not a sound came from the earth. The period of vibration was of the order to ten to twenty seconds or more. The shaking lasted fully three and a half minutes and was followed for the next hour by other shocks, all having a slow, rolling motion. . . . In the Region of the Lakes . . . the movement began smoothly and continued for some two minutes, just as in other localities, when suddenly, a loud subterranean noise was heard followed by a sharp jarring motion and a more rapid, less regular vibration of the earth. Similar reports were obtained at other points to the east of the Lakes, and it seems from these that another earthquake took place here . . . while the ground was still shaking from the first shock" (St. Amand, 1961).

Aftershocks continued to be felt for months. St. Amand records 119 shocks, from the main one in May 1960, until June 1961. While none was as severe as the main quake, which had a Richter magnitude (see p. 437) of 8.4, thirty-two of the aftershocks were of a magnitude of 5 to 6. An earthquake of this size in California floods the police switchboards with phone calls and merits several columns on the front page of the local newspaper. Twenty-one of the Chilean aftershocks had magnitudes between 6 and 7, and three had magnitudes over 7. It is obvious from these figures that southern Chile suffered from a most impressive earthquake swarm.

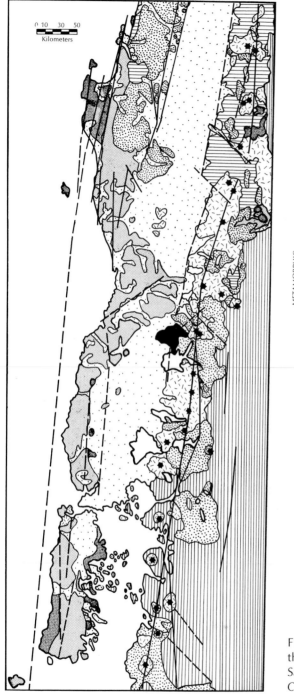

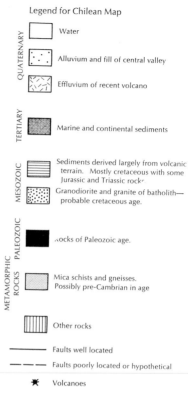

Legend for Chilean Map

QUATERNARY

Water

Alluvium and fill of central valley

Effluvium of recent volcano

TERTIARY

Marine and continental sediments

MESOZOIC

Sediments derived largely from volcanic terrain. Mostly cretaceous with some Jurassic and Triassic rocks

Granodiorite and granite of batholith— probable cretaceous age.

METAMORPHIC ROCKS

PALEOZOIC

Rocks of Paleozoic age.

Mica schists and gneisses. Possibly pre-Cambrian in age

Other rocks

Faults well located

Faults poorly located or hypothetical

Volcanoes

Fig. 16-7   Geologic map of the region of the Chilean earthquake. (After Pierre Saint-Amand, *Los Terremotos de Mayo-Chile,* 1960.)

Effects of the Chilean Earthquakes on Land

Because the active area was so large, 160 kilometers (100 miles) wide and almost 1600 kilometers (1000 miles) long, and therefore includes a wide variety of land features and surface conditions, this earthquake showed almost all the effects that such motions can have, with one notable exception. Even though a thorough search was made by investigators, both on the ground and from the air, no places were found which showed large offsets of the surface along a fault trace. This is in sharp contrast to earthquakes in other regions, such as our own San Andreas fault in California where lateral movement is obvious, or the earthquake which took place at Hebgen Lake, Montana, in 1959 which showed vertical movement of as much as 9 meters (30 feet) (Figs. 16-8, 16-9).

Other effects on the land surface, however, were numerous. Landsliding was one of the most common, where the earthquake triggered the sudden movement of unstable material. Cracks in the ground were caused by the settling of fill and the subsidence of areas underlain by soil which had liquefied. Liquefaction also caused huge earth flows, probably because of the presence of quick clays. "In the harbor of Puerto Montt a motor ship was caught in a

Fig. 16-8 Fresh fault scarp formed along the mountain east of Hebgen Lake, Montana, during the earthquake of August 1959. (Courtesy of Montana Highway Dept.)

Fig. 16-9   Close-up of escarpment showing vertical movement of as much as 30 feet. Hebgen Lake, Montana. (Courtesy of Montana Highway Dept.)

current of sand and mud that flowed from a nearby dock area into the bay, thus creating for itself the unique distinction of being the first ship to go aground in a landslide. Unable to move the ship, but undaunted, the owners converted it into a hotel" (St. Amand, 1961).

In the area of the most intense shaking, some trees were simply snapped off, while others were uprooted and fell. In the case of dry trees, the broken branches formed a circular pile of debris on the ground around the trunk.

Flooding resulted from changes in the level of the land. While some areas, particularly near the coast, were left 1.5 to 2 meters (4 to 6 feet) above sea level, in others the last of the great sea waves associated with the earthquake simply never receded completely, leaving those coastal regions permanently inundated. The rivers, too, produced flooding. In places the severe shaking compacted the poorly consolidated or unconsolidated material of the river banks, so that these banks were lower than they had been before. The motion of the earth also actually shook water out of the ground so that the rivers were unusually full. Almost continuous rain added to the problem.

Two days after the main shock, the volcano Puyehue started an eruption that continued for several weeks. Steam and ash issued from a fissure about

300 meters (984 feet) long and from several smaller openings. The last stage of the eruption was characterized by a number of flows of viscous lava.

"The local newspapers, in an unparalleled burst of enthusiasm, reported that 12 volcanoes had exploded and that two new ones had been formed. Lava was reported to be flowing down the sides of several of these volcanoes, and towns were said to have been buried. Because of the bad weather and poor visibility in the central valley, the inhabitants of the valley towns all believed that the volcanoes were, indeed, erupting and were concerned by the situation for a period of several weeks. Newspapers and news magazines all over the world repeated and enlarged on these stories" (St. Amand, 1961). This shows the importance of responsible journalism and the value of restoring communications as rapidly as possible after a disaster.

Other reports of new volcanoes proved to be sand-blows and mud volcanoes. These features are caused by the effects of soil compaction and vibration on ground water.

Effects on Bodies of Water

The Chilean Region of the Lakes lies inland from the coast but is still well within the area affected by the earthquakes. The lakes exhibited a feature

Fig. 16-10 Slippage of a section of highway into Hebgen Lake, Montana, after the earthquake of August, 1959. (Courtesy of Montana Highway Dept.)

Fig. 16-11　Fence rails shortened by surface slippage accompanying the Hebgen Lake Earthquake of 1959, near West Yellowstone, Montana. (Photograph by A. M. Bassett.)

known as a *seiche*. The motion of the quake sets up in each lake a wave which oscillates back and forth in the lake basin. The water sloshes, much like the familiar wave in a bathtub. Seiches in the Chilean lakes were small compared to that in Hebgen Lake, Montana, during the 1959 earthquake there. Hebgen Lake was an artificial body of water held in place by a dam. The earthquake suddenly lowered the bottom of the lake. "An eyewitness standing in the moonlight on Hebgen Dam and looking down its sloping face could not see the surface of the water, so far had it receded. Then with a roar it returned, climbing up the face of the dam until it overflowed the top, and poured over it for a matter of minutes. Then the water receded again, to become invisible in the moonlit night. The fluctuation was repeated over and over, with a period of about seventeen minutes; only the first four oscillations poured water over the top of the dam, but appreciable motion was still noted after eleven hours" (Hodgson, 1964).

A similar withdrawal of the water was noticed in the Chilean quake in the ocean itself. Shortly after the main shock, the coastal waters withdrew in a jumbled and confused state well below the lowest low tide line. The inhabitants knew from experience what to expect and fled from their homes to nearby hills to wait and watch for the *tsunami*, often erroneously called tidal wave. They had from fifteen to thirty minutes to wait and to complete their evacuation, whereupon the sea returned in a mighty wave that reached 6 meters (20 feet) in height at some places and extended as much as 3.2 kilometers (2 miles) inland. The waves continued for the rest of the afternoon, generally diminishing in height, although the highest wave is reported to have been the third or fourth rather than the first. In some areas as the water retreated it clawed great grooves in the coastal vegetation, grooves that exposed the soil beneath and that measured 1 meter (3.2 feet) in width and 20 meters (65.6 feet) in length.

The tsunami is believed to have been the result of a rapid lowering of the sea floor near the coast. Other tsunamis probably result from other kinds of land or sea-floor movement during earthquakes. Whatever their cause, they are particularly destructive, and create havoc not only near their place of origin but also many miles away. The Chilean tsunami crossed the Pacific Ocean at remarkable speeds, up to 644 kilometers (400 miles) per hour, and smashed into the Hawaiian Islands. In Hilo its force was concentrated by the shape of Hilo Bay, and the water surged over the sea wall, causing severe damage to the waterfront buildings. The wave continued its relentless journey west and resulted in extensive destruction to parts of the coast of Japan.

Measuring Earthquakes

The amount of damage done by an earthquake is expressed on the *Mercalli Intensity Scale* (Fig. 16-12) and is given in Roman numerals. It is based primarily on the effects on people and structures. Naturally the intensity of an earthquake will vary from area to area, depending on the character of the ground, the type of building construction, and the distance from the *focus*, the actual spot at which the rocks ruptured. A point on the surface directly above the focus is called the *epicenter*. An earthquake, then, does not have just one intensity, but several, according to where the observations are made.

This is in contrast to the *Richter Scale of Magnitude* which is based on the total amount of energy released by the earthquake and is expressed in Arabic numerals. This quantity is determined from the *seismogram*, a picture made by the *seismograph*, an instrument which records the waves set up in the earth by an earthquake. The Richter Scale is logarithmic rather than arithmetic (Fig. 16-13), which means that an earthquake of magnitude 8 releases not twice as much energy as a shock of magnitude 4, but more than one million times as much. There is only one magnitude for each earthquake.

Earthquakes—A Geologic Hazard

Forces which cause such sudden and severe changes in the land and the sea are naturally also extremely damaging to man and man-made structures. The actual destruction of buildings by virbration alone is awesome, and was particularly frightening in the earthquake which struck Long Beach, California, in 1933. Here a spectacular difference was noted between damage to buildings

Fig. 16-12   Modified Mercalli Intensity Scale.

I. Not felt except by a very few under especially favorable circumstances.

II. Felt only by a few persons at rest, especially on upper floors of buildings. Delicately suspended objects may swing.

III. Felt quite noticeably indoors, especially on upper floors of buildings, but many people do not recognize it as an earthquake. Standing motor cars may rock slightly. Vibration like passing truck. Duration estimated.

IV. During the day felt indoors by many, outdoors by few. At night some awakened. Dishes, windows, doors disturbed; walls make creaking sound. Sensation like heavy truck striking building. Standing motor cars rocked noticeably.

V. Felt by nearly everyone; many awakened. Some dishes, windows, etc., broken; a few instances of cracked plaster; unstable objects overturned. Disturbances of trees, poles, and other tall objects sometimes noticed. Pendulum clocks may stop.

VI. Felt by all; many frightened and run outdoors. Some heavy furniture moved; a few instances of fallen plaster or damaged chimneys. Damage slight.

VII. Everybody runs outdoors. Damage negligible in buildings of good design and construction; slight to moderate in well-built ordinary structures; considerable in poorly built or badly designed structures; some chimneys broken. Noticed by persons driving motor cars.

VIII. Damage slight in specially designed structures; considerable in ordinary substantial buildings, with partial collapse; great in poorly built structures. Panel walls thrown out of frame structures. Fall of chimneys, factory stacks, columns, monuments, walls. Heavy furniture overturned. Sand and mud ejected in small amounts. Changes in well water. Disturbs persons driving motor cars.

IX. Damage considerable in specially designed structures; well-designed frame structures thrown out of plumb; great in substantial buildings, with partial collapse. Buildings shifted off foundations. Ground cracked conspicuously. Underground pipes broken.

X. Some well-built wooden structures destroyed; most masonry and frame structures destroyed with foundations; ground badly cracked. Rails bent. Landslides considerable from river banks and steep slopes. Shifted sand and mud. Water splashed over banks.

XI. Few, if any, (masonry) structures remain standing. Bridges destroyed. Broad fissures in ground. Underground pipelines completely out of service. Earth slumps and land slips in soft ground. Rails bent greatly.

XII. Damage total. Waves seen on ground surface. Lines of sight and level distorted. Objects thrown upward into the air.

that had been well designed and those which were poorly designed. The schools especially suffered heavily; 85 schools in Long Beach were estimated to have sustained 75 per cent damage and some collapsed completely. Had the quake occurred during school hours, the loss of life among children would have indeed have been appalling. Fortunately this was not the case, but the possibility of such a disaster in earthquake-prone California led to the passage of the Field Act by the state legislature. This act makes mandatory the use of earthquake-resistant construction in all school buildings.

The value of this move was amply demonstrated in 1952, when a severe earthquake shook parts of Kern County, California, near Bakersfield. Almost all damage to school buildings was limited to those built before the Field Act. One post Field Act school showed slight structural damage, but it would not have resulted in any injuries had the building been occupied at the time of the quake. Investigations conducted after this earthquake confirmed the previous findings: buildings with good design and good workmanship suffer little or no damage, while shoddy workmanship and poor design result in extensive damage, if not total destruction.

Fire and landslide are two of the most destructive side effects of earthquakes on man's structures. As we have seen in the San Francisco quake, fires often break out in heavily populated areas soon after an earthquake because of broken gas mains, overturned stoves, and short circuits in electrical wiring. In the 1923 quake in Japan, fire demolished great portions of Yokohama and Tokyo. It occurred just as the noon meal was being prepared over thousands of individual open fires. In addition, fire-fighting equipment was either destroyed by the quake or hampered in its movements by narrow streets. In both cities there was insufficient water to fight the fires anyway because the water mains had been ruptured.

The Alaska earthquake of 1964 pointed up, as never before, the dangers of building on unstable or potentially unstable material in earthquake-prone areas. Much of the city of Anchorage is built upon Bootlegger Cove clay, one of the quick clays which are notorious for their habit of liquefying at the

Fig. 16-13   Richter Scale of Magnitude.

Earthquake magnitude	Approximate energy released
1.0	6 ounces T.N.T.
1.5	2 pounds T.N.T.
2.0	13 pounds T.N.T.
2.5	63 pounds T.N.T.
3.0	397 pounds T.N.T.
3.5	1,990 pounds T.N.T.
4.0	6 tons T.N.T.
4.5	32 tons T.N.T.
5.0	199 tons T.N.T.
5.5	1,000 tons T.N.T.
6.0	6,270 tons T.N.T.
6.5	31,550 tons T.N.T.
7.0	199,000 tons T.N.T.
7.5	1,000,000 tons T.N.T.
8.0	6,270,000 tons T.N.T.
8.5	31,550,000 tons T.N.T.
9.0	199,000,000 tons T.N.T.

least provocation. The 1964 quake was certainly provocation enough, and the Bootlegger Cove clay failed conspicuously. Sliding was not confined to the expensive residential area of Turnagain-by-the-Sea, but was, in fact, responsible for most of the damage in Anchorage.

Living with Earthquakes

Earthquakes, like many other natural phenomena, cannot be prevented; we will have to learn to coexist with them. There are, however, measures that we can take to limit the amount of damage they do.

First, we should discover what factors determine the amount of damage that will occur, for earthquakes sometimes show extraordinarily freakish effects. One factor, of course, is the population density in the affected area. The Long Beach earthquake, for example, was not of very great magnitude, but it caused severe damage, partly because it occurred in a densely populated area, and partly because many buildings had substandard construction. In addition, much of the town was built on alluvium and gravel which always react badly under earthquake stresses, while damage to structures on bedrock is usually much less.

The type and extent of damage also depend upon the depth at which faulting occurs. If the focus is deep, as it was in Chile, the damage will be widespread, whereas if the focus is shallow, damage will be confined to a much smaller area, but it may be much more severe.

Prevention of Damage

Obviously we can have no control over the depth at which earthquakes will occur. We can, however, do something about the design and construction of buildings. Normally buildings are constructed to withstand vertical stresses, and they generally stand up, even if their design is poor and their workmanship shoddy. The forces in an earthquake are quite different, however, and are mainly horizontal. Engineers have decided that earthquake-resistant construction means that a building must be able to withstand a horizontal force of 15 per cent of the total weight of the building and its contents. In addition, the parts of the building must be tied together so that they vibrate together, which will prevent the separation of roofs from walls from floors from the foundation.

Since the type of soil or rock beneath the foundation is important, care should be taken not to build on ground that is unstable or may become so. What should be done is plain; getting the necessary building codes and zoning laws enacted and enforcing them is quite another matter. It requires a knowledgeable, concerned citizenry and co-operation among officials, industry, and scientists.

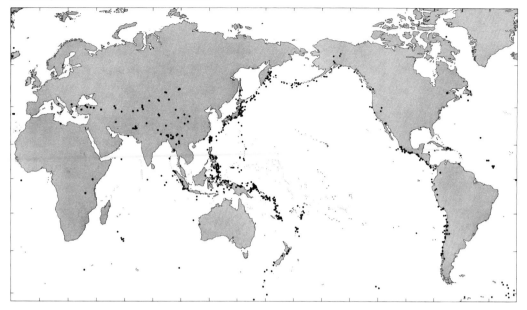

Fig. 16-14   The distribution of major earthquakes 1904 to 1952. (After B. Gutenberg and C. F. Richter, *Seismicity of the Earth and Associated Phenomena,* copyright © 1954, Princeton University Press. By permission.)

Prediction in Space and Time

It would be foolish indeed to build every building in the world so that it could withstand earthquakes. It is necessary, then, to know where earthquakes are likely to occur (Fig. 16-14). While we cannot yet predict earthquakes in time with much accuracy, we do know that there are areas of the earth's surface where earthquakes have occurred with some frequency, and we assume that they will continue to be active. The majority of the world's earthquakes occur where the earth's crust is being deformed. This means that earthquakes have a notable concentration around the borders of the Pacific Ocean and westerly from its rim along the line of the Himalayas and the mountains of the Middle East through to the Mediterranean. Japan among the industrialized nations is almost certainly the uncontested champion with regard to earthquake frequency and intensity. Not far behind, however, are the islands of Indonesia, as well as New Zealand, the Philippines, and the west coasts of North and South America.

Some regions, such as the great shield areas of every continent, are virtually free from earthquakes. The north European plain is a good example of a relatively earthquake-free area, and here the most telling evidence of their absence in past centuries is the heaven-aspiring steeple of the Gothic cathedral. It would be difficult to devise a worse structure for earthquake country.

On a much smaller scale, geologists can often tell us exactly where active faults are located. The state of California, for example, is now making strip

Fig. 16-15 Aerial view of the Daly City, California, area, showing the trace of the San Andreas fault. (U.S. Geological Survey.)

maps of faults suspected of recent activity. Geologists search every inch of the fault trace for signs of recent movement. Some faults are moving constantly, but only a very small amount. This movement, or *creep*, is of great interest to geologists. Some believe that it may help to predict not only where, but even when, an earthquake will occur.

The forces that deform the earth are probably acting continuously, but some rocks are hard to break, even along faults, and the force must accumulate for some time before rupture can take place. Other faults, however, show creep, and geologists do not yet know whether a creeping fault is relieving those forces a little at a time so that no major movement may be expected, or whether creeping indicates an unusually active fault which may give way at any time. It occasionally appears that increased creeping precedes a major break, while at other times faults are unusually quiet just before a large earthquake.

The San Andreas fault in California differs along its trace. Some portions of the fault are "locked," or show no creep. These portions appear to have the most severe quakes at wide intervals in time. Other parts of the fault do creep and generally have less severe earthquakes at more frequent intervals. Both San Francisco and Los Angeles are on locked portions of the fault, and there has been extensive construction directly on the trace in spite of the hazards (Fig. 16-15).

Japan, as well as the United States, is conducting intensive research which, it is hoped, will lead to accurate prediction of earthquakes. The Japanese have set up large nets or arrays of instruments as Matsushiro and have even predicted periods of earthquake activity in this area with some success. They discovered that swarms of very small quakes occur several months before a larger shock and indicate the epicentral region. Changes in the level of the ground were recorded with tiltmeters, and fluctuations in the earth's magnetic field were also observed shortly before stronger activity.

Some geologists (Pakiser et al., 1969) believe that we may soon be able to predict the occurrence of earthquakes within a few days or even hours. To do this would require constant monitoring of siesmic activity along an active fault, as well as continuous recording of tilt and magnetic fluctuations in the surrounding rocks.

Causes of Earthquakes

While it is apparent that earthquakes are caused by rupturing and movement of rocks along a fault, the causes of instability in the earth's interior remains a mystery. Some geologists believe that large chunks of the earth's crust are drifting around on the mantle. We will discuss this further in Chapter 18. Nobody, however, has come up with an entirely satisfactory theory to explain why these chunks drift. Given the instability of the earth, there remains another puzzle—the triggering mechanism of earthquakes. What pushes the strain in the rocks past the breaking point?

Some geophysicists have pointed out a correlation between solar activity, or sunspots, and large earthquakes. Others believe that they are related to earth tides and thus to the moon. Another relation exists between major earthquakes and changes in the position of the earth's pole of rotation. This pole has several motions, one of which is called the Chandler Wobble, whose driving force has long been a puzzle. Now it seems that earthquakes may cause the wobble.

Man, in his eternal attempt to "conquer" nature, has discovered accidentally that he can trigger earthquakes. In 1961 the U.S. Army drilled a deep well at the Rocky Mountain Arsenal northeast of Denver, Colorado. It was drilled 3671 meters (12,040 feet) through sedimentary rocks into Precambrian igneous and metamorphic rocks. Its purpose was the disposal of waste fluids from the manufacture of chemicals, and from March 1962 through September 1963 an average of 21 liters (22 quarts) per month were focibly injected into it. Just one month after injection began, the Denver area started to experience earthquakes; they had magnitudes of 3 to 4 and were all centered within 8 kilometers (5 miles) of the well. When a relationship between the well and the earthquakes was suspected, injection of waste fluids was changed to simple gravity flow and finally was stopped altogether in 1966. Special seismographs were set up to study the area. Still the quakes continued, even after use of the well ceased, and they reached magnitudes of 5 to 5.5 in 1967. The epicenters continued to be located around the well, and the foci were 4 to 6 kilometers (2.4 to 3.7 miles) deep, just below the bottom of the well.

In this case, old faults were present in the bedrock and the introduced fluid increased the pore pressure in the surrounding rocks, lessening their frictional resistance to faulting. Although the Army started a slow withdrawal of the liquid in 1968, earthquakes have continued, although they are smaller in magnitude and less frequent. Until an equilibrium is reached, more earthquakes can be expected. Other cases of this type have since been discovered— in particular, in the Rangely Oil Field in northwestern Colorado. Here the forcible injection of water in order to recover petroleum from otherwise unproductive wells caused a series of small earthquakes.

As the earth's population increases and finding space for waste disposal becomes an ever more serious problem, the temptation to get rid of it underground will also increase. In some regions this may prove a satisfactory answer, but before it is used in any area, the geology must be thoroughly investigated in order to prevent undesirable side-effects. These would include earthquake swarms, as well as contamination of ground water and underground water storage areas.

Another of man's activities which triggers earthquakes is the construction of large reservoirs. Apparently the load of water in the reservoir can reactivate long-dead faults if they happen to be present. The filling of Lake Mead behind Boulder Dam caused earthquakes of magnitudes 4 and 5, while quakes up to 5.8 occurred at Lake Kariba in Zambia, and a 6.5 magnitude quake at Koyna, India, caused extensive damage and loss of life.

Underground nuclear explosions, too, can trigger earthquakes, as geologists have found at the Atomic Energy Commission's Nevada Test Site. There thousands of aftershocks were set off by a test explosion, some of them as much as 13 kilometers (8 miles) from ground zero. This is what makes the use of Amchitka Island, one of the Aleutians, as an underground testing area so controversial. Its location in the circum-Pacific area means that it is in one of the most seismically active regions on earth, and, in fact, the tests are scheduled to take place only 30 to 100 kilometers (18 to 62 miles) from the known Aleutian thrust fault. The first test shot did not set off any major earthquake, but it was a small one, and the remaining tests will involve devices which are too large to be detonated in Nevada. Thus many geologists feel that much more should be learned about earthquake mechanisms before more tests are undertaken. If they can be conducted safely, however, there is a possibility that underground nuclear explosions might be used to release stresses on a locked fault, so that the forces causing very large and damaging earthquakes could be dissipated in a series of smaller shocks. Such control might also be achieved by the injection of fluids into deep wells, although this, too, will require much study before it can be attempted.

Seismographs and Seismograms

The founder of modern seismology—a word that comes from the Greek, *seismos*, for earthquake—was an English mining engineer, John Milne. He went to Japan in 1875 as one of the dedicated group of men who were responsible for bringing the technology and educational systems of the West to the Japanese people under the Meiji restoration. By 1880, he had had sufficient stimulus from a nearly continuous exposure to earthquakes to emerge as the first full-fledged seismologist, interested, active, and effective in every facet of that many-sided science. Through his efforts the seismograph became a precise instrument, and Milne-Shaw horizontal pendulum seismographs were installed at approximately fifty co-operating stations by 1892. From that modest beginning seismology has advanced until today there are nearly a thousand stations in operation throughout the world. In the United States data from all these stations are received and processed at the National Earthquake Information Center, ESSA Coast and Geodetic Survey in Rockville, Maryland. This organization publishes a bi-monthly magazine, *Earthquake Information Bulletin*, whose objective is "to translate into understandable terms the techniques used in investigating and describing earthquakes and related phenomena, and to present the results of past and continuing studies in seismology." An international summary of the world's earthquakes is issued by the Bureau Central Internationale Seismologique in Strasbourg—a program supported by UNESCO.

As a consequence of the Alaska earthquake in 1964, the U.S. Geological Survey established the National Center for Earthquake Research in Menlo Park, California. The Survey also works closely with its Japanese counterpart on earthquake research programs. Seismology, since it is concerned with the earth as a whole rather than some political subdivision of its surface, is a truly bright star in the firmament of international scientific co-operation.

The Seismograph

A seismograph is the instrument used to record the vibrations set in motion by an earthquake. The perfection of this instrument is the result of an immense amount of inspired and patient work over the many years since the first one was contrived by L. Palmieri in Italy in 1855. The problem that had to be solved was how to isolate an object on the earth's surface so that when every-thing surrounding it was set in motion by earthquake waves it remained sta-tionary, or as nearly stationary as it could be kept. Then the motion of the earth's crust relative to the immobile mass could be measured.

The most successful device for meeting this difficult requirement is the pendulum. Its inertia tends to hold it at rest in space while the ground beneath and the support holding it up are in motion. Then, if some method can be worked out so that a stylus, or pen point, can be attached to the inert mass, it will inscribe a record showing the amount of motion of the ground relative to the mass standing still in space. This is roughly comparable to the illusion we experience when we are sitting in a stationary passenger train and suddenly a freight train starts to move on the adjacent track. For a moment we think that we are moving and that it is the other train that is standing still.

The first primitive seismographs traced their record in smoothed loose sand or on a piece of smoked paper, and could give little information on direction or even the characteristics of the vibration they recorded. Milne made a vast im-provement over the limitations of such crude instruments so that it was then possible to determine the characteristics of earthquake waves traveling in a given direction. A completely equipped seismological station in the early part of this century customarily would have had three instruments: two horizontal pendulums, one mounted on an east-west axis, one mounted on a north-south axis, both to record horizontal ground movement, and the third pendulum sus-pended on a sensitive spring to measure the vertical component of motion.

One vexatious problem is that the amplitude of waves recorded directly by a seismograph is likely to be too small to enable us to study them profitably. This means magnification is necessary, and if this is to be achieved mechani-cally, then operating problems arise which seriously affect the accuracy of the result.

The desired magnification can be achieved photographically by using light rays rather than springs or levers and by recording the results on photographic paper rather than by dragging a stylus or pen over the surface of a sheet of paper. The torsion seismograph (Fig. 16-16) is such an instrument. Basically, it consists of an 20- or 22-centimeter (8- or 9-inch) tungsten wire with a pencil-thin weight mounted on it and suspended between the poles of a magnet. A tiny mirror is mounted on the weight, which in effect is the pendulum. The mir-ror reflects light from a source back through a set of prisms and lenses onto the surface of a rotating drum on which is mounted a sheet of light-sensitive paper. On this paper the light will appear as a minute, but very sharply defined,

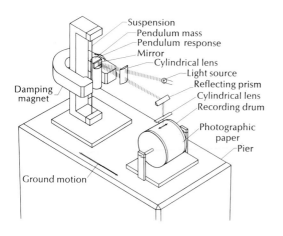

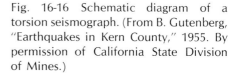

Fig. 16-16 Schematic diagram of a torsion seismograph. (From B. Gutenberg, "Earthquakes in Kern County," 1955. By permission of California State Division of Mines.)

square spot. It no shocks are recorded, the spot leaves a perfectly straight line behind it on the rotating sheet of paper, which shows up when it is developed. A typical instrument of this sort for recording rapid vibrations may have a pendulum with a free period of 0.8 seconds and a magnification of 2800 times —which is the ratio of light-spot displacement to ground displacement. For waves from slower, more distant shocks, a pendulum with a free period of 8 seconds and a magnification of 800 would be representative.

Instruments of far greater subtlety and accuracy than the ones described above are in existence today. Many of them were created through the efforts of Dr. Hugo Benioff and his associates at the Seismological Laboratory of the California Institute of Technology. In one of these the movement of the pendulum generates power electromagnetically, and this may be used to drive recording galvanometers with differing periods and magnification. This instrument bears the imposing name of a variable reluctance electromagnetic seismometer.

The pulsating record that a seismograph traces on a moving roll of paper not only tells us much about the earthquake itself, but reveals more of the nature of the earth's interior than any other device known.

Reading the Seismogram

To the uninitiated, the squiggles on a seismograph record seem wholly without meaning and have no more significance than a random series of not very well executed doodles. They are wave trains left by a variety of different kinds of waves. Those that follow paths that lie within the earth are *body* waves, while those that follow paths in the outer crust of the earth are *surface* or *long* waves.

The long, or *L*, waves are the ones we feel during an earthquake and they are the ones that cause destruction. The *L* waves travel at slower velocities than the body waves, and are the ones that write the largest squiggles on a seis-

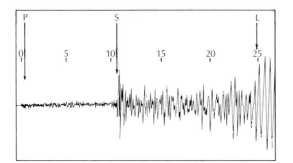

Fig. 16-17 A typical seismogram. The horizontal scale represents the number of minutes after the arrival of the first *P* wave. *S* waves began arriving at time *S,* and *L* waves at time *L.*

mogram (Fig. 16-17), but their effect diminishes rapidly with distance. The *L* waves are limited to the *crust.*

To greatly simplify a very complex story, there are two major classes of body waves, and these are (1) *primary* and (2) *secondary* waves. Since they are customarily abbreviated as *P* waves, perhaps the easiest way to remember them is as *push* waves. This not only is a good mnemonic device but also the word, push, gives a clue as to their behavior, as we shall see in a moment. Secondary waves are abbreviated as *S* waves, and one way to keep them in mind is to call them *shake* waves. (Fig. 16-18).

Primary waves are akin to sound waves, and thus produce alternate compression and rarefaction in the medium through which they travel—much like the waves that spread out through the air in all directions from a tuning fork. Each particle moves to and fro in the same direction that the wave is traveling. This is a behavior pattern somewhat like the slam-bang pulse that travels the length of a long freight train when the engine takes up the slack, especially if it were possible for all the couplings to be elastic.

In secondary waves, the particles in the rock through which the wave is traveling vibrate at right angles, or transversely, to the direction of propagation. The waves that travel down a stretched clothesline, or a garden hose, or a throw rug when one end is shaken vigorously are familiar examples. This sort of wave is called a shear wave because the particles within a rock *shear,* or glide, past one another much like the individual cards in a deck of cards can be made to slide past one another.

The direction and the speed at which waves travel is their *velocity.* The velocity of *P* waves is very nearly twice that of *S* waves, and in general the velocity of both increases with depth in the earth. For example, at a depth of 40 kilometers (25 miles) *P* waves travel at 8 kilometers (5 miles) per second and *S* waves at 4. At a depth of 2896 kilometers (1800 miles) the velocity of *P* waves has increased to 13 kilometers (8.5 miles) per second and the velocity of *S* waves has increased to 7 kilometers (4.6 miles) per second. This immediately raises the question: why this seemingly strange behavior of body waves within the earth? Why should the velocity of *P* and *S* waves increase with increased depth in the earth? Laboratory experiments under conditions of temperature and pressure at the surface of the earth made to determine the speeds of such waves in solids show that the velocity of these body waves increases with the rigidity and decreases with the density of the medium through which they are moving. Both rigidity and density increase with pressure, and if anything is certain about the interior of the earth, it is that pressure increases toward the center, finally to reach values incomprehensible to us on the surface.

The important thing here is that the effect of the increasing pressure on the rigidity of materials within the earth is greater than the increase in density with regard to body waves, and therefore their velocity increases in general down to a depth of 2896 kilometers (1800 miles). Below this depth there is a sharp decrease in velocity, and we shall return to the explanation offered for this strange behavior farther on in this discussion.

The unequal speed at which P and S waves travel provides us with an effective means of determining the distance between a seismograph station and the starting point of an earthquake, or its focus.

The problem here is much like a horse race. Both the P and S waves start out together, a great deal like two horses at the instant the barrier is raised. The P waves are a sure thing, however, since they always run about twice as fast as the S waves. The longer the distance the two waves travel, the wider the gap separating them becomes. The diagram (Fig. 16-17) shows the pattern of an idealized and greatly simplified seismogram, and on it the pulses announcing the arrival of first the P and then the S waves can be accurately timed. Then the seismologist can turn to his travel-time curves, of which Figure 16-19 is a simplified illustration. We can see from it that if the difference in arrival time of the P and S waves is about 5 minutes, then the epicenter is about 2000 miles away; if the difference is 10 minutes then it is about 5500 miles away.

This reading of the seismograph record tells us only the distance; it gives no clue as to the direction of the source. All we can do is draw a circle on a map with the location of the seismograph station as the center and the radius of the circle equal to the distance between the seismograph station and the *epicenter*

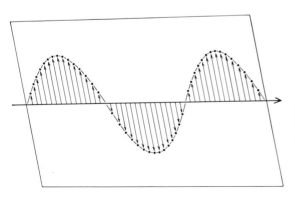

Fig. 16-18 The movement of particles in S waves. The particles move up and down while the wave moves horizontally. (Reprinted with permission. Copyright © 1962, Scientific American, Inc. All rights reserved.)

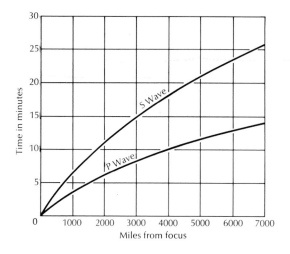

Fig. 16-19 Travel-time curves for *P* and *S* waves. The time lag between first arrivals of the *P* and *S* waves is used as a measure of the distance to the epicenter. (From J. H. Zumberge—after Bullen, *Seismology,* 1954—*Elements of Geology,* 1958. By permission of John Wiley & Sons, Inc. and Methuen & Co., Ltd.)

or geographic location of the earthquake source. If we use the example shown in Figure 16-20, and consider first of all the record received at Pasadena, then all we know is that the epicenter is somewhere on the circumference of the circle drawn with Pasadena at the center. When Pasadena exchanges its records with Berkeley, then two locations are possible, because two circles can intersect at two points. When the third station, in this case Reno, is heard from then the epicenter can be located with considerable accuracy because three nonconcentric circles intersect only at a single point.

In practice, the actual determination is usually more complex because, for one thing, many earthquakes do not originate at a point but may commence simultaneously along a fault line. The San Francisco earthquake of 1906 is a representative example since it originated along a 4345-kilometer (270-mile) failure of the San Andreas fault.

P and *S* waves traveling on the shortest route are the ones recorded by the seismograph, and when their arrival times are converted to distance, this will prove equal to the shortest distance between the seismograph (S) and focus (F). Should the earthquake prove to be the variety known as a *deep-focus earthquake,* then a circle with a radius of SF drawn on a map will overshoot the epicenter, and it will prove impossible to get the three circles to intersect at a point. In other words, the method we have been describing works well for shallow-focus earthquakes, but different travel-time tables must be used for those that originate at great depths below the crust.

The existence of deep-focus earthquakes was first discussed in 1922 by H. H. Turner and after debate and discussion during the following decade their reality was widely accepted. Part of the evidence for their existence is that they write a different sort of seismogram from shallow-focus earthquakes. For one thing, the *L* waves are lacking, or are likely to be ambiguous. More critically, though, the *P* waves that pass through the earth arrive sooner than they would have had the focus been nearer the surface. They should because they have the advantage of a considerable head start.

To the initial surprise of many geologists, deep-focus earthquakes were demonstrated to have originated at depths as great as 644 kilometers (400 miles) (Fig. 16-21). This is taken as an indication that at depths as great as this, rocks rupture in about the same way they do at shallower depths to produce a surface shock. According to Richter (1958):

> All this evidence indicates that deep-focus earthquakes originate in a process involving shear and elastic rebound and of the same general nature as that causing shallow shocks. The problem of plastic flow at great depth must then be faced; the apparent contradiction is resolved by appealing to a time parameter. Slowly accumulating strains will be relieved by flow before they can arrive at fracture, but rapidly accumulating strains may progress until fracture is reached. The behavior may be compared with that of a block of wax, which flows gradually under pressure or even under its own weight but fractures sharply if struck with a hammer.

A map of the distribution of deep-focus earthquakes would show that the locations of their epicenters correspond to some extent with those of surface quakes and yet differ in some significant respects. The coincidence is closest perhaps on the western side of the Pacific; on the eastern side the most noteworthy thing is their absence under the North American continent. There is also a significant concentration of deep-focus earthquakes far below the Himalayas and under the Andes—much farther inland in South America than the

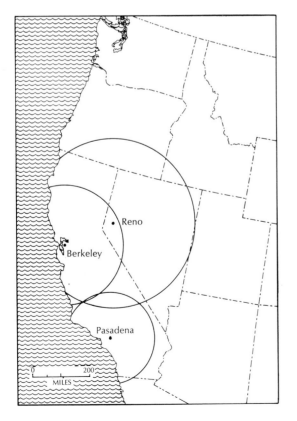

Fig. 16-20 Determination of epicenter of earthquake by knowing its distance from the seismographs. The three independently determined circles intersect at a common point.

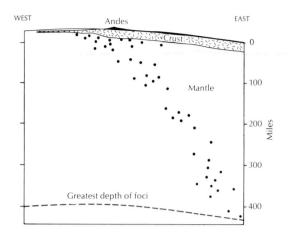

Fig. 16-21 The position of earthquake foci under western South America. (After Jacobs, *Physics and Geology*, copyright © 1959, McGraw-Hill Book Co., Inc. By permission.)

shallow-focus earthquakes clustered along the Pacific coast. Much the same distribution characterizes the epicenters of deep-focus earthquakes below the island arcs of the western Pacific. The same pattern characterizes the Andean deep-focus earthquakes, with the deepest which are also the farthest inland starting from foci far below the Bolivian plateau.

This curious linear distribution of deep earthquake foci along the surface of what appears to be a dipping plane has suggested to some geologists that these may represent points of failure where strains have built up along fractures akin to thrust faults inclined inland away from the Pacific basin. We will meet them again when we come to Chapter 18.

Seismic Clues to the Earth's Interior

We have seen how man-made explosions from a ship can tell us something about the rocks just beneath the ocean floor. This technique is called *wave-reflection* because the sound waves are reflected from the boundaries between types of rocks. With two ships we can use *wave-refraction*, which gives us information about conditions at greater depth. One ship maintains a stationary position and "listens"; the other cruises a straight line some distance away and sets off dynamite charges at regular intervals. Some of the waves produced are reflected off the bottom; some penetrate the crust and travel rapidly through it; some go deeper still and travel more slowly through another layer called the *mantle*. The boundary between the two is known as the *Mohorovičić discontinuity*. This Moho, as it is familiarly called, is a zone of change from a less dense to a more dense substance, a change which takes place within the relatively narrow space of about one kilometer.

Explosions can be detonated on land, also, and these, along with other methods such as gravity measurements, can tell us a good deal about the crust, the uppermost layer of the earth's interior. The crust itself has two layers, the SIAL, a light upper layer of primarily granitic composition, and a heavier, denser layer, the SIMA, which is mostly basalt. The crust is much thicker under the continents and this is because the thick SIAL layer is present there. The SIAL disappears at the edge of the continents, and the oceans are underlain only by SIMA (Fig. 16-22).

The lighter, less dense continents which rise above sea level have correspondingly deep roots in order to stay afloat on the mantle. This state of equilibrium is known as *isostasy*, from the Greek *isos*, equal, and the Latin *stare*, to stand. We have met the effects of isostasy before in glaciation. The weight of the continental ice sheets depressed the land area so that it bulged down on the mantle beneath it. When the ice melts, the land slowly rises as the mantle rebounds.

If puny man-made explosions can tell us so much about the crust, then nature's greater shocks should be able to tell us about the mantle and what lies beneath it. As was mentioned before, seismology is an international science, and quite early in its history a number of perplexing patterns became apparent in the recording of body waves transmitted through the earth. When seismological observatories in different parts of the world compared records, a puzzling pattern emerged with regard to the disappearance of S waves. Both P and S waves are recorded by seismograph stations within a distance of about 103° of arc or 10,900 kilometers (6800 miles) from the epicenter. Beyond this point both the P and S waves fade away and there is a wide band, approximately 5100 kilometers (3,200 miles) across, the so-called *shadow zone*, in which no body waves are received. Then at a distance at 143°, or 16,000 kilometers (10,900 + 5100) from the epicenter the P wave is recorded, usually with quite a strong signal, but S waves appear no more.

Another curious behavior pattern of the P waves that follow a path through the center of the earth is their apparent delay en route. Were they to travel with the same velocity they have at a depth of 2896 kilometers (1800 miles), they should go completely through the earth in about 16 minutes, whereas actually it takes them 20 minutes to make the trip. Obviously something has happened in the deep interior to cancel out the S waves and to retard the P waves by approximately 4 minutes.

An explanation for this anomalous behavior was given in 1913 in Europe by Beno Gutenberg, who for many years was director of the Seismological Laboratory at the California Institute of Technology. His widely accepted theory is that this wave behavior can be explained by the presence of a profound discontinuity within the earth at a depth of approximately 2896 kilometers (1800 miles). The part of the earth above this discontinuity is called the *man-*

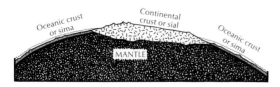

Fig. 16-22 Relation of SIAL to SIMA in the earth's crust.

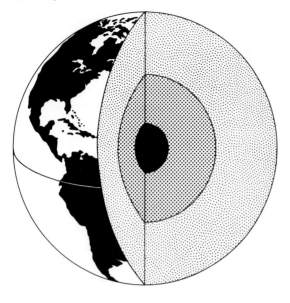

Fig. 16-23   The earth's interior, showing the relative size of the mantle (light stippling), outer core (dark stippling), and inner core (black). The crust is approximately the thickness of the line on the outer surface of the earth.

tle; the inner 3380 or 3540 kilometers (2100 or 2200 miles) of the earth is the *core* (Fig. 16-23).

Seismic observations made over the past half-century have established with reasonable certainty that the velocity of *P* waves drops abruptly at the Gutenberg discontinuity from about 13 kilometers (8.5 miles) per second to around 8 kilometers (5 miles) per second, or back to very nearly the same velocity they had at the base of the crust close to the surface (Fig. 16-24). The graph also shows the disappearance of the *S* waves at a depth of 2896 kilometers (1800 miles) as well as the fall-off and then gradual increase of velocity of *P* waves within the core.

What is the explanation for this behavior of the two waves? All we can do is make a number of surmises about the physical state of the core based on what we know of the behavior of matter at the earth's surface. We do know that *S* waves at the earth's surface are *not* transmitted through liquids. They are shear, or distortional, waves, and since fluids do not have distortional elasticity the inference drawn from this is that under the temperatures and pressures prevailing there, the core has more of the physical attributes of a liquid than it does a solid.

The fact that *P* waves cross the core does not help us settle this specific problem of the physical state of the earth's central zone because they are comparable to sound waves and thus can be transmitted through either solids or liquids. The abrupt reduction in velocity at the boundary between the mantle and the core is a puzzler, because not only does the velocity diminish almost by half, but it picks up again and increases slowly toward the center of the earth (Fig. 16-23), until it reaches about 11 kilometers (7 miles) per second.

Reasoning by analogy from what we know of the compressional or longitudinal waves in the surface layers of the earth, their velocity is affected by (1) the density and (2) the rigidity of the medium through which they are traveling. The more rigid and incompressible the rock, the higher the velocity. On the other hand, increasing density brings about a decrease in velocity.

The big drop in velocity of *P* waves at the Gutenberg discontinuity can be interpreted to mean that this boundary marks a change in the physical state of material in the earth's interior at a depth of 2896 kilometers (1800 miles). This is a change involving a marked loss in rigidity and at the same time an increase in density.

This supposition fits well with the blanking out of *S* waves mentioned in the paragraph preceding—that the Gutenberg discontinuity is the boundary surface separating a solid mantle above from a liquid core below. In a seismological sense a substance is a solid if it has both incompressibility and rigidity. It is a liquid if its rigidity is insignificant compared to its incompressibility, rather than whether or not we can drink or wash in it. Water, incidentally, has negligible rigidity, or resistance to shear, but it is a relatively incompressible substance.

This is not to say that the deep interior of the earth is like water on the surface, or even like lava, because among other things its rigidity, after the initial drop-off at the upper margin of the core, picks up again until it reaches a value of perhaps as much as two to four times that of steel. This is interpreted by many seismologists to mean that there may be a smaller inner core, perhaps 2570 kilometers (1600 miles) in diameter, that is solid, thus giving the earth a structure consisting of a solid mantle, liquid outer core, and possibly a small, dense, solid inner core.

That the core itself is denser than the mantle is demonstrated by the peculiar

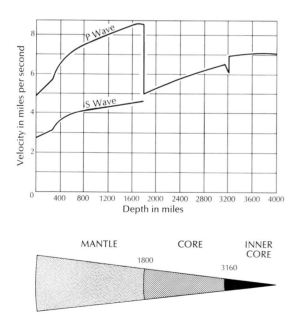

Fig. 16-24 Discontinuities in the earth's interior as inferred from changes in the velocity of *P* and *S* waves. (From J. H. Zumberge—after P. Byerly, *Seismology,* 1942—*Elements of Geology,* 1958. By permission of John Wiley & Sons, Inc. and Prentice-Hall, Inc.)

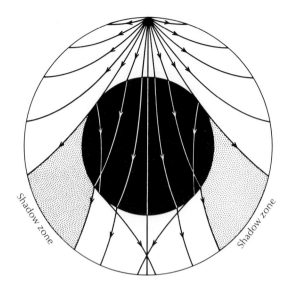

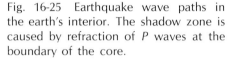

Fig. 16-25 Earthquake wave paths in the earth's interior. The shadow zone is caused by refraction of P waves at the boundary of the core.

behavior of the *P* and *S* waves where they fade out and disappear in the shadow zone. Waves between the epicenter and a point 1094 kilometers (6800 miles) away travel on direct, but slightly curved, paths until they emerge at the surface (Fig. 16-25). On the other hand, waves that penetrate the interface at the top of the core have their paths bent or deflected. This is called *refraction*, and is familiar to all of us in the way that light waves are bent on passing from air into water. The same thing happens, too, when light waves are bent on passing from air into a glass lens. The light rays converge toward the center as their speed is slowed down in the denser medium. This is the principle that is used with such telling effect in a magnifying glass when all the light rays falling on its surface are concentrated into a spot which is intense enough to start a fire. The cross section of the earth (Fig. 16-25) shows how the paths of *P* waves are deflected toward the center of the earth when they enter the denser material of the core and then away from it when they leave the core on the opposite side and start out into the lighter material of the mantle. The diagram shows how this bending, or refraction, of the *P* waves—especially of the ones that just penetrate the discontinuity at the top of the core—is responsible for the shadow zone on the far side of the earth. Either increased density or decreased rigidity can account for this effect. The important thing is that anything that decreases the velocity can achieve this result.

Much of our picture of the true nature of the earth's interior is veiled. We do know that pressure increases with depth until, if our assumptions are correct, it may reach the enormous figure of 20,000,000 pounds to the square inch.

We believe, too, that with this rise of pressure there is a corresponding increase in density. In Chapter 2 we learned that the earth has a density, or more properly a specific gravity, 5.519 times greater than an equal volume of water. This high value for the earth's over-all density poses an interesting problem because the average for rocks visible at the surface is about 2.7. To arrive at an average figure of nearly twice as much as that, somewhere within the earth there must be a concentration of heavier material. Logically, this would be in the inner core, and density expectedly would increase from the surface of the earth to the center. In a general way it very likely does, but a debate is still being conducted over how materials of different composition and density are distributed within the earth.

There is no practical way yet known for us to reach the center of the earth, or even to obtain a sample of it. All our knowledge comes from indirect evidence, such as the cryptic squiggles written by the waves from far-distant earthquakes on a seismograph drum. Here the evidence we have discussed in the preceding paragraphs makes it seem clear that a great increase in density occurs at the interface between the mantle and the core—perhaps from 4.7-5.7 to as much as 9.4-11.5.

We have no direct knowledge of what the density may be in the inner core, but according to Bullen, an Australian geophysicist (Fig. 16-26), it may rise to as much as 14.5 to 18 times as much as an equal amount of water. Bullen's diagram also shows that this is not an increase at a constant rate from the surface to the center of the earth, but indicates graphically the large increase in density at the top of the core. What is the core made of that it should be so heavy? No one knows, of course, but many geologists believe that it is composed of iron, or an alloy of iron and nickel.

From the limited information obtainable from studies of deep wells and mines we know that temperatures in the uppermost layers of the earth increase with depth. We do not, however, have much of a base from which to extrapolate because the deepest mines go down to only 3000 meters (9800 feet) and oil wells have been drilled to slightly more than 7600 meters (25,000 feet). In these probes which seem so deep to us, but still at the very most are only

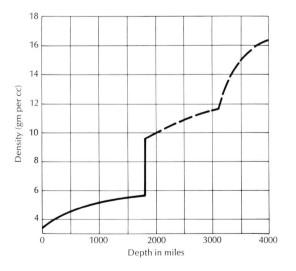

Fig. 16-26 Densities at various depths in the earth. (After K. E. Bullen, *An Introduction to the Theory of Seismology*, 1953. By permission of Cambridge University Press.)

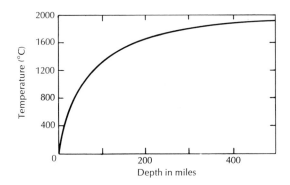

Fig. 16-27 Internal temperature of the earth. (After B. F. Howell, Jr., *Introduction to Geophysics,* copyright © 1959, McGraw-Hill Book Co., Inc. By permission.)

1:1000 part of the distance to the center, the temperature increases with increasing depth. The amount of this increase varies from place to place, and is greater in the vicinity of hot springs and geysers than in less thermally stimulated regions. A rough average for the outermost skin of the earth accessible to us would be about 1°F. for every 18 meters (60 feet). If this rate of increase is projected all the way to the center of the earth, remarkably high figures result, perhaps as much as 100,000° C.

Very likely it has been such straight-line projections as this which have led to the picturesque interpretation that the earth has a flaming, molten core enclosed by a shallow envelope through which volcanoes have forced their way to pour out their fiery streams.

A second concept that has probably influenced our thinking about the internal temperatures of the earth is the belief that if the earth solidified from a once-molten sphere, then high internal temperatures may be a residue, as it were, of this cosmic event.

With the recognition of the importance of radioactivity in the rock-forming minerals, a new source of heat within the earth could be proposed. Careful study through the years has shown that most of the radioactivity is concentrated in such elements as thorium, radium, and potassium. Since these are relatively more abundant in the minerals of granite than they are in those of basalt, it seems reasonable to expect that temperatures will be higher in the first rock and lower in the second. A great variety of evidence, much of which is internally consistent, demonstrates with reasonable certainty that rocks of granitic composition are limited to the earth's crust and that those that are related to basalt dominate at greater depth. Therefore, much of the heat supply available within the earth is concentrated within its higher levels rather than in the depths below as the molten core hypothesis calls for.

With this revision in our thinking, there unfortunately is almost no certainty as to what the actual temperatures in the inner recesses of the earth may be. The consensus today appears to be that temperature does not increase directly with depth, but that the greatest *rate* of increase is near the surface, and of course the highest actual temperature is near the center, possibly something like the pattern shown in Figure 16-27. What this temperature will be is sheer speculation, but in all acceptable theories today the amount is thought to be not over 10,000° C. and, according to J. Verhoogen of the University of California, it is probably not less than 2000° C. Possibly 3000° C. or 4000° C. may be closer to the answer. Most recent figures for temperatures in the mantle are even lower—900° C. at the top and 2900° C. at the base.

Confining pressures within the earth are high enough that probably nowhere are there large volumes of molten rock. Solid rocks must expand to melt, and pressures in the deeper parts of the earth are too great to allow this expansion. Very likely melting through a drop in pressure occurs only locally where a sudden release of stress occurs, as in regions of active stress or crustal activity.

Summary

From what has gone before, we learn that earthquakes result from the failure of rocks in the earth's crust when stresses have built up to the point that rupture occurs. Most often such a rupture takes place along a plane of weakness, such as a fault, and of these, the San Andreas fault in California is one of the world's pre-eminent examples.

Such failure sets in motion a variety of waves; surface or long waves which are responsible for the destruction of cities and buildings, and body waves, such as longitudinal (or compressional) waves and transverse (or shear) waves that move through the earth.

Patient unraveling of the tangled skein of the records written by seismographs at observatories all over the world shows that the earth appears to be arranged in concentric shells. At the surface the thin layer of the crust is divided into a light upper layer which underlies the continents but is usually absent beneath the oceans, and a lower heavier layer. The crust is separated from the mantle by the Mohorovičić discontinuity. Recent work done at the California Institute of Technology's seismological laboratory has shown that the mantle itself may be divided into as many as ten layers, and that its upper portion, perhaps 48 to 145 kilometers (30 to 90 miles) down, may be partly molten. This can be explained by the fact that temperature increases faster with depth than pressure does. The mantle is separated from the core by the Gutenberg discontinuity, and the core has an inner and outer portion marked by the Lehman discontinuity. Information about these shells of the earth's interior are summarized in Figure 16-28.

Much more is to be learned about the earth's interior than is known today. The gulf of our ignorance is both deep and wide. In many ways the realm beneath our feet is as mysterious as interstellar space. It is true that we have come a long way down the road of understanding, but the goal of full knowledge still is far distant. It is a strange commentary, though, that the pendulum, seemingly the simplest of all instruments—even with all the subtleties of the modern seismograph added to it—should be the means by which we have gained the insight we possess.

Fig. 16-28   Summary of the composition of the earth.

Depth in kilometers	Discontinuity	Layer	Possible chemical composition	Velocity of primary waves in km/sec	State	Density	Temperature in °c
		Upper crust	Granite	5.5–6.1	Solid	2.7	
17–25	Conrad						400
		Lower crust	Basalt? Gabbro?	6.4–7.2	Solid	3.0	
32–38	Mohorovičić						600–1000
		Mantle	Peridotite?	8.0–8.2 13.6	Solid	3.3 5.3–6.7	
2,900	Gutenberg						1,500–5,500
		Outer core	Ferro-nickel?	8.1	Liquid	9.0–10.5	
5,000–	Lehmann			9.4–10.4		11.5	
5,200				11.2–11.7		11.17	
							1,900–6,000
		Inner core	Ferro-nickel		Solid		
5,371				11.2–11.7		10.18	1,900–10,000

After André Cailleux, *Anatomy of the Earth.*

Selected References

Healy, J. H., Rubey, W. W., Griggs, D. T., Raleigh, C. B., 1968, The Denver earthquakes, Science, vol. 161, no. 3848, pp. 1301–10.

Hodgson, John H., 1964, Earthquakes and earth structure, Prentice-Hall, Inc., Englewood Cliffs, N.J.

Iacopi, R., 1964, Earthquake country, Lane Book Co., Menlo Park, Calif.

Pakiser, L. C., Eaton, J. P., Healy, J. H., Raleigh, C. B., 1969, Earthquake prediction and control, Science, vol. 166, no. 3912, pp. 1467-74.

Phillips, O. M., 1968, The heart of the earth, Freeman, Cooper & Co., San Francisco.

Richter, C. F., 1958, Elementary seismology, W. H. Freeman and Co., San Francisco.

Saint-Amand, P., 1961, Los terremotos de Mayo—Chile, 1960, Technical Article 14, U.S. Naval Ordnance Test Station, China Lake, Calif.

Witkind, I. J., 1962, The night the earth shook: a guide to the Madison River Canyon earthquake area, Department of Agriculture, U.S. Forest Service, Miscellaneous Publication No. 907.

Fig. 17-1   Mount Gardner, Ellsworth Mountains, Antarctica. (Courtesy of *Science*; photograph by Thomas Bastien.)

17
Mountains and Shields

Of all the landforms on the earth's surface, none, surely, is closer to the heart of geology than mountains. The greatest variety of rocks is visible in their valleys and on their ridges and peaks. Many of the more dramatic aspects of erosion are concentrated in them. Landslides and other kinds of mass movement have their maximum development; streams are more powerful in mountains where their gradients are steeper; frost-riving efficiently subdues the higher alpine summits; lastly, it is in high mountains that glaciers add the final touch to make montane scenery a source of joy and inspiration to many of us dwellers of the lowlands.

That mountains can be a place of solace and of beauty is a relatively new point of view. Mountains were greatly feared by travelers in medieval times, and rightly so. Roads crossing them were few, and almost all were rough and tedious. Inns were incredibly crude by our standards and distances between them in terms of travel time were great. Dangers from landslides, cold, snow, and armed robbers were very real. Certainly few people would have been deranged enough to climb a mountain just for the sake of climbing it.

By the eighteenth century a change set in. Not only had climbers started Alpine ascents, but there was an awakening intellectual curiosity as to the nature of the mountain world. One of the leaders in this emerging inquiry was a Swiss, Horace Benedict de Saussure (1740–99), who, in addition to making a first ascent of Mont Blanc, wrote a four-volume work, *Voyages dans les Alpes*, in which, among descriptions of birds, flowers, trees, etc., is a first account of the complex structure of the rocks within these famous mountains.

Mountains have been explored scientifically through the years since the middle of the eighteenth century, until today we know vastly more than our predecessors about the rocks and structures found within them. We still are far from understanding how mountains are formed, and what processes operate within the earth's crust to raise some of them, such as the Andes and Himalayas, to their imposing heights.

Fortunately, some mountains are relatively simple compared with others; these we shall consider first. Like many other natural phenomena mountains are difficult to divide into classifications that have real meaning in the natural world. We are prone to make rigid, pigeonhole categories into which to fit features that may have had more than one kind of origin, or that tend to merge with one another rather than to have sharp boundaries. The very abbreviated classification below serves to differentiate the more distinctive types and yet not be so arbitrary as to be inflexible.

VOLCANIC MOUNTAINS   These are built up of an accumulation of magmatic material, such as ash, pumice, bombs, and lava flows.

BLOCK MOUNTAINS   These owe their elevation to differential movement along faults, so that some parts of the crust are raised and others lowered relative to one another.

FOLDED MOUNTAINS   These commonly are made up of folded sedimentary rocks.

COMPLEX MOUNTAINS   This composite class consists generally of igneous and strongly deformed sedimentary and metamorphic rocks, but commonly this category grades into the preceding class of folded mountains.

Volcanic Mountains

Some of the world's best-loved and most scenic peaks are volcanoes. Among the more familiar are Fujiyama, Mt. Rainier, Mt. Etna, Mauna Loa, and the lofty Andean summits, such as El Misti and Aconcagua. Volcanic mountains may range in size from cinder cones to such a monolithic edifice as Mauna Loa, which, counting the submerged as well as the visible part, rises 9100 meters (30,000 feet) from a base 145 kilometers (90 miles) in diameter on the sea floor; it is the world's largest isolated mountain mass.

Volcanoes constitute a fairly high percentage of the world's mountains; this is apparent on the map showing the distribution of active craters (Fig. 4-33). Add to these the dormant centers, such as the Cascade volcanoes of the western United States, and the total number is large. Could the waters of the sea be rolled away, we should be doubly impressed, because then the peaks of many of the volcanic islands would loom above the surrounding abyssal plain, and we should also see for the first time the hidden slopes of seamounts, all of which are volcanic.

However, volcanic mountains differ fundamentally from all the others in our classification since they are accumulations of material piled up on the earth's surface. In Chapter 4 we discussed the great diversity of form that volcanic mountains may show; there is no need to repeat the description here. The chief point to be emphasized is that volcanoes are heaps of pyroclastic material or of lava or of both. Although they may be grouped in clusters, or even chains, they do not form the long and nearly continuous ridges so characteristic of folded or complex mountains, such as the Alps, the Himalayas, or the Cordillera de los Andes. Characteristically, volcanoes rise as conical, or dome-like, mountains above their surroundings.

Block Mountains

To the emigrants making their way with immense travail westward in the Gold Rush days, as well as during the post Civil War expansion, few elements of the landscape of the arid West were more taxing than the succession of isolated mountain ranges separated from one another by broad and desolate basins, many of which contained saline playas surrounded by gravel-carpeted wastes. This region, now known as the Basin and Range Province, is shown in Figure 17-2.

To the geologists who accompanied the early exploring expeditions these mountains were a challenge. Obviously their structure was wholly unlike that of the central Appalachians with nearly continuous ridges broken at long intervals by water gaps such as those of the Cumberland, Potomac, and Susquehanna rivers. By the 1840's the outlines of the geologic structure of the Pennsylvania portion of the Appalachians had been deciphered through the geological surveys of the Rogers brothers, who pointed out that the internal arrangement of the mountains was a succession of synclines and anticlines whose crests and troughs much resembled a train of waves.

Although men looked for a similar geometry in the Great Basin, its discovery eluded the nineteenth-century geologists who were attached to the various railway surveying parties or to military expeditions. Instead of the relative simplicity of the central Appalachians, the internal structure of many of the Great Basin ranges is wonderfully complicated with intricate patterns of thrust faults, folded stratified rocks, and complex hierarchies of igneous intrusions. Certainly there was no wave-like progression of rhythmically folded strata.

It remained for one of America's greatest geologists, Grove Karl Gilbert (1843–1918), to solve the riddle of the origin of these mountain ranges. This was a difficult scientific feat indeed, for scientifically next to nothing was known of the remote and arid Southwest. No adequate maps existed to show even the location and extent of many of the desert ranges. As a young man of twenty-eight, Gilbert accompanied an exploring party of the Corps of Engi-

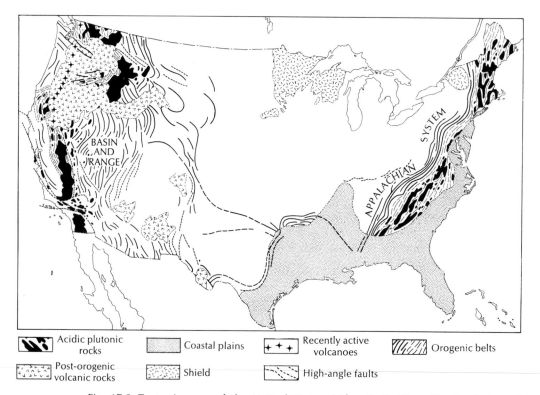

Acidic plutonic rocks

Post-orogenic volcanic rocks

Coastal plains

Shield

Recently active volcanoes

High-angle faults

Orogenic belts

Fig. 17-2 Tectonic map of the United States. (After P. B. King, *The Evolution of North America,* Copyright © 1959 Princeton University Press. By permission.)

neers, U.S. Army. In addition to the difficulties imposed by the hostile terrain, Gilbert, a civilian, had to contend with the arbitrary decisions governing the movements and route of a militarily oriented topographic surveying detachment. That he accomplished so much in so short a time, while still an untried and largely self-taught geologist, earned him an enduring place among the pioneers of scientific exploration.

In simplified terms, Gilbert saw that the unusual topographic form of these generally straight-margined mountains, separated from one another by broad, gravel-floored basins, was most plausibly explained by their having one or more faults along their margins by which they had been uplifted (Fig. 17-3). For its time (1872) this was a bold generalization, but investigations of these and other mountains of the Far West, as well as of comparable ones around the globe in the years since then, support Gilbert's prescience in making a sound generalization from the limited data available to him.

As is often true with scientific discoveries, later work demonstrates that an originally simple concept becomes increasingly complex as more and more information comes to light. No one as yet has devised an explanation for these fault-bordered mountain ranges that is completely satisfactory to geologists.

There appears to be a general consensus on the nature of the boundary faults — they are relatively straight along their strike; their dip is steep, perhaps 60-70°; and they appear to be more akin to normal faults than to other types. Very possibly the same general kind of deformation is responsible for long, down-dropped wedges of the earth's crust such as the Owens Valley, east of the Sierra Nevada in California (Fig. 17-4), the Rhine graben of western Germany, and the system of Rift Valleys of eastern Africa.

Clearly, vertical movements in the earth's crust have been responsible for the origin of these fault-bordered mountains and troughs. There is no evidence here of compression, or shortening of the earth's crust, the process that commonly is believed necessary for the formation of structures such as anticlines, synclines, and thrust faults, among others. Whatever the cause may be, its discovery eludes us but remains a challenge for some future generation to solve. From our limited understanding, the forces operating to uplift or depress these

Fig. 17-3   The base of the straight steep escarpment of these block mountains near Lakeview, Oregon, is marked out by a fault. (Courtesy of Oregon State Highway Dept.)

Fig. 17-4 East side of the Sierra Nevada and the floor of Owens Valley. (Photograph by Roland von Heune.)

blocks of the earth's crust appear to be unlike the ones responsible for the elevation of complex mountain chains such as the Himalayas, the Andes, the Rockies, and the Alps. Such cordilleras as these belong in the next two categories of folded and complex mountains, which are combined in order to simplify their discussion.

Folded and Complex Mountains

Geologists knew long ago that in many of the world's mountain chains the leading kinds of rocks were sedimentary, or if metamorphosed, were once sedimentary. Even for some of the highest ranges, such as the Himalayas, this generalization seems to hold, for on the upper slopes of Mount Everest, which rises more than 8800 meters (29,000 feet) above the sea, there are Tertiary rocks containing fossils that once accumulated on the ocean floor.

Not only are sedimentary rocks likely to be one of the leading components of a mountain range, they commonly are much thicker than their counterparts underlying the neighboring lowlands. This paradoxical relationship was commented on in 1859 by James Hall (1811–98), a young and energetic member of the pioneering New York Geological Survey, at a time when much of upstate New York was still a frontier. He found that Paleozoic strata underlying the Mississippi Valley states, such as Iowa and Illinois, were much thinner than those of identical age underlying the Allegheny Plateau of New York and Pennsylvania. From this observation, as well as from the knowledge he acquired through reading, he reasoned that many of the world's mountains had somehow been formed in regions with abnormal thicknesses of sedimentary rocks.

A contemporary of his, James Dwight Dana (1813–95), had done his stint of wandering as a geologist attached to the vessels of the U.S. Exploring Expedition (1838–42). This was a round-the-world investigation that has disappeared into unmerited oblivion in the scientific memory of this country but deserves to stand with the much better known voyages of the *Beagle* and the *Challenger*. Dana was greatly impressed by Hall's discovery of the abnormal thickness of sedimentary rocks exposed in the Appalachians, and, recognizing the rather unusual circumstances under which such sediments accumulated, coined the name *geosynclinal*, now shortened to *geosycline*, for the trough in which they were laid down.

Although the two men agreed upon the physical relationships of the rocks and structures in the Appalachians, they differed strongly on how they got that way. Hall believed that the earth's crust was bowed down as a consequence of the load imposed upon it by a localized accumulation of sediment. The more sediment that was deposited in such a wedge, the deeper the crust would be arched downward. Obviously this made for a rather uncertain equilibrium. The comparatively weak material of these sediments could be crumpled readily and thrown into contorted folds by inward movement of the vise-like margins of the trough. Fracturing of these infolded sediments also provided a means by which magma could invade shallower zones in the earth's crust, or possibly even reach the surface.

Dana did not believe the weight of such accumulated sediments was capable by itself of bending the crust downward. Without the knowledge that seismology has given us today of varying densities of the material in the earth's crust and in the mantle, he recognized the difficulty of having relatively light, water-soaked, and unconsolidated or only recently consolidated sediments displace subcrustal material whose density was higher. Dana was one of the earlier geologists to voice the opinion that the earth was slowly contracting through time, and this contraction of the interior (very likely, he believed, through loss of heat) caused the crust to be thrown into folds, or mountains, much like the wrinkles on the skin of a drying apple.

In the century since then, both points of view have continued to find their adherents, and we still are far from achieving a theory of mountain building which has universal acceptance. Fortunately, we have learned a number of additional facts since the days of Hall and Dana, and as a consequence competing theories nowadays are more sophisticated.

There is general agreement that the Appalachians, for example, progressed through an evolutionary sequence somewhat as follows:

1. A lens of sediment perhaps 9000 to 12,000 meters (30,000 to 40,000 feet) thick and 3200 kilometers (2000 miles) long and 480 kilometers (300 miles) broad, accumulated slowly through the Paleozoic Era—roughly from around 500 million to about 185 million years ago. Whatever the sediment source may have been, one certainty from the record of the rocks that mark its former site is that it was a region where considerable volcanic activity was concentrated.

Such a thick accumulation of sediments does not mean that the first grain of sand to accumulate settled down through water 6 or 7 miles deep, to be followed by other grains until the geosyncline filled to the brim. In fact, the evidence of the physical and paleontological record is that the water probably was fairly shallow during the 300,000,000 years the trough was in existence. During part of its history, the time the coal beds of West Virginia and Pennsylvania were being laid down, some of the geosyncline was a broad marshland above sea level. At other times it may have been an expansive delta surface, perhaps much resembling the Mississippi Delta of today.

During the long passage of years, much more happened than a peaceful, nearly imperceptible sinking of the geosynclinal floor. Times of deformation were interspersed with times of slow subsidence. The disturbed intervals are recorded in the rocks by unconformities, and by structures such as anticlines, synclines, and faults. There were occasional volcanic episodes, and a succession of alternating incursions and retreats of the sea.

2. Finally this long depositional episode in earth history, in what is now the eastern United States, ended at the close of the Paleozoic Era. Then, the thousands of meters of patiently accumulated strata were thrown into folds, or were broken by great, low-angle thrust faults—the latter chiefly in the southern part of the range.

3. Omitting some minor episodes of deformation and volcanism, the mountain range created by this disturbance of the earth's crust was worn away, and in its place a wide-spread peneplain beveled the upturned edges of the rocks which had been folded and faulted in the late-Paleozoic deformation.

4. Later still, the barricade of the Appalachians from New England to Alabama was upwarped to its present altitude. As a consequence of this elevation, streams which had wandered sluggishly over a gentle terrain of low relief were rejuvenated and became entrenched in such deep defiles as those now forming the ridge-crossing segments of the Susquehanna, Delaware, and Potomac rivers.

This highly condensed version of the history of the Appalachians embodies what might be regarded as the life cycle of the typical mountain range. From the paragraphs immediately preceding, we have learned that this cycle reduced to its simplest terms consists of: (1) a period of *geosynclinal deposition*, (2) a time of intense *deformation* involving folding, faulting, and possibly igneous activity, (3) an interval of quiescence and widespread *erosion*, and (4) a final episode of vertical *uplift* to carry the range to the height that it now stands. All these steps required something like 500 million years.

In science it often seems that no sooner is a generalization advanced than exceptions leap to the fore. Essentially the same sequence of events that typifies the Appalachians occurred in other mountains, but was enormously compressed in time. The southern Coast Ranges of California are a good illustration. There, within the short limits of the Pleistocene Epoch, strata a mile thick were deposited; then they, as well as the older rocks of the region, were intensely deformed, eroded to make a landscape of moderate relief, and finally were uplifted at least a quarter of a mile above the sea—the last episode is demonstrated by a flight of marine terraces which marches like a cyclopean stairway up the seaward slope of some parts of the range.

Shields

If we look at the distribution of these folded and complex mountains throughout the world (Fig. 17-2), we can see that they generally occur at the margins of continents. It is almost as if they had been pushed up against an unyielding center part of each continent. These central areas are known as *shields*, or *cratons*, and they have several interesting characteristics. For one thing, they are extremely old, being almost exclusively Precambrian in age. Radiometric dating, in fact, shows them to be over 1700 million years old. They consist of highly contorted metamorphic and plutonic rocks which have undergone extensive erosion to form some of the most widespread erosional surfaces in the world. One of the most fascinating things about them is that they have been so stable since Precambrian time, i.e. they have not been involved in any orogeny for about 600 million years.

They have not always been at the same level, however, but have been subjected to slow, minor vertical uplifting and subsidence. When they were low, they may have been completely or partially covered by shallow seas which left layers of sedimentary rocks. Hudson Bay in the Candian shield is an example of such a shallow sea. There is no evidence that any shield areas were ever covered by deep oceans. Some of the shallow marine deposits have since been removed by erosion, but others are represented by Paleozoic strata that overlap the edges of the Precambrian mass. These Paleozoic sedimentary rocks extend from the Canadian shield into the Arctic and into the central United States where they form vast expanses which, like the shield, also hqve not undergone orogeny. Therefore deep drilling would probably reveal the shield underlying them. Similar stable regions commonly are adjacent to shield areas on other continents.

Development of Mountains

Before constructing a theory to explain the building of mountains, and assuming that the Appalachians are typical, we should remember that such a theory must include the basic elements of geosynclinal deposition with accompanying volcanism, and also deformation with the intrusion of granite and related rocks, as well as the uplift to form higher parts of the earth's crust. The observations which James Hall made in the late nineteenth century form the basis of the classic theory of orogeny, a modification of which is presented here.

Using uniformitarianism as an approach, we should look for present-day evidence of the formation of geosynclines. In fact, we must look for two kinds of geosynclines, for the record of the past shows two parts to many folded and complex mountains. Again we may use the Appalachians as an example. The part nearest the shield contains no volcanics, but is composed entirely of sedimentary layers which thicken as they are traced eastward and which are shallow ocean deposits. They have not been so complexly deformed as the eastern part of the mountains, but are primarily a series of gently folded, plunging anticlines and synclines. This is called the *miogeosyncline*.

The eastern part of the mountain chain contains volcanics interbedded with sedimentary strata, some of which appear to have been deposited in the deep ocean. These rocks have been extensively deformed, with folds and faults, including thrust faults, and have often been intruded by large granite batholiths, or they grade into granite-type rocks in the mixed-up form called migmatite. This more active part is called the *eugeosyncline*.

For a long time geologists were puzzled because they could find no present-day example of a geosyncline. The vast underwater delta of the Mississippi River in the Gulf of Mexico was cited by some as an incipient geosyncline, but deltaic deposits differ in several respects from the sediments found in mountains, and so this was not entirely satisfactory. The deep trenches associated with island arcs on the perimeter of the Pacific ocean and elsewhere seemed a likely beginning of a geosyncline, but they too have their difficulties. The main one is that they are almost free of sedimentation, and what is there, of course, is characteristic of deep-ocean conditions only, since the trenches are the deepest parts of the earth's crust.

It remained for advances in mapping the ocean floor to show a better possible geosyncline, both miogeosyncline (the quiet one) and eugeosyncline (the active one). According to Robert Dietz, geologist and oceanographer, the present continental shelves are miogeosynclines, while the continental rises are eugeosynclines. As we saw in Chapter 11, the continental shelf consists of a great wedge of sediments derived from continental erosion and deposited at relatively shallow depths. Where there is any offshore barrier forming a basin, this wedge then develops on top of the basin filling. Some sediment is carried over the shelf break and some moves down submarine canyons; this sediment

DEVELOPMENT OF MOUNTAINS

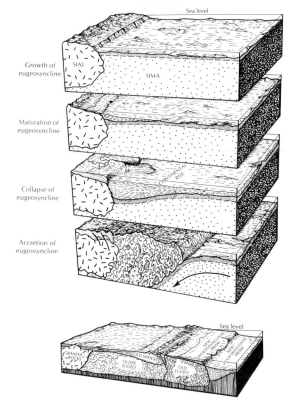

Fig. 17-5 Development of mountains from the continental rise. (After R. S. Dietz and J. C. Holden, "Deep-sea Deposits in but Not on the Continents," *Bull. Amer. Assoc. Petrol. geol.*, Vol. 50, No. 2.)

does not stay on the continental slope, but is deposited in the deep ocean at the base of the slope, forming the continental rise. The existence of the rise was unsuspected until echo-sounding techniques advanced to the stage where many soundings could be easily obtained. Figure 17-5 shows the subsequent development of the geosynclines and the mountain ranges. As the rise builds up, the increased load depresses the top of the mantle, and resulting fractures allow volcanic material to rise to the surface. This subsidence under the rise carries with it the continental shelf, enabling the shelf deposits to thicken outward, even though they remain shallow-water deposits. Eventually, through some unknown mechanism, the continent and the offshore deposits cease to act as a unit; they become uncoupled and act independently. The rise collapses and is forced against the continental mass, causing folding of the shelf sediments. Some of the sediments are thrust up and over the continental areas forming the new mountain range. Some are forced deeper into the crust and upper mantle where they may undergo granitization and even actual melting. The edge of this orogeny forms a new continental slope, and wave action forms the beginning of a new continental shelf.

If this theory is true, the continents have not always had their same general size and shape, but have been gradually added onto at their margins and have been growing larger and larger outward from their stable shield interiors. This is very pretty as far as it goes and fits many of the facts as observed very well, but it also raises other questions. Why are the interiors of the continents so stable; why do they act as a barrier against which other rocks crumple; why do these processes take place only at continental margins; and perhaps the most monumental question of all—what is this mysterious force that shoves tens of thousands of meters of rocks against the continental margins with strength enough to fold and break them and thrust them up over the margins themselves?

It might help in seeking answers to these questions if we stop thinking of the origin of mountain ranges as if it were an isolated problem. Actually, of course, mountains are only a part of the earth's crust. We shall see in the next chapter that when geologists start looking at the earth as a whole, many exciting new ideas are generated. Such "mega-thinking," as it is sometimes half-facetiously called, can be compared to the reconstruction of jigsaw puzzle. Until recently many pieces of the puzzle were missing; many probably still are, but now geologists have some very important new data to try to fit into the puzzle. This study of the broad structural features of the earth and their causes is called *global tectonics* and makes geology one of the most stimulating and rapidly developing sciences today.

Selected References

Dietz, R. S., 1963, Collapsing continental rises: an actualistic concept of geosynclines and mountain building, Journal of Geology, vol. 71, no. 3, pp. 314–33.

Eardley, A. J., 1951, Structural geology of North America, Harper and Brothers, New York.

Gilluly, J., 1949, The distribution of mountain building in geologic time, Geological Society of America Bulletin, vol. 60, pp. 561–90.

Kay, M., 1951, North American geosynclines, Geological Society of America Memoir No. 48.

Fig. 18-1   The rift valley of the Red Sea is at left, the Gulf of Oman at right. The
Afar Triangle bounds the Gulf on the south. (NASA)

18

Global Tectonics

One of the first modern scientists to develop a theory about earth history on a global scale was Alfred Wegener. In 1910, while he was a teacher of meteorology in Germany, he was struck by the similarity in shape of the east coast of South America and the west coast of Africa (Fig. 18-2). The coasts looked as though they might fit together if there were no South Atlantic Ocean. Other persons had undoubtedly noticed the same thing and had discarded the thought as fantastic, or at least merely coincidental. So did Wegener, for about a year. Then he accidentally came across some evidence from fossils of a former land connection between the two continents. This inspired him to look into the matter more carefully and, on the basis of what he found, to develop a theory of continental drift.

Wegener's Hypothesis

Wegener suggested that all the land in the world at the end of the Carboniferous formed one great continent which he called Pangea (Fig. 18-3a). This land mass began to split apart during the Jurassic, until in the mid-Tertiary it looked like Figure 18-3b. The continents, he believed, continued to drift apart, until in the early Quaternary they resembled Figure 18-3c.

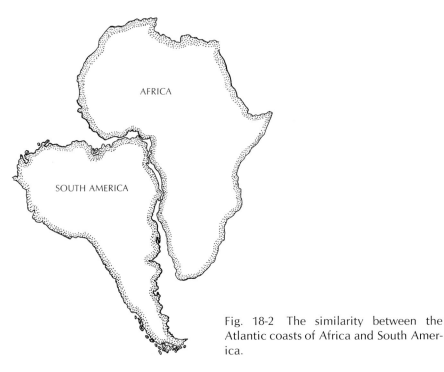

Fig. 18-2 The similarity between the Atlantic coasts of Africa and South America.

Paleontological Evidence

What evidence did Wegener discover that led him to formulate his theory? We have already mentioned that paleontology provided some of the evidence. Marine plants and animals, of course, can range throughout the world's seas, but terrestrial plants and animals are confined to the land and are stopped in their expansionist tracks by large bodies of water. Paleontologists studying in the Southern Hemisphere had known for some time that certain identical or very similar species had lived on both Africa and South America, as well as India and Australia. This evidence for a land connection between them had been explained by the existence of now extinct land bridges. We know today that there is no indication of submerged continental crustal material in the Atlantic and Indian Oceans, but even in Wegener's day there was no supporting evidence for a land bridge other than the fossils. Wegener reasoned that continental drift explained the paleontological facts at least as well, and, in addition, he had other suggestive clues.

Evidence from Isostasy

Vertical movements of the continents was an accepted fact by that time. The amount of uplift in Scandinavia could be measured and was known to be caused by the melting of the heavy continental ice sheet which had depressed the peninsula. Although the magnitude of movement and the direction in such isostatic adjustment was vastly different from what was proposed in continental drift, it did suggest that the continents were not completely fixed in place, but that some motion was possible.

Evidence from Stratigraphic and Structural Features

Wegener also cited in support of his theory the "torn newspaper" concept. If a page of newsprint is torn into several pieces and then the pieces are reassem-

Fig. 18-3a, b, c   Reconstruction of the map of the earth according to Wegener. The dark areas are oceans, the dotted areas shallow seas, and the white areas land. (After Alfred Wegener, *The Origin of Continents and Oceans.*)

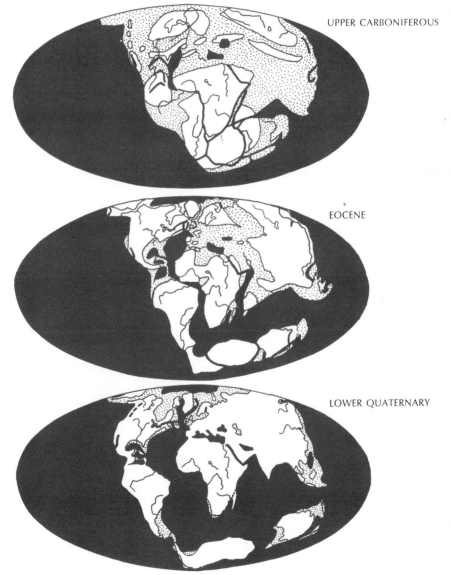

UPPER CARBONIFEROUS

EOCENE

LOWER QUATERNARY

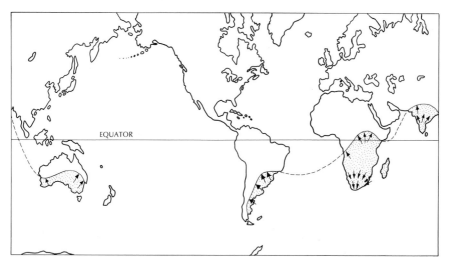

EQUATOR

NORTH AMERICA

EURASIA

AFRICA

SOUTH AMERICA

INDIA

AUSTRALIA

ANTARCTICA

Fig. 18-4 (a) Distribution of permo-carboniferous glaciation on a present-day map. (b) Permo-carboniferous glaciation on continents arranged according to Wegener. (Both modified after Takeuchi, Uyeda, and Kanamori, *Debate About the Earth.*)

bled, the lines of print will continue across the fractures; they will match, in other words. If Africa and South America, and Europe and North America were indeed one continent, then structural features on one side of the Atlantic should have their continuations on the other. This Wegener could show to be true. The Cape Mountains in South Africa, for example, appear to be the same type of Permian folded mountains with the same types of rocks as occur south of Buenos Aires. There are other instances of agreement between sedimentary strata and igneous rocks on both sides of the South Atlantic, but none younger than the Cretaceous. This suggests that the split occurred during the lower to middle Cretaceous periods, a conclusion borne out by the fossil evidence.

Matching rocks and structual features can be found on both sides of the North Atlantic, too, but here the date of rifting appears to be much later, so that the Atlantic rift seems to have started in the south and gradually extended northward, as if the original continent had unzipped.

Evidence from Glaciation

In Chapter 12 we mentioned some of the perplexities connected with the Permian-Carboniferous glaciation in the Southern Hemisphere. One of them is that the glaciated areas are so widespread, including even regions near the equator. Also there is no corresponding glaciation in the Northern Hemisphere which seems to have been tropical at that time. In Wegener's placement of the continents, the glaciated areas are all arranged in one small group (Fig. 18-4). Assuming that the Permian glaciation was centered around the south pole, the north pole would then be somewhere in the central Pacific Ocean where there was no land to be glaciated.

Evidence from Paleoclimates

This reasoning assumes that the earth's poles also have no fixed position. By Wegener's time polar wandering had been discussed for some years. It was indicated by the study of fossil climates, or *paleoclimates*. A region which has a frigid climate today often shows evidence of having had a subtropical climate in a past geologic era, and since climate is largely controlled by latitude, past latitudes of continental masses may be inferred by studying these paleoclimates. This scientists had tried to do by Wegener's time, and, although some type of polar wandering was definitely required, it remained vague—until Wegener reassembled the continents. Combining drifting continents with wandering poles proved to be a workable solution.

Controversy and Decline

Wegener's hypothesis seemed, then, to fit much of the known and suspected information about the earth. Yet it could not be proved; methods for testing the theory simply did not exist. This did not prevent considerable discussion, argu-

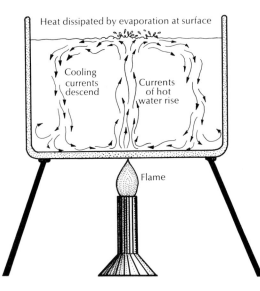

Heat dissipated by evaporation at surface

Cooling currents descend

Currents of hot water rise

Flame

Fig. 18-5   Convection currents in boiling water.

ment, and modification, however, which continued for more than a decade. DuToit, for example, believed that there were two large land masses instead of Wegener's one: "Laurasia," consisting of Europe, Asia, North America, and Greenland, and "Gondwana," comprising South America, India, Australia, Africa, and Antarctica. These two supercontinents, he believed, were separated by a geosynclinal ocean which he called the "Tethys." Even though current thought calls for one large land mass, the name Gondwana is preserved in the *Gondwana succession*, a distinctive series of strata containing tillites and diagnostic fossils, which is found in many places in the Southern Hemisphere. It gets its name from the location in India where it was first described.

In 1928 the American Association of Petroleum Geologists held a symposium on continental drift theory. Supporting evidence was presented, objections were raised, refutations were argued more or less satisfactorily. The fourteen speakers divided evenly as to whether they supported or rejected the theory.

One of the principal objections was that there was no satisfactory way to explain why or how the continents drifted. Wegener's idea that the tidal forces of the sun and moon on the earth were the driving mechanism was shown to be incorrect because the forces involved are so small and are not confined to one direction. Thus it seemed that nothing further could be accomplished by argument or testing, and so the theory faded from the scene, generally disregarded in the New World, but still considered by some in the Old World as a possibility.

Convection Proposed

At about the same time geologists were also arguing about the cause of mountain-building, especially the sequence of events leading to the formation of the belts of folded and complex mountains. It was realized that a shrinking, cooling earth could not account for the amount of crustal shortening considered necessary. In addition, the discovery of radioactivity in the earth's crust indicated warming rather than cooling with time. Arthur Holmes suggested that this radioactive heating might set up convection currents in the mantle, currents of the same type as are set up in a kettle of boiling water (Fig. 18-5). Even though the mantle was believed to be solid, given sufficient time, and heat, it could perhaps flow enough to produce convection currents. The heated material would rise within the mantle (Fig. 18-6). As the currents turned downward, they would crumple the crust above into piles of folded rocks. When the heat was dissipated, the convection current would stop, and the pile of crustal material in the folds, being light, would rise by isostasy and form mountains. Meanwhile, radioactive heat would be accumulating in another place in the mantle.

Obviously, it was not long before thermal convection was called upon to explain drifting continents. If the rising heated material reached the top of the mantle at a spot beneath a continent, the spreading motion of the mantle material would split the continent (Fig. 18-6). The continental masses would then be carried along on the moving mantle. In areas where the mantle material was descending, the bottom of the continent would be dragged into folds as described above, and would rise into mountains when the convection ceased. Meanwhile, the rising mantle material would be heated enough to melt at surface pressures (or lack of them) and would then form new ocean floors.

Ingenious as these ideas were, even their authors and proponents realized that they could not be proven until some kind of independent evidence was produced.

Independent Evidence

Without methods of testing the hypothesis of continental drift, it gradually faded away from most geological considerations. After World War II, however, independent evidence began to appear. Refined geophysical techniques that were being used in quite different studies started to produce data that led to the resurrection of the drift hypothesis.

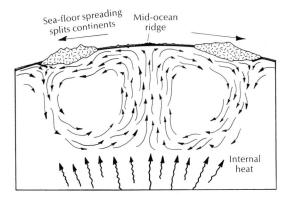

Fig. 18-6 Convection currents in the mantle as suggested by some geologists.

Paleomagnetism

One of these studies concerned paleomagnetism, which was discussed in Chapter 2. There it was seen that when igneous rocks cool, they preserve, in the orientation of some of their minerals, the direction of the earth's magnetic field at that point in time. With the new and sensitive instruments now available, it was possible to measure the much weaker geomagnetism in sedimentary rocks as well. This was extensively done in Britain where there are sedimentary strata representing all geologic periods from the Precambrian to the Quaternary. The fact that these strata are relatively undeformed was also important. A study of the geomagnetism in the Triassic rocks of Britain showed that there was indeed a change in direction from that of the present—not a reversal in polarity, but an effect that could have been produced had the island been rotated some 30° since then. This rotation was also supported by paleoclimatic evidence.

As additional evidence of these changes of direction of paleomagnetism became available, the tide of excitement began to rise. Was it time to revive Wegener's old hypothesis? No, since thus far all these changes could easily be accounted for by wandering poles.

As data was accumulated from other continents, however, there appeared to be something very wrong. The path of the pole as determined from evidence in North America (Fig. 18-7a) was quite different from that obtained from Europe, which was different again from that from Africa. When all the polar paths were plotted on a single map (Fig. 18-7b) a great deal remained to be explained. Polar wandering is not sufficient to account for even the most reliable paleomagnetic data. Some migration of the land masses, and not just migration but rotation as well, is necessary. Paleomagnetic studies, even though they are continuing and are being refined, cannot tell us everything we want to know.

The Spreading Sea Floor

As we saw in Chapter 11, increased exploration of the ocean floors in the last two decades with newly developed instruments and techniques has revealed some extraordinary features. One of the most astonishing has been the discovery of the mid-ocean ridges which virtually encircle the globe. Detailed studies show that the crests of many of the ridges are marked by exceedingly rugged terrain and a great central rift or crack paralleling the trend of the ridge.

These rifts are obviously active areas. They are the location of extensive earthquake and volcanic activity and have what geophysicists call *high heat flow*. This means simply that more heat is escaping from the earth in the vicinity of the rifts than from other areas of the earth's surface. When geophysicists tried to find an explanation, they thought naturally of convection. Are the rifts places where the heated portion of the mantle is rising to the surface? Is this where the sea floor begins spreading, thus causing drifting of the continents? Possibly, but again independent evidence is required.

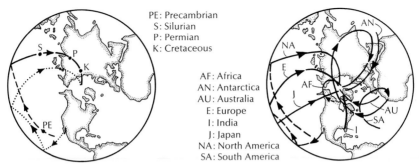

Fig. 18-7 (a) Polar wandering as determined from evidence in North America. (Modified after Takeuchi, Uyeda, and Kanamori, *Debate About the Earth.*) (b) Different paths of polar wandering based on evidence from different continents. (Modified after Takeuchi, Uyeda, and Kanamori, *Debate About the Earth.*)

What is the nature of the evidence other than high heat flow? Seismology, with its techniques for probing deep into the earth, has produced some. Earthquakes are common on the crests of the ridges, but these are shallow quakes. As we saw in Chapter 16, deep earthquakes are found only under the edges of continents, where they become deeper and deeper as the distance from the ocean increases (Fig. 16-9). This plane of deep earthquakes is called the *Benioff Zone.* Perhaps the Benioff Zone marks the place where the convection currents drag the spreading ocean floor down into the mantle. An engaging hypothesis, but not evidence.

Recent seismographic work has shown a low density layer in the upper mantle, called the *asthenosphere* in contrast to the crust or lithosphere. What does this layer mean? It seems to indicate that there is a portion of the mantle which, while not exactly molten as we know it, is capable of fluid-type motion. In other words, it could be the transporting medium for continental drift.

Going back now to those shallow earthquakes on the mid-ocean ridges, it has been found that they are not actually on the ridge itself, but only on the offsets of the ridge. On a map, the ridges appear to have been laterally displaced by a series of strike-slip faults. Seismologists have discovered a disturbing thing about these faults, however. They have made *first-motion* studies of the earthquakes there; these determine, as the name implies, the direction of the movement which takes place at the very beginning of the tremor, not the total displacement. First-motion studies show that these ridge faults do not act like strike-slip faults at all. In a strike-slip fault, the ridge crests would become farther apart with time (Fig. 18-8a). This does not happen, however, with the faults on the mid-ocean ridges; the ridge crests maintain their apparent offset with time (Fig. 18-8b). The motion and geometry of these special faults, called *transform faults,* can be explained, if it is assumed that spreading of the sea floor is taking place and that the ridge crests are the centers of the spreading (Fig. 18-9). This figure shows what happens at a transform fault with time. The ridge crests do not change their position; the ocean floor does, however,

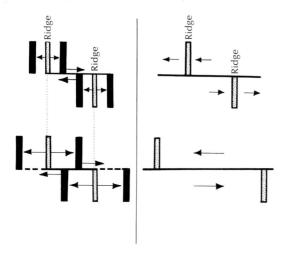

Fig. 18-8a and b   Contrast in the apparent offset with time between transform faults (left) and strike-slip faults (right). (After J. Tuzo Wilson, "Transform Faults, Oceanic Ridges, and Magnetic Anomalies Southwest of Vancouver Island," *Science*, 22 Oct. 1965.)

move, and where the two segments of sea floor move in opposite directions, line A-B, the appearance of a strike-slip fault is produced, and earthquakes occur.

If the ocean floor is actually being created at the ridge crests, the rocks farthest from the crests should be older than those at the crests. Dating of cores, especially those taken recently by the *Glomar Challenger*, has shown this to be true. Moreover, there seem to be no rocks in the ocean floor older than Jurassic. If the ocean basins are indeed no older than 160 million years, this would explain the lack of vast thicknesses of sedimentary rocks on the sea floor; they simply have not had time to accumulate.

Paleomagnetism confirms the notion of sea-floor spreading. As we saw in Chapter 2, the reversals of polarity (from N-S to S-N) of the earth show a definite pattern, and, combined with radioactive dating, extend our knowledge of the age of the ocean floor. The most fascinating discovery is that the pattern of reversals is identical on each side of many of the mid-ocean ridges (Fig. 18-10). This means that there are matching ages on each side of the ridges, the youngest ages being near the crests. For most geologists this piece of the puzzle, added to the others, indicates strongly that spreading of the sea floor does actually occur. These motions of the ocean floor, of course, make drifting of the continents ever more plausible.

Paleontology

To evidence from paleomagnetism and oceanography must be added that from the paleontology of Antarctica. Most of the fossil evidence that can be found for the joining of the continents in the Southern Hemisphere has involved plants and invertebrate animals. Even though they were all land dwellers, the objection has always been, from Wegener's day until very recently, that the seeds and spores of such plants could be transported great distances over open oceans by drifting on air currents, and that invertebrates could be carried by water currents, possibly in a larval form. In 1969, however, a group of geologists discovered in Antarctica bones of extinct vertebrates. Furthermore, these vertebrates were labyrinthodonts and lystrosaurian reptiles which cannot tolerate salt water. Similar remains have been found in sedimentary beds of the same age (Triassic) in Africa, Madagascar, and Australia. Thus, Antarctica was probably part of Gondwana, if such a land mass ever existed separately, or of Pangea if there was only one supercontinent.

A Unifying Theory

The fondest hope of scientists is always for one grand hypothesis that brings all the puzzle pieces together. This has the benefit of being psychologically rewarding, but it has even more important consequences. It shows where the gaps are, and where more work needs to be done. It is always a tremendously exciting time when a science reaches the stage where such a synthesis is possible. That is where geology is now, and this is the picture that is beginning to emerge.

As long ago as the Permian, all the continental mass was one large piece. Sometime during the Jurassic the land mass started to split up along rift zones. As new material from the mantle rose to the surface in these rifts, the continents were pushed apart and even rotated. The rifts, as they progressed, broke the land mass into approximately six large plates and a few smaller ones (Fig. 18-11). The boundaries of the plates are the active areas. When a continent is "locked" to its adjoining section of the sea floor, as the Atlantic coasts of North and South America seem to be, the deposition of eroded terrestrial materials can build up thick wedges of sedimentary strata in the form of continental shelves and continental rises. If, for some as yet unknown reason, a zone of weakness develops at the juncture of the continent and ocean, the spreading ocean-floor material will be thrust beneath the edge of the continent, forming deep trenches in some cases. As this relatively cool material which is stable at surface temperatures and pressures comes in contact with the mantle material,

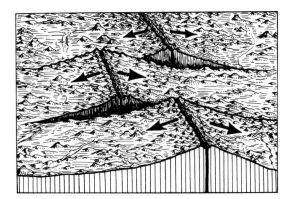

Fig. 18-9 Ridge crests as the centers of sea-floor spreading. (After H. W. Menard, "The Deep-ocean Floor," *Scientific America,* Sept. 1969.)

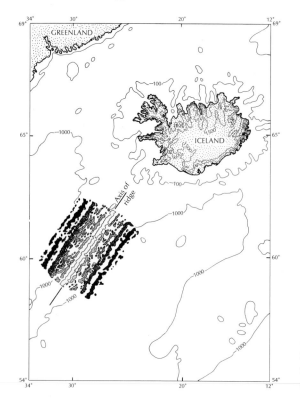

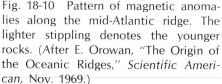

Fig. 18-10 Pattern of magnetic anomalies along the mid-Atlantic ridge. The lighter stippling denotes the younger rocks. (After E. Orowan, "The Origin of the Oceanic Ridges," *Scientific American*, Nov. 1969.)

stresses will be created which will cause deep-seated earthquakes such as are characteristic of the Benioff Zone. At the same time, sediments of the continental margins will also be dragged under, and when they become hot enough, will combine with the oceanic basalt, and will erupt in volcanoes of a composition intermediate between the two. This rock type is andesite, and, interestingly enough, andesitic volcanoes are found all around the periphery of the Pacific Ocean. The boundary between andesitic eruption and basaltic eruptions is known as the *andesite line*.

Not all of the ocean-floor sediments will be dragged down, but some may be scraped off on the edge of the continent. These deep-ocean sediments will be involved with the continental-margin sediments, and all will be compressed, complexly folded, and even thrust against the continental interior. When the continent-ocean boundary again becomes stable, the folded sediments will rise through isostatic processes and form mountains, and another period of continental rise and shelf accumulation will start. When two continental areas are shoved against each other, their margin sediments will also form mountains. This may be the explanation of the Himalayas and the Urals, mountain ranges which are deep within present continental masses rather than at the edges of continents. Thus, in the light of new data concerning ocean-floor spreading and other evidence for continental drift, our classical concept of the geosynclinal cycle of orogeny may have to be considerably modified. As with most other geologic subjects, increased knowledge means increased complexity and diversity. While the mountains of the world may have some characteristics in common, it appears timely to look instead at their differences. These differences will probably indicate a different history for each individual mountain system. "Perhaps like granites, there are mountains and mountains. . . . Periods of rifting would further complicate evolutionary histories. These possibilities inherent in coupled, uncoupled, rifting, and colliding plates, added to varying motions along transform boundaries and intraplate transcurrent [strike-slip] shifts, and irregularities in shapes of colliding margins, could explain why no single model fits all mountain systems. . . . It would seem one of the objectives now is to determine what the variants are in these complex histories as well as the invariants so that we may approach a clearer understanding of the tectonic evolution of mountain systems."

Pre-drift History

The theory of continental drift as presented above starts in the Jurassic, at least the actively rifting parts seem to have started then. What about the tectonic history prior to the Jurassic? The Permian glaciation in the Southern Hemisphere suggests that the one large land mass existed as such at least as long ago as 250 million years. How much longer it had that form we cannot say.

Examination of the oldest rocks on earth, the Precambrian, provide conflicting evidence. They are, in general, so complexly metamorphosed that they must have had a very complicated history. Even though they now form cratons, and appear not to have been deformed to any great extent since the Precambrian, still they give much evidence of having been involved in great

Fig. 18-11 The major crustal plates and their motions. (After Bullard, "The Origin of the Oceans," *Sicentific American,* Sept. 1969.)

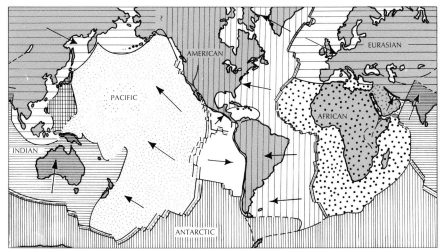

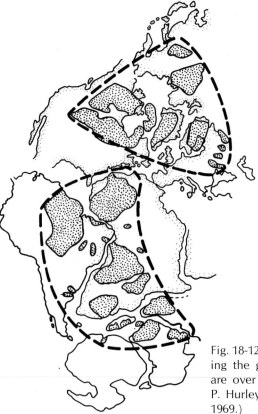

Fig. 18-12   Pre-drift reconstruction showing the grouping of areas whose rocks
are over 1700 million years old. (After
P. Hurley and J. Rand, *Science,* 13 June
1969.)

epochs of mountain-building during very ancient times, and, if so, they may
well have suffered rifting and drifting.

On the other hand, when rocks over 1700 million years old, well within the
Precambrian, are plotted on a pre-drift reconstruction (Fig. 18-12) an interesting pattern emerges. These continental nuclei fall into two distinct groups.
The geologists who noticed this grouping have inferred that these very ancient
rocks were in this position as far back as the Precambrian and that there was
only one great period of rifting—that which started in the Jurassic. Obviously
much work needs to be done to determine even the barest outline of the earth's
pre-drift history.

Present-day Rifting and Drifting

Now that the theory has been stated, it is pertinent to ask if we can see any
stages of these processes going on today. The answer seems to be yes: in the
Red Sea and in the Gulf of California. The map of the mid-oceanic ridges
(Fig. 11-13) shows these to be places where the ridge appears to go beneath
the continents.

The Red Sea

Examination of the floor of the Red Sea shows it to be in fact oceanic-type crust and not an ordinary shallow inland sea on continental crust. It also has a mid-ridge, an active rift, and unusually high heat flow. Furthermore, the edges of the Red Sea and of the Gulf of Aden fit together remarkably well (Fig. 18-13). Such a reconstruction calls attention to the Afar Triangle, where the two land areas overlap, as a critical area. Although it is exceedingly difficult to maneuver there, some geologists have managed to do investigating with the aid of a helicopter. The area surrounding the Afar Triangle is definitely continental in type, as would be expected, but the triangle itself exhibits all the characteristics of oceanic crust. It is actually a part of the deep-ocean floor which has been temporarily faulted above sea level. The Red Sea then appears to be an incipient ocean. Furthermore, it seems likely that this rifting is closely related to the long belt of the East African rift valleys, which may represent a zone of weakness where separation is about to occur.

The Gulf of California

Very similar conditions obtain in the Gulf of California, where the peninsula of Baja California is apparently being rifted away from the mainland. Very recent work indicates that sedimentary and volcanic strata match across the Gulf. The intriguing question arises as to where the East Pacific Rise goes. It has been suggested that it underlies the Basin and Range Province of the southwestern United States and may be responsible for their block-fault type of mountains and for the uplift of the Colorado Plateau. This much activity deep within a continent which ought to be stable is difficult to explain in any other way. The idea was put forward by H. W. Menard, an oceanographer, who says, "This digression from discussion of the familiar ocean basins to the mysterious continents may serve to emphasize that large elevated regions of the continents and the ocean basins may be produced by the same bulges of the mantle. The origin of rises [ridges] may be determined by studying plateaus. Unfortunately we know even less about plateaus than about rises."

The coastal region of the western United States shows two other interesting peculiarities in this connection. For one thing, there is very little continental deposition in the Pacific; the continental shelf is so narrow as to be almost non-existent. The second strange thing is that in this area there is only one side to the East Pacific Rise. There is only the west side of the pattern of magnetic reversals.

How can these facts be explained? One way is as follows. The American continent and the Atlantic Ocean are now locked together to form the American Plate. Acting as a single unit, they are drifting westward and overriding the Pacific Ocean floor. They have slid over the East Pacific Rise and hidden the eastern limb of it, and are also overriding the Benioff Zone and any ter-

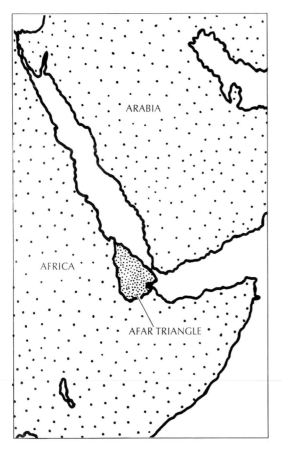

Fig. 18-13 The coasts of Arabia and Africa, showing the Afar Triangle where they would overlap if pushed closer together. (After Bullard, "The Origin of the Oceans," *Scientific American*, Sept. 1969.)

restrial sedimentary deposits that form at the continental margin. It has also been suggested that the San Andreas fault and other large strike-slip faults bordering the Pacific Ocean are not strike-slip faults at all, but are the laterally moving parts of large transform faults between ridges (Fig. 18-14).

Mechanisms
Convection

This makes a very neat picture, but nothing has yet been said about the mechanism that drives these processes. That was the major problem with Wegener's original hypothesis, and it still plagues geologists today. As the "new" global tectonics is described above, it might seem as if some convection current model would fit, but unfortunately there are serious objections to all of them. In the deep-convection model (Fig. 18-15a), which is essentially what was proposed in the late 1920's and early 1930's, the heated mantle rises under the oceanic ridges, separates, and flows in opposite directions until it descends to be heated and recirculated once again. The continents are carried along on the backs, so to speak, of the convection cells. Seismology has indicated, however, that the lower parts of the mantle are probably much too rigid for convection to take place at that depth.

In the shallow-convection model (Fig. 18-15b), convection takes place only in the upper part of the mantle which has a known low-density layer, the asthenosphere, where flow can occur. Again the mantle carries the continents along. The problem here is that the two parts of the mantle's convection cell

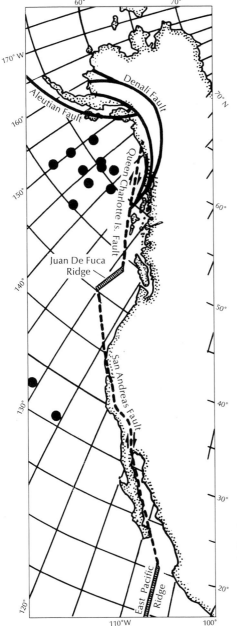

Fig. 18-14 The San Andreas as part of a transform fault. Black dots are locations of guyots. (After J. Tuzo Wilson, "Transform Faults, Oceanic Ridges, and Magnetic Anomalies Southwest of Vancouver Island," Science, 22 Oct. 1965.)

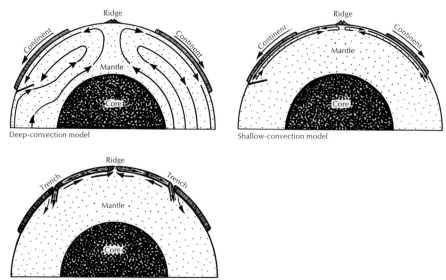

Fig. 18-15 Convection models. (After E. Orowan, "The Origin of the Oceanic Ridges," *Scientific American*, Nov. 1969.)

going in opposite directions are assumed to be only about 300 kilometers (200 miles) apart in a vertical direction, much less than the horizontal dimension of each plate. It seems to many that under these conditions several thin convection cells would form rather than one large thin one.

Transvection

Another model for continental drift and sea-floor spreading is called *transvection*. In this, an unknown force drives the crustal plate itself, either forcing it from behind, i.e. the ridge area, or causing *subduction*, or dragging down, of the leading edge (Fig. 18-15c). The viscous layer of the mantle then flows toward the mid-ocean ridge and produces new floor material. Thus no thermal convection is necessary. This may explain *how* continental drift occurs, but still leaves us with that unknown force that begins the whole process. A mechanical engineer, applying the results of laboratory experiments, has suggested that local heating of both the rigid and the viscous portions of the mantle causes the rigid portion to move, as a unit, away from the heated area. The viscous layer remains in place, but is subjected to tensional stress. This stress results in a fracture in the oceanic crust. Water and other fluids migrate toward and into the resulting fractures and combine with the crustal minerals to form lighter, more hydrous minerals which take up more space and thus build up a ridge. Simple gravity flow moves the lighter material down the sides of the ridge. New mantle material rises to fill the gap and it in turn reacts with the fluids. The swelling of the ridge sets up shear stresses which result in the transform faults that appear to offset the mid-ocean ridges (Fig. 18-16). Once started, such a process could continue without large convection cells.

Others have postulated that as the hydrated oceanic crust dips beneath the continental edge into the mantle, the temperature and pressure increase until it loses its water and other lightweight substances which rise to the crustal area. The denser, heavier component, well on its way to becoming mantle material once again, is heavy enough to drag down more oceanic crust in its wake. This hypothesis makes subduction the driving force in a continuing process, but still does not account for its beginning.

Global Expansion

Since we can be reasonably sure that the earth is not contracting, might it not be expanding, and might not this be the cause of rifting of the crust? It is a hypothesis that appeals to some geologists, although at the moment they are in the minority. One geologist worked for twenty years on a spherical map table with all available data trying to reassemble the present continental land masses into one supercontinent. No matter how he juggled them, there were still gaps, but he succeeded in fitting them together nicely on a globe three-fourths the diameter of the present one. In addition, his work showed that the perimeter of the Pacific Ocean is constantly getting larger. This is difficult to explain when the surrounding plates are all encroaching upon it—unless the earth is expanding.

A geophysicist approached the problem of an expanding earth from a different point—a study of paleogeographic maps which show the distribution of land and water during past geologic periods. He determined that the amount of land covered by water has steadily decreased. In other words, there is more land above sea level now than there has ever been. This, of course, is quite apart from temporary fluctuations in sea level caused by climatic changes and local warping. His findings are the opposite of what would be expected, since the amount of water has actually increased by some 4 per cent through additions from volcanoes. It would be reasonable, however, if the earth were actually getting larger.

Evidence has come from other quarters also, especially evidence that shows that our days are becoming longer. This is thought to be due to a slowing down in the earth's rotation which, in turn, could be caused by an increase in its

Fig. 18-16 A possible mechanism for the development of mid-ocean ridges. (After E. Orowan, "The Origin of the Oceanic Ridges," *Scientific American,* Nov. 1969.)

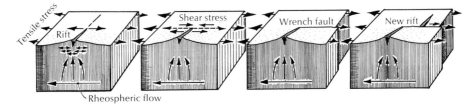

radius. On the other hand, there is some question whether the amount of enlargement is sufficient to account for the growth of the ocean basins. Furthermore, any reason for global expansion must necessarily be highly speculative in our present state of knowledge of the earth's interior.

There is no reason to adopt an either/or attitude toward the mechanisms suggested for continental drift. It is quite possible that expansion of the earth and some form of convection or transvection are proceeding simultaneously. In all probability, we will discover that a number of processes are at work in varying degrees of effectiveness. "Geologists have a new game of chess to play, using a spherical board and strange new rules" (Hurley, 1968).

Caution

Such play is enormous fun, of course, and it has a very definite purpose. The object is to come up with hypotheses that are not only speculations, however well based they may be, but with hypotheses that can be tested. The danger, of course, is that once a unifying theory such as the "new global tectonics," or "plate tectonics," is stated, then it is a great temptation to find facts that fit into it and to ignore facts that do not. As Ashley Montague says, "it is notorious that theorizers often become so enamored of their theories that, like their prejudices, they begin to mistake them for the Laws of Nature."

It is probably reasonable to look at the new global tectonic theory as a framework, some of whose supports are quite sturdy, but others of which are definitely shaky. The shoring up and the filling in of the framework will take many years, and its appearance will no doubt change, sometimes drastically, from year to year. Sir Edward Bullard (1969) has remarked, "What we have is a sketch of the outlines of a history; a mass of detail needs to be filled in and many major features are quite uncertain. Nonetheless, there is a stage in the development of a theory when it is most attractive to study and easiest to explain, that is while it is still simple and successful and before too many details and difficulties have been uncovered. This is the interesting stage at which plate theory now stands."

Selected References

Bullard, Sir Edward, 1969, The origin of the oceans, Scientific American, vol. 221, no. 3, pp. 66–75.

Heirtzler, J. R., 1968, Sea-floor spreading, Scientific American, vol. 219, no. 6, pp. 60–70.

Hurley, P., 1968, The confirmation of continental drift, Scientific American, vol. 218, no. 10, p. 52.

Orowan, E., 1969, The origin of the oceanic ridges, Scientific American, vol. 221, no. 5, pp. 102–119.

Takeuchi, H., Uyeda, S., Kanemori, H., 1970, Debate about the earth, revised edition, Freeman, Cooper & Co., San Francisco.

Wilson, J. T., 1963, Continental drift, Scientific American, vol. 208, no. 4, pp. 86–100.

Fig. 19-1   Panning for gold. Boulder Creek, Alaska, 1900. (Photograph by A. H. Brooks, U.S. Geological Survey.)

19
Mineral Resources and the Future

Resources are our most important product. Without natural resources not only our civilization but even life itself would be impossible; raw energy is not enough. There are many kinds of natural resources: agricultural resources, human resources, water resources, recreational resources, to name only a few. Those resources which are properly the concern of geology are the mineral resources, which are generally considered to include water, coal, crude oil, and natural gas, even though these are not minerals in the strict sense.

One way of classifying all mineral resources is according to their use (Fig. 19-2). The metallic elements can be subdivided into those which are common, those that are rare, and those that are intermediate. The nonmetallics, subdivided according to use, include (1) those minerals which are chemically valuable either as fertilizer or as chemicals for industrial and other processes; (2) rock and mineral materials used for construction purposes; (3) resources used to supply our energy, the lifeblood of our industrial civilization, primarily the fossil fuels, coal, oil, and natural gas, but uranium as well; and (4) water. Even though water is probably our most vital natural resource, we will not consider it further here, as its geological aspects have been discussed in previous chapters.

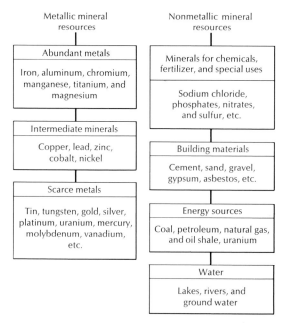

Fig. 19-2 Classification of some mineral resources. (Modified from B. J. Skinner, *Earth Resources.*)

Distribution of Mineral Resources

How different history would have been if all the minerals we use were evenly distributed throughout the world! A glance at Figure 19-3 will show that however much order and organization we may find in nature, it is certainly no respecter of national boundaries. Or perhaps that is putting it backwards—man has not drawn his boundaries with regard to mineral resources. He sometimes, however, tries to redraw them in order to favor a particular country, a practice that frequently leads to war. In peaceful times, the inequitable distribution of mineral resources leads to trade, and is, in fact, the basis of all trade. The search for scarce mineral commodities also, in part, spurred the exploration of our planet, the land areas in the past and the oceans today. This search may in very small measure inspire the exploration of space.

The role played by shortages of mineral resources and the extent to which they influence national policies is not always realized. A. M. Bateman, an economic geologist, says, "Germany, Italy, and Japan are notably lacking in most critically important minerals. They are . . . embarrassingly short of oil, iron ore, copper, tin, bauxite, iron-alloy metals, gold, and asbestos; and Italy also lacks coal. It is no coincidence that they became aggressor nations and challenged the unequal distribution of minerals." Now, under more favorable economic and political conditions, they are able to remedy their deficiencies through extensive trading activity. So do most of the nations in the world today, although there persists the doubt in the minds of many whether the United States is fighting in southeast Asia to protect democracy or to protect certain American economic considerations. Lovering (1943) states, "When ethics rather than expediency determine national policy, when both markets and sources of raw materials are equitably administered for all, when our natural resources come to be regarded as property in which all people of the world own a joint interest, regardless of their location, it may become possible to effect a mental disarmament based on mutual confidence, friendship, and tolerance which alone can prevent future wars."

Geologic Occurrence

Metals and Some Nonmetals

Minute amounts of many valuable minerals are distributed throughout common rocks. An ore deposit, however, is a concentration of one or more of these minerals in a relatively restricted area. Furthermore, such a concentration is

Fig. 19-3   Some metals: their ore minerals and occurrence.

1. Iron	Hematite Fe_2O_3 Magnetite Fe_3O_4	U.S., Brazil, U.S.S.R.
2. Aluminum	Bauxite $Al_2O_3 \cdot nH_2O$	the Guianas, U.S., France
3. Chromium	Chromite $Fe(Cr,Fe)_2O_4$	U.S.S.R., Rhodesia, Turkey
4. Manganese	Pyrolusite MnO_2	U.S.S.R., India, South Africa
5. Titanium	Ilmenite $FeTiO_3$	India, Canada, Norway
6. Magnesium	Magnesite $MgCO_3$	Austria, Manchuria, Greece
7. Copper	Chalcocite Cu_2S Bornite Cu_5FeS_4 Enargite Cu_3AsA_4	U.S., Chile, Peru
8. Lead	Galena PbS	U.S., Germany, England
9. Zinc	Sphalerite ZnS	U.S., Spain
10. Cobalt	Cobaltite $CoAsS$	Norway, Canada
11. Nickel	Pentlandite $(Fe,Ni)_9S_8$	Canada, U.S.S.R.
12. Tin	Cassiterite SnO_2	Indonesia, Bolivia, Congo
13. Tungsten	Wolframite $(Fe,Mn)WO_4$	China, Malay Peninsula, Burma
14. Gold	native gold Au	South Africa, Siberia, Canada
15. Silver	Argentite Ag_2S	Norway, Germany, Mexico
16. Platinum	native platinum Pt	U.S.S.R., South Africa, Canada
17. Mercury	Cinnabar HgS	Spain, Italy
18. Molybdenum	Molybdenite MoS_2	U.S.
19. Vanadium	Carnotite $K_2(VO_2)_2(VO_4)_2 \cdot 3H_2O$	U.S.

Fig. 19-4 The aluminum ore bauxite. (Photograph by William Estavillo.)

not technically an ore deposit unless the ore can be extracted with profit. Obviously, many operations shift from being an ore deposit to being a marginal deposit, and back again, when commodity prices fluctuate.

It is clearly of considerable economic interest to discover how such concentrations come about. Four geologic processes are responsible for the formation of ore deposits. Sometimes more than one process takes place at any given ore deposit, operating simultaneously or at different times. In addition, the same ore minerals may be formed by different processes in different areas. Thus we can see that the concentration of ores into economic deposits is as complex as most other geologic events.

The four most important processes are (1) igneous activity, (2) sedimentation; (3) weathering; and (4) metamorphism. While volcanoes are the source of numerous gases besides water vapor, these gases usually are not produced in economic quantity. Most igneous activity that produces ore deposits is plutonic. As a body of magma crystallizes, the metallic elements may be segregated at any time. Some separate out early and, if they are heavier, sink to the bottom of the magma chamber and are concentrated in this way. Others are late to crystallize and are thus separated from the already solid components of the magma. These occasionally solidify in place, or they may be injected into the surrounding country rock, as intrusives called *pegmatites*. It is especially easy for these residual liquids to migrate, since they are commonly enriched in gases and water which makes them highly mobile. The very latest substances often contain much water, either from the magma or, as they approach the surface, from mingling with ground water. These hot, aqueous solutions fill any cracks and cavities that they can find, and solidify to form *veins*. They may react with or replace some of the surrounding rocks. Not all of these fluids are rich in valuable elements, but many are.

At times a plutonic rock may have more than the usual amount of valuable mineral substances, but they may be so dispersed as to be uneconomic. In such cases weathering may be the concentrating agent. Weathering forms ore deposits in several ways. It may remove by solution or other means all the uneconomic components, leaving the valuable ores concentrated in place. Mechanical weathering can also break down a rock into particles which can then be transported. During such transportation, the heavier substances will tend to be sorted out from the lighter ones and so may be deposited in one place. Parts of an original rock may be dissolved and carried to ocean basins, inland seas, or lakes, where they are concentrated by evaporation when the body of water dries up. Chemical weathering in place may change an element from an uneconomic form to a valuable one, and concentrate it by leaching it from an upper level and reprecipitating it at a lower one.

Metamorphism, too, can change the form of mineral compounds by the effects of heat and pressure and sometimes with the addition of small quantities of water.

Building Materials

Building stone, crushed rock, sand, and gravel can consist of almost any type of rock, so that the geologic occurrence of these materials is extremely varied. Since they are heavy, transportation costs usually result in the development of quarries as near to the site of use as possible.

The manufacture of portland cement rests heavily on the use of limestone of the proper composition. Thus the location of cement plants is usually governed by the occurrence of such a limestone, and the other necessary materials, of which less is required, are shipped in from elsewhere.

Clays which form the basis of many widely used ceramic materials are products of weathering, and so their occurrence is almost entirely confined to surface portions of the earth's crust.

Coal

As we saw in Chapter 5, coal is a sedimentary rock composed of partially decomposed remains of land plants. As the plants lose their hydrogen, they progress from peat to lignite (brown coal) to bituminous (soft coal) to anthracite (hard coal) and finally, after much metamorphism, to graphite (pure carbon). As the coal proceeds along this path, up to and including anthracite, the amount of heat that is produced on burning increases. Coal beds generally occur in ancient basins containing fresh-water sediments. A large quantity of our coal supply was formed in the extensive swamps of the Pennsylvanian and Mississippian periods. In fact, these two periods together are sometimes referred to as the Carboniferous for just this reason.

Petroleum: Crude Oil and Natural Gas

Every conceivable method was used by the oil pioneers, except science, in their search for oil in the hills and dales of Pennsylvania. According to P. H. Price, who wrote a history of oil prospecting, "It was a popular saying among early-day oil men that 'geology never filled a tank' and one prominent producer

Fig. 19-5   The vessel *Atlas*, shown at San Francisco, California, in the year 1907, carried packaged petroleum products to other Pacific ports. (Courtesy of Standard Oil Company of California.)

remarked that if he wanted to make sure of a dry hole he would employ a geologist to make the location."

In spite of such slurs, the geologists of that time were not disheartened, and, although ignored, they persisted in their theorizing. Oddly, because most of us look on it as a largely United States enterprise, the first explicit statement of close relationship between the occurrence of oil and the structure of the enclosing rocks came from Canada. The key relationship between oil and anticlines was rediscovered in the United States by two geologists, one of whom staked his repuation in 1889 on the drilling of the Mannington field in West Virginia, at least 40 kilometers (25 miles) from the nearest producing well. His theory was abundantly vindicated, but he had to battle for at least eight years more to convince the skeptics, both professional geologists and practical oil-seekers, that geology had some relevance to the occurrence of this fugitive substance.

One of the first geologists actually hired by an oil company to look for oil was employed by the Union Oil Company of California in 1899. By 1900, grudging approval had generally been won throughout the industry, and with increasing success came increasing acceptance until, today, the percentage of wildcat wells located with geological advice has risen to 83 per cent. That not all of them are crowned with success, however, is brought out in the chart (Fig. 19-7).

Fig. 19-6 Modern tankers carry petroleum in enormous quantities over the seven seas. (Courtesy of Standard Oil Company of California.)

A question that may very well be asked here is just what has the geologist learned about oil in the sixty years he has been seeking it, during which time his employers have grown to become America's second largest industry. The answer is, a great deal, but there is still much more to be learned—possibly more than all the knowledge acquired up to now.

Oil clearly is related to sedimentary rocks, and its accumulation appears to be part of the normal sedimentary process, rather than some freakish event. Most oil seemingly has accumulated in sedimentary rocks deposited on the sea floor, and these apparently were laid down in shallow to moderate depths, rather than in the abyss. Although an organic versus an inorganic origin for petroleum was once vigorously debated, the fires of this controversy are quenched, and there seems to be little doubt that most petroleum started out as an organic accumulation within sediments on the sea floor.

Beyond this simple statement there is little agreement and much uncertainty.

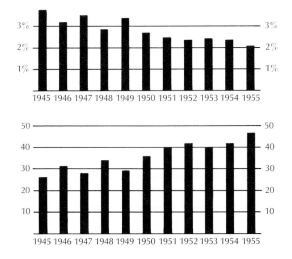

Fig. 19-7 Chart A shows, for the 17 principal oil-producing states, the percentage of all the exploratory wells that discovered an oil field with more than one million barrels in reserve. If the figure one million barrels seems large, remember that the consumption of petroleum in the United States is more than 7 million barrels per day. Chart B shows, for the same 17 states, the number of exploratory wells drilled for each well that discovered a new oil field with more than 1 million barrels in reserve. (After J. B. Carsey and M. S. Roberts, *Bull. Amer. Assoc. Petrol. Geol.*, Vol. 46, 1962. By permission.)

Fig. 19-8 Modern portable drilling rig on the California coast near Ventura. The well is curved or slanted beneath the ground surface, so that the bottom of the well is more than a mile away under the ocean floor. (Courtesy of Standard Oil Company of California.)

One of the essential requirements is a large and continuing supply of marine life—close to the surface of the sea. Then there should be rapid accumulation of their dead remains on the sea floor, followed by burial that is quick enough so that decay is inhibited and the natural distillation of the organic litter can commence. An additional factor that seems helpful is the accumulation of organic debris in a nearly enclosed basin where circulation of water is at a minimum and oxidation is retarded. Examples of such basins today are the Persian Gulf, the Red and Black Seas, and the bottoms of some of the Norwegian fjords.

Nearly all sedimentary rocks, with the exception of some unusual types, such as red beds, contain significant amounts of organic matter. Oddly enough, the amount present seemingly bears little or no relationship to the color. Formerly it was believed that black or dark-colored rocks are the only ones rich in oil, but this need not be true. At most the organic content of typical marine sediments is low, seldom does it exceed 5 or 10 per cent. How this organic matter is converted into the liquid hydrocarbon, petroleum, is still unknown, but the evidence is strong that burial is required, and perhaps not too great a lapse of time, although some geologists are convinced that at least a million years are necessary.

More than source sediments are required to make a successful oil field, because as a rule their organic content it too slight and, since many of them are shales, their permeability is too low to permit oil to flow freely and relatively rapidly—an essential requirement for a producing oil well. What is needed next is a reservoir rock. Nearly 60 per cent of the world's petroleum reserves are in sandstone, the remaining 40 per cent are in limestone, dolomite, etc., and perhaps 1 per cent are contained in other rocks which are sufficiently fractured to permit oil to accumulate. This means that some time after its formation in source rocks, oil has migrated slowly to be concentrated in more permeable strata, such as sandstone and limestone.

Oil might well escape from its reservoir unless there is some sort of lid, and this is the *cap rock*. This may be nothing more elaborate than a stratum of fine-grained shale overlying the more permeable reservoir rock. All it need be is a layer of such low permeability that it acts as a diaphragm, preventing the upward escape of oil.

The third requirement is some sort of *trap*. This usually means some kind of geological structure that (1) retards the free migration of oil and (2) concentrates the oil in a limited space. The most common of such traps is the anticline, and as pointed out earlier, this is the structural control that was deduced in 1885 by I. C. White. The basic factors needed to make such a trap operate are shown by the cross section of the Kettleman Hills, California (Fig. 19-9). The reservoir rock which crops out in Reef Ridge picks up water at the surface of the ground, and as this water travels through the permeable reservoir rock it does two things: (1) builds up an increasing hydrostatic pressure with depth, and (2) carries the oil along with it.

When the natural gas-oil-water mixture reaches the anticlinal crest, the three constituents have an opportunity to separate from one another on the basis of their density differences. Most of the natural gas rises to the crest of

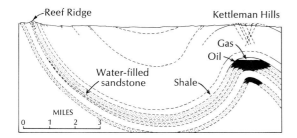

Fig. 19-9 Cross section from Reef Ridge on the west to the Kettleman Hills on the east showing the distribution of gas, oil, and water in porous sandstone. The oil is held in the anticlinal trap by the overlying cap rock of shale.

Fig. 19-10   The famous Lakeview No. 1 gusher, Kern County, California. The figures in the foreground are reflected in a pool of oil. (Courtesy of Union Oil Company of California.)

the fold, and since oil is lighter than water it rises to the next higher level in the structure. Actually these three constituents do not separate quite so tidily; a considerable amount of water may be mixed with the oil as an emulsion, just as a good deal of gas will be dissolved in the oil. This gas is of the greatest importance in the productive life of an oil field. Gas pressure can be utilized to drive oil to the surface in the early life of a well drilled near the top of a structure. Eventually, the pressure declines, and the well is no longer free flowing, but has to be pumped. As the oil, in the main part of the anticlinal structure is withdrawn, water at the edge of the fold and at the oil-water interface moves into the space once occupied by the oil and then this part of the field is through.

Few more spectacular examples could be cited of the efficacy of expanding gas in driving oil to the surface than the Lakeview gusher in California (Fig. 19-10). This potential one-well oil field was taken over by the Union Oil Company after the four original drillers gave up at a depth of 550 meters (1800 feet) following a succession of setbacks, financial and technical. When Union took over the drilling, no one could possibly have forecast what was to come. On March 14, 1910, Lakeview No. 1 blew in totally out of control with a column of oil and gas that rose hundreds of meters into the air to tower above the floor of the San Joaquin Valley. The whole derrick collapsed into the crater that had been blasted out of the ground. The flow in the first twenty-four hours was estimated at 125,000 barrels and for months afterward was around 50,000 barrels per day. No available storage capacity could hope to hold the tide—at least 600 men, with scores of them hauled out of bars and hobo jungles all over California, were sent to dig ditches and build reservoirs to stem the flood. By building earth dams across the nearby creeks, a series of immense oil lakes was created which impounded at least 9,000,000 barrels of oil before the gusher subsided after eighteen months of uncontrolled flow. The economic effect of such an inundation was to flood the limited market of that day and to drive the price of oil down to thirty cents a barrel.

The world is not likely to see such a spectacle again. Today, with half a century of experience and a whole aresenal of technology, gushers are virtually extinct. We can only lament the passing of a more colorful age.

Another gusher that made history blew in on January 10, 1901, when Captain A. F. Lucas, whose determination was only equaled by his ignroance of geology, drilled a well on an unobstrusive little hill at Spindletop, near Beaumont, Texas. It blew in as an uncontrolled gas-propelled fury in the same way as Lakeview did nearly a decade later, and as a consequence the Texas Gulf Coast was swept up in a speculative frenzy that was the equal of the Pennsylvania scramble of a generation before. Without knowing how on earth it had happened, the dazed captain had discovered a second major type of structural trap—the *salt dome*. Oil can be trapped against the impervious salt, especially where strata are bent up around its margin, as at Avery Island, Louisiana (Fig. 19-11). Scores of similar structures are found along the coastal margin of Texas and Louisiana, as well as offshore in the Gulf of Mexico. Their discovery has required the expenditure of millions of dollars and the utilization of the full resources of geology and geophysics, since a surface indication of these inscrutable structures is a rarity.

The East Texas field, the largest single oil field in the Western Hemisphere and one of the most prolific in this country's history, with a total production of around 3,000,000,000 barrels, could scarcely have selected a worse time for its debut. In 1931, the nadir of the depression, it loosed a flood of oil upon an unwilling market, resulting in a drop in price to ten cents a barrel.

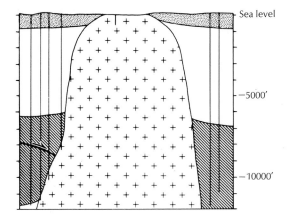

Fig. 19-11 A cross section of the Avery Island salt dome, in south-central Louisiana. The vertical lines indicate drill holes. Salt (crosses) is shown piercing Cenozoic sedimentary rocks. (After P. B. King, *The Evolution of North America*, copyright © 1959, Princeton University Press. By permission.)

This sea of oil is concentrated, not along an anticlinal axis, but in a stellar example of a third type of accumulation—the so-called *stratigraphic trap*. The oil is trapped under an unconformity where the gently dipping reservoir rock wedges out beneath the overlying impervious layer; actually, the field is at the intersection of two unconformity planes.

An unusual type of trap because of the exotic imagery it provides is a buried, or fossil, *organic reef*, such as a former coral reef. These are analogous in their form to that of living reefs, except, of course, for their burial under later sediments. Perhaps the best known of such structural traps are the oil fields of Canada, near Leduc, Alberta, and those in north central Texas.

The nature of such a structure is shown in cross section in Figure 19-12, which is that of the so-called Golden Lane of Mexico. The main thing to notice is that sediments, which once were muds, terminate abruptly against limestone, which once was the main body of the reef. Limestone, if cavernous, holds oil, and once this narrow, buried limestone ridge near Tampico, nearly 80 kilometers (50 miles) ling but mostly less than a mile wide, held more than a billion barrels. The all-time wonder well of oil history, Potrero del Llano No. 4, came in as a gusher with a daily production of 260,000 barrels, and a total output of around 60 million barrels, until one awful night when it failed completely and went to salt water.

There is a wide variety of other structural traps, which make interesting challenges for the geologist but quantitatively are not very important. Among these are accumulations of petroleum along faults. These may act as an impervious membrane against which oil collects when its flow is halted along a reservoir bed. The relative economic importance of a variety of structural traps described in this chapter are shown in Figure 19-13.

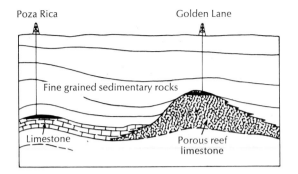

Poza Rica   Golden Lane

Fine grained sedimentary rocks

Limestone

Porous reef limestone

Fig. 19-12 Cross section of the Golden Lane, Mexico. (Courtesy of Petroleos Mexicanos.) After D. W. Rockwell and A. G. Rojas, *Bull. Amer. Assoc. Petrol. Geol.*, Vol. 37, 1953. By permission.

Anticline

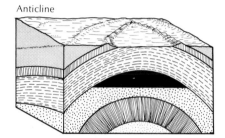

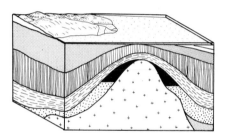

Fault trap

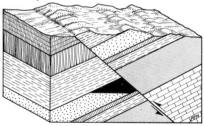

Stratigraphic trap

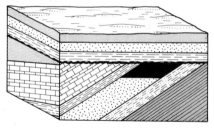

Organic reef

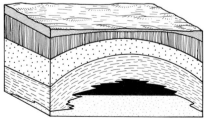

Fig. 19-13 Types of oil traps shown in their order of economic importance. Anticlines and salt domes account for about 58 per cent of the oil fields and 80 per cent of the world's total oil production. Faults, organic reefs, stratigraphic combination, and other traps are far less important.

Fig. 19-14   Saudi Arab drilling crew at work in Arabia. (Courtesy of Aramco.)

Conservation of Mineral Resources

Many of the earth's resources are *renewable;* i.e. there is more where that came from—indefinitely. Agricultural crops are an example; they can be grown and harvested again and again, their only limitations being lack of space, water, and fertilizers. Trees, too, are a renewable resource, but time is here an additional restriction, for it takes years for a seedling tree to reach maturity.

In contrast, most mineral resources are *nonrenewable;* there is only a definite quantity of each one present in the earth, and there can never be any more.

Fig. 19-15 Kennecot Copper pit, Utah. (Courtesy of Salt Lake Area Chamber of Commerce.)

Fig. 19-16 Water and mud spout from a "shot hole" as an explosive charge is set off to create an artificial earthquake in seismic exploration for oil. (Photograph by Anthony E. L. Morris.)

Some resources are more nonrenewable than others, however. When an aluminum can is discarded, it can be reprocessed and the aluminum used again. Resources of this type are nonrenewable, but they are *reusable*. On the other hand, when petroleum products are burned they are gone forever.

Mineral Fuels

Crude oil, natural gas, and uranium are the three major mineral sources of the energy that is required in tremendous amounts to maintain our western type of civilization. A recent report on energy resources published by the National Academy of Sciences-National Research Council, estimates that the United States will have reached its peak production of crude oil near the end of the 1960's and its peak production of natural gas ten years later. After those dates the production of each will start to decline. World production of petroleum will start to decline in about thirty years, and within sixty-five years 90 per cent of the earth's crude oil and natural gas will have been used. Clearly it is time to look carefully at other energy sources.

The ultimate source of all energy is, of course, the sun, which will provide a limitless supply for millions of years to come. The direct use of solar radiation for producing electrical energy is fraught with difficulty. King Hubbert (1969) has calculated that a square area of land 6.5 kilometers (3.7 miles) on a side would be required to gather enough solar energy to power a plant of a size comparable to our present power plants. He goes on to say, "There is no question that it *is* physically possible to cover such an area with energy-collecting devices, and to transmit, store, and ultimately transform the energy so collected into conventional electric power. However, the complexity of such a process, and its cost in terms of the metals and physical, chemical, and electrical equipment required . . . renders such an undertaking to be of questionable practicality." He concluded that man's only use of solar energy will be extremely small-scale.

Fig. 19-17   Exploration for oil has moved off the land and out to sea. The picture shows a movable drilling barge at work in the Gulf of Mexico. A helicopter, on the landing platform at the left, transports the drilling crew back and forth. A permanent type of drilling platform is shown in the upper right section of the picture. (Courtesy of Humble Oil and Refining Company.)

Fig. 19-18 Hydraulic mining for gold in gravels at Cherokee Flat, California. (Photograph by J. S. Diller, U.S. Geological Survey.)

Some of our present power comes from hydroelectric power plants. If all the hydroelectric potential in the world were to be developed, it might be just possible to maintain our present level of industrialization without any other energy source. This would allow for no increase in our energy requirements and in any case would not last for very long. Eventually the reservoirs behind the dams become filled with sediment brought to them by rivers and streams. In addition, it would mean sacrificing much of the earth's most beautiful and spectacular scenery.

Utilizing the power of the tides is an appealing thought, and is, in fact, in operation at la Rance estuary in France. The Russians are reported to have a small experimental plant on the Barents Sea. The number of places where such plants are feasible is, of course, very limited, and even if all of them were to be developed, they would account for less than 1 per cent of the world's power needs.

The same type of limitations will restrict the use of geothermal energy (heat from the depths of the earth, volcanic steam, hot springs, and steam wells). Many small power plants are possible, but in very few regions, and their total power potential is 'a very small percentage of what is and will be required.

Everyone's great hope is, of course, nuclear energy. This supposedly endless resource is not without its difficulties, too. As an aid to understanding them, let us classify the ways in which nuclear energy may be produced.

FISSION
   Burner Reactors
      Fuel: uranium-235
   Breeder Reactors
      Fuel: uranium-238 and thorium-232
FUSION
      Fuel: hydrogen (deuterium and tritium)

Fig. 19-19   The results of hydraulic mining. Sierra County, California. (Photograph by H. W. Turner, U.S. Geological Survey.)

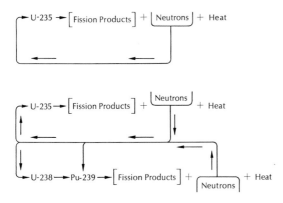

Fig. 19-20 Schematic diagram of the operation of a burner reactor for uranium-235.

Fig. 19-21 Schematic representation of breeder reaction for uranium-238.

In a burner reactor, the uranium isotope uranium-235 is converted into energy as shown in Figure 19-20. The amount of heat produced is great; one gram of uranium-235 produces as much heat as is obtained from burning 2.7 metric tons of coal, or 13.7 barrels of crude oil. The difficulty lies in the fact that uranium-235 is a very rare isotope, comprising only $\frac{1}{140}$th of the total amount of uranium.

There is, however, a way out of this difficulty. Uranium-238 and thorium-232 are both much more available. They are themselves not fissionable as uranium-235 is, but they can be converted by neutron bombardment into fissionable isotopes: uranium-238 becomes plutonium-239, while thorium-232 becomes uranium-233. If conversion is accomplished in a special way, the rare isotope, uranium-235, is needed only as a starter material. After the reaction once gets going, more fissionable material is produced during the process than is used up. This is why only a limited amount of uranium-235 is necessary at the beginning and why such a plant is called a *breeder reactor*. Its operation is diagrammed in Figure 19-21.

All the nuclear reactors now in existence are of the burner type, or have such low conversion ratios that they are essentially burners. The demand for more nuclear power is increasing rapidly, and all the reactors now being planned are also burners. Meanwhile the supplies of uranium-235 are quickly being exhausted. Needless to say, the Atomic Energy Commission has speeded up the development of breeder reactors, and the first large sodium-cooled fast breeder has been constructed in Michigan—the Enrico Fermi Atomic Power Plant. This is still an experimental plant, and the AEC. is trying to achieve commercial plants before the end of the century. The persistent fear is that by that time there will not be enough uranium-235 to get them started; that it will all have been consumed in the burners. "The energy potentially obtainable by breeder reactors from rocks occurring at minable depths in the United States and containing 50 grams or more of uranium and thorium combined per metric ton is hundreds or thousands of times larger than that of all of the fossil fuels combined. It is clear, therefore, that by the transition to a complete breeder-reactor program before the initial supply of uranium-235 is exhausted, very much larger supplies of energy can be made available than now exist. Failure to make this transition would constitute one of the major disastors in human history" (Hubbert, 1969).

Obtaining energy from controlled nuclear fusion (as opposed to fission) is appealing for many reasons. Imagine a source of power that (1) uses water as a fuel and is therefore cheap and abundant; (2) produces no radioactive waste, but instead two saleable commodities—helium and tritium; (3) cannot explode; (4) if there is a breakdown in the cooling system, will not overheat and vaporize radioactive materials; (5) has only small quantities of radioactive materials at the reactor site, and those relatively harmless to human beings; and (6) can work with 60 per cent to 90 per cent efficiency with very little waste heat.

This ideal power source operates by fusing one or more of the isotopes of hydrogen (hydrogen-1, deuterium-2, tritium-3) to make helium-3 or helium-4. Fusion reactors will not be a reality for some time. Some scientists do not foresee their possibility ever. Others firmly believe that it can be done, but have grave doubts that power can be produced economically in this way. All agree that it will be at least 1980 before we can know with any certainty if it will be possible.

Other Minerals

Even assuming an inexhaustible, cheap power source, man cannot live on energy alone. His civilizations, especially the highly developed, western-type ones, depend for their lives on a continuing supply of mineral resources. It is not sufficient to find enough of these mineral resources to maintain our present level of use; the demand for them will increase and at an increasing rate. "The entire metal production of the globe before the start of World War II was about equal to what has been consumed since" (Lovering, 1969).

Even if the goal of zero population growth is finally achieved, it will not happen next year or even before the end of the century. Meanwhile it is inevitable that there will be more and more human beings demanding their share of the earth's resources. Another increase in the demand will arise from advances in technology; new users are being found for mineral products all the time. A third factor will be the insistence of many of the underdeveloped nations on achieving a western-type, industrialized, mineral-consuming society as rapidly as possible.

As we saw from our classification chart, some industrially important minerals are abundant, iron and aluminum, for example. But even these are nonrenewable; there is still only a limited supply available in the world. Others are extremely scarce. Whereas aluminum constitutes about 8 per cent of most igneous rocks and shales, mercury is at most only .04 per cent of all the rocks in the crust. Along with mercury in this category are tungsten, tantalum, silver, tin, and molybdenum. Still other metals are intermediate between the abundant and the extremely scarce; these include copper, cobalt, nickel, vanadium, zinc, and lead.

The fact is that we are rapidly running out of a supply of our scarcer metals, while the intermediate ones will not last much longer. Let us take mercury as an example. High-grade mercury mines have mostly been depleted, and very little mercury is produced as a by-product from ore-processing for other metals. Yet in the United States the average use of mercury has increased 3 per cent per year for the last twenty years. At the same time the price of mercury has gone up 500 per cent, because it costs more to process low-grade ore. If the price of mercury is $200 per flask, the total reserves of the United States has been estimated at 140,000 flasks. Now, if the price is raised to $1000 per flask, the reserves jump to 1,287,000 flasks. *But*, at our present rate of increase in consumption, 3 per cent, those reserves will last only fifteen years. The world picture is equally unreassuring. Similar figures could be quoted for all the scarce metals, and reserves for the intermediate ones are not much less disturbing.

It is imperative, then, that globally oriented thinking be brought to bear upon the problem of the wisest uses of the materials of this planet. The Committee on Resources and Man of the National Academy of Sciences and the National Research Council has recommended as a first step "that there be a large increase in the effort directed toward a comprehensive geochemical census of the crustal rocks of the nation, the continent, and the earth, including those parts beneath the sea." We cannot determine how best to use what we have, until we know how much there is and where it is. At the same time, we will need a watchdog group to revise and update constantly the mineral reserves of the country and to point out imminent shortages and make recommendations for getting around them before the situation reaches crisis proportions. One might also add that there should be a similar group charged with the same responsibilities for the earth as a whole.

A realistic view of the situation will show that advances in technology will result in our making the most of our mineral resources. We must improve our methods of exploration. Most of the easily found and easily worked deposits of the world have already been discovered. New and improved techniques are required to find the remaining deposits, most of which are well hidden deep within the crust. We must also have new methods of extracting the ore when it is found so that all of it can be used. It will no longer be acceptable to leave pillars of high-grade ore, for example, to hold up the ceilings of underground workings. We already know quite a bit about recharging oil wells with gas or water to flush out the last stubborn drops of crude oil after the natural gas pressure of the well is gone. In addition, we must find more effective ways of separating minerals from the rocks in which they occur, and economical ways of treating low-grade ores so that they may become useful.

At the same time that improved technology is helping us to exploit the resources of the earth, a constant guard is necessary to insure that crimes against the environment are not permitted, as they have been in the past. Scientific research must be put to work here also for the benefit of man, in preventing pollution of the air and the waters of the earth, and assuring that extraction operations damage the landscape as little as possible. In this connection it would be well to study the actions of the American Metal Climax Company (AMAX) in setting up a new molybdenum mine in the Rocky Mountains of Colorado. AMAX engineers have worked closely with conservationists to see that as little damage to the surrounding area as possible will be done. For example, 300 million tons of finely ground rock tailings must be put somewhere. The company has agreed to move the tailings pond site from beside a highway, which was the original plan, to a spot 21 kilometers (13 miles) away, out of public view, even though this means putting a nine-mile tunnel beneath the continental divide. They also will recycle the water from the ponds for factory use, and will build canals so that water run-off will by-pass the ponds and not pollute the valley below. The original landscape will be preserved as much as possible and reclaimed when the mine is worked out. In addition, much of the land will be opened for public use. Unfortunately such co-operation between industry and conservationists is all too rare.

It will be important to find substitutes for minerals which are in short supply. While substitutes may not exist for all uses of a particular mineral, some will be possible, leaving more of the scarce mineral to be used where it is irreplaceable.

The recycling of many minerals will extend the supply of those minerals for years. Products should be designed to last far longer than they do at present, and their design should also make easy the separation of their mineral components as an aid to recycling. This process will help to relieve some of the waste-disposal problems as well. A past director of the United States Bureau of Mines has written, "Even as we bury metal in one place, we are looking elsewhere for ores that may well be leaner than our sanitary landfills. . . . The 34 million metric tons of municipal refuse incinerated annually in this country contain more than 2.8 million metric tons of iron and some 180,000 metric tons of aluminum, zinc, copper, lead, and tin." Mine dumps are another source of reusable mineral resources. Mining operators should separate worthless country rock from piles of low-grade ore rock which may someday become profitable and necessary.

The Energy Myth

It has been suggested that we really do not need to worry about shortages of minerals vital to our economy because cheap, abundant energy will solve all our problems. With cheap, abundant energy, it is reasoned, we can afford to mine ordinary rocks and extract their widely disseminated metals and other minerals. Every common rock would become an ore deposit.

Metals exist in ordinary granite in a ratio of about 1 in 2000, which means that for every pound of metal extracted, a ton of crushed rock must be put somewhere. Put it back where it came from? Rock when crushed increases some 20 per cent to 40 per cent in volume, so you could fill all your holes and still have problem quantities left on your hands.

This same difficulty, unfortunately, is also encountered in getting uranium and thorium for breeder reactors. It is quite true, for example, that the Devonian Chatanooga Shale, a low-grade source of uranium, underlies a great part of Tennessee, Kentucky, Ohio, Indiana, and Illinois. One 4.5-meter (15-foot) thick member of this formation would provide per square meter of surface area the energy of 10,000 barrels of crude oil. The Conway Granite of New Hampshire has been cited as a good source of thorium. It is exposed over an area of 750 square kilometers (300 square miles), and if mined to a depth of 100 meters (330 feet) would provide the fuel equivalent of 150×10^{12} barrels of crude oil, 750 times the present reserves. But did anybody ever stop to imagine what large areas of Tennessee, Kentucky, Ohio, Indiana, and Illinois would look like during this mining, even if the land could be reclaimed afterward? Or what appearance New Hampshire would present after 750 square kilometers had been quarried to a depth of 100 meters?

Clearly, even the breeder reactors are only a stopgap measure and the research and development necessary for nuclear fusion energy must proceed. Even so, we are still faced with the shortage of many metals. Lovering (1969) puts it this way:

> To ensure both the ecologically nondestructive procurement of mineral supplies and a more equitable distribution of their beneficial results must surely be among the foremost objectives of a successful economic system. During the next century adequate supplies and equitable distribution will not be achieved merely by recycling scrap metal nor by processing dozens of cubic kilometers of common rock to supply the metal needs of each major industrial nation. When the time comes for living in a society dependent on scrap for high grade metal and on common rocks for commercial ore, the affluent society will be much overworked to maintain a standard of living equal to that of a century ago. Only our best efforts in all phases of resource management and population control can defer that day.

The Ocean Myth

It has also been suggested that anxiety about mineral resources is uncalled-for because we have not even begun to tap the riches of the oceans. There are three possible sources of minerals in the sea: (1) sea water; (2) the continental margins; and (3) the deep-ocean floor. Sea water can provide all we need of a few substances, notably magnesium, bromine, table salt, potassium, iodine, and maybe strontium and boron.

The difficulty of getting others from sea water is best explained by Preston Cloud (1969). "Take zinc, high among the metals sought by industry. A modest operation aimed at grossing $120,000 per year at 1968 values before costs for salaries, operations, and investment would require the complete stripping of zinc from 9000 billion gallons (nearly nine cubic miles) of ordinary sea water annually—a volume equivalent to the combined average annual flows of the Hudson and Delaware rivers. The production from this operation, however, would be only about 400 tons of zinc; a trickle compared with the 122,400 tons used industrially in the United States in the same year. The practicality of such an operation is not impressive."

The continental shelves, slopes, and rises, on the other hand, may be expected to have mineral resources approximately equal to those of a comparable area of land above sea level. These will be notably crude oil and natural gas, truly nonrenewable and nonreusable resources. In addition, the difficulty and expense of exploration and extraction are many times greater for underwater deposits. While such activities should continue, if they can be done without damage to the environment, the total area of continental margins is quite small, and their resources, while considerable, will not long rescue us from shortages on land.

Geologically, the deep-ocean floor is not promising. Added to the tremendous difficulties encountered in prospecting these areas and in mining in great depths of water is the fact that all the ocean basins appear to be floored with basalt. Basalt does not contain a great variety of minerals to begin with; it does not generally favor the initial concentration of minerals, and the ocean-floor basalts have not been exposed to the enriching processes that can be accomplished by weathering.

Manganese nodules on the ocean floor have been highly touted as an example of the wealth of the ocean realm. The truth of the matter is that (1) the nodules appear to be a thin film and do not continue at depth; (2) there is at present no technology for recovering the nodules from the sea floor; (3) there is no technology for extracting manganese or other metals from the silica-rich nodules, and, if there were, extensive land deposits would then become economic; and (4) they appear not to form as rapidly as had been hoped.

Cloud (1969) sums up the potential of the oceans as a source of mineral substances as follows: "A 'mineral cornucopia' beneath the sea thus exists only in hyperbole. What is actually won from it will be the result of persistent, imaginative research, inspired invention, bold and skillful experiment, and intelligent application and management—and resources found will come mostly from the submerged continental shelves, slopes, and rises. Whether they will be large or small is not known. It is a fair guess that they will be respectably large; but if present conceptions of earth structure and of sea floor composition and history are even approximately correct, minerals from the seabed are not likely to compare in volume or in value with those yet to be taken from the emerged lands. As for seawater itself, despite its large volume and the huge quantities of salts it contains, it can supply few of the substances considered essential to modern industry."

The Science Myth

We have seen that inspired research and improved technology will be of great help in the wise use and conservation of our natural resources. But we would be as wrong to think that "the scientists will always find a solution" as we would to count on cheap, abundant energy or the mineral potential of the oceans.

It has been said that people are our problem, and as far as quantity goes, this is frighteningly true, people *are* our greatest problem, but people are our only solution as well. Left alone the earth might heal itself of the scars mankind has inflicted, but as long as man continues to exist, people working *with* the earth, instead of trying to conquer it, are the only ones who can save both it and us.

It may well be that in order for the human race to survive in an environment worth surviving in, we will have to modify our ways of thinking and living. We will have to stop equating "progress" and "growth," beginning with the population. We will have to substitute global views of our problems for narrow nationalistic views. We will have to change our ways of living and moving about in order to use our mineral resources in a way that provides the most benefit for the most people. The present inequitable distribution of mineral wealth must be remedied. Hibbard (1968) states: "With only about 9 per cent of the Free World population in 1965, the United States consumed between 30 and 40 per cent of the Free World's mineral supply." William Sloane Coffin, Jr. puts it this way: "Exercise your imagination and reduce the planet's population of 3 billion people to a town of, let's say, one thousand. Proportionally, sixty of those people would be Americans and nine hundred and forty would be all the rest of the world's population; those sixty Americans, 6 per cent of the total, would enjoy 50 per cent of the resources of that town." He was defining "violence" and concluded that that situation qualifies. Conservation must become a personal as well as a global way of life.

Hubbert (1969) sums it up: "It now appears that the period of rapid population and industrial growth that has prevailed during the last few centuries, instead of being the normal order of things and capable of continuance into the indefinite future, is actually one of the most abnormal phases of human history. It represents only a brief transitional episode between two very much longer periods, each characterized by rates of change so slow as to be regarded essentially as a period of nongrowth. It is paradoxical that although the forthcoming period of nongrowth poses no insuperable physical or biological problems, it will entail a fundamental revision of those aspects of our current economic and social thinking which stem from the assumption that the growth rates which have characterized this temporary period can be permanent."

Selected References

Cloud, P., 1969, Mineral resources from the sea, *in* Resources and man, W. H. Freeman and Co., San Francisco.

Hibbard, W., 1968, Mineral resources: challenge or threat? Science, vol. 160, pp. 143–50.

Hubbert, M. K., 1969, Energy resources, *in* Resources and man, W. H. Freeman and Co., San Francisco.

Lovering, T. S., 1943, Minerals in world affairs, Prentice-Hall, Inc., New York.

———, 1969, Mineral resources from the land, *in* Resources and man, W. H. Freeman and Co., San Francisco.

Skinner, B. J., 1969, Earth resources, Prentice-Hall, Englewood Cliffs, N.J.

Fig. A-1. A crater 33 meters in diameter located approximately 60 meters east of the Apollo XI lunar module. Note the pavement of blocky fragments in the crater floor. (From Mutch, *Geology of the Moon,* Princeton University Press, 1971.)

Appendix **A**

Geology and Space

The Greeks had a word for it: geology—"geos" meaning "earth" and "logos" meaning "description." Therefore it may seem erroneous to discuss the "geology" of other bodies of the solar system. Fortunately Thomas A. Mutch in *Geology of the Moon* has redefined the word as "that science which deals with the history of a planet revealed in its rocks." He defends his redefinition by arguing that "geologic principles worked out from examples on the Earth can be applied equally well on the Moon. It is as if we have been allowed to practice on one planet, freely drawing and erasing patterns as our knowledge and technical skills increase. Then, for a final version, we are presented a second planet virtually unblemished by the pencil marks of previous geologists."

Mutch is using a method of reasoning often employed by geologists—analogy. If two features appear to be identical or nearly so, it is possible that they may have had similar origins. As we will see, this has been widely applied in the study of the geology of the moon. It has helped to make possible a geologic map of the moon prior to man's actual landing there.

Space Bits and Pieces on Earth

The return of samples of the lunar surface from that first landing was, of special interest to geologists. Before that time, however, scientists were not without samples of material from outer space, most notably, meteorites.

Meteorites

Chunks of solid material called *meteoroids* roam the solar system, often in wildly eccentric orbits. When these orbits intersect our own, meteoroids may enter the earth's atmosphere where the friction causes them to heat up until they are luminous. At this stage they may become visible to the naked eye and are called *meteors* or shooting stars. Most are completely burned up and never reach the earth's surface, but exceptionally large ones may retain enough substance to persist as solid matter and collide with our planet. They are then called *meteorites.*

Most meteorites are fragmented by their impact with the earth; after all, they enter our atmosphere at speeds between 12 and 72 kilometers (8 and 45 miles) per second, and, while friction slows them down, they still have velocities many times greater than that of a rifle bullet. Most fragments are so small that they are never noticed, but others have been found that weigh from a fraction of an ounce to the 50 tons of the Hoba West meteorite which was discovered in South-West Africa.

In 1964 the Smithsonian Astrophysical Observatory set up the sixteen-station Prairie network with the express object of recovering freshly fallen meteorites. Each station is equipped with an automatic camera so that the area of the fall can be fairly accurately determined. Surprisingly enough, it took six years to "capture" a fresh meteorite on film. This one fell in Oklahoma and was picked up only six days afterward. Although it weighed only 10 kilograms (22 pounds), scientists estimate that it weighed nearly a ton when it entered our atmosphere. The reason for obtaining fresh meteorites is to study those products of cosmic-ray bombardment which have short half-lives and which have essentially disappeared in meteorites that have been lying around for some time.

Meteorites have been classified according to their composition: (1) those containing mostly iron with considerable amounts of nickel, (2) those composed of stony nodules in a groundmass of iron, and (3) those which are almost entirely stony with only a little iron. When the predominantly metallic ones are cut open and the cut surface polished and etched with acid, a distinctive lattice-like pattern emerges. This pattern, called Widmanstätten figures, shows the intergrowth of nickel-poor "alpha" iron which forms at low temperatures, and nickel-rich "gamma" iron which forms at high temperatures.

Stony meteorites are made up of some of the same minerals found on earth: olivine, pyroxene, feldspar, pyrite, and graphite, to name a few of the most common. A few meteorites, among them some of the fragments from Meteor Crater in Arizona, contain diamonds. These diamonds apparently formed from graphite as an effect of the tremendous shock waves. Some diamond-producing shock waves originated from terrestrial impact, but others must have occurred before the meteoritic material entered our atmosphere, presumably from a fragmentation in space. Studies of the composition and other features of meteorites are enabling scientists to reconstruct their histories.

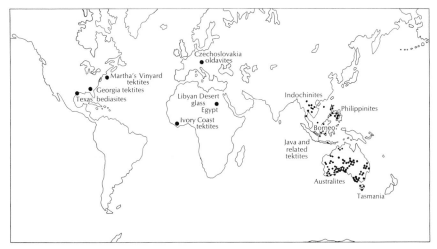

Fig. A-2   Distribution of tektite strewn fields. (After J. A. O'Keefe, "Tektites and Impact Fragments from the Moon," *Scientific American,* Feb. 1964.

Tektites

Tektites are small glassy objects, usually weighing only a few grams, which are found scattered in several areas of the earth's surface. The largest, and youngest, of these *strewn fields*, as they are called, covers much of Australia, Indonesia, Indochina, and the Philippines, and is called the Australasian strewn field. Others are on the Ivory Coast of Africa, in Czechoslovakia, in North America, and in Egypt (Fig. A-2). Their origin has been a knotty problem ever since they were first discovered in 1787.

Some scientists believe that they were formed on earth of terrestrial material, but there are major objections to this theory. For one thing, they contain spherules of an iron-nickel compound which is unknown elsewhere on earth, but which is found in meteorites. For another, the shape of many indicates that they have undergone remelting, such as would occur if they had entered our atmosphere at a velocity over 7 kilometers per second.

Their composition also indicates that they did not travel very far through space, for they do not have any aluminum-26 which is produced by cosmic-ray bombardment. This has led to the hypothesis that they may have come from the moon, possibly as a result of a great meteorite impact there. It has also been suggested that they are the result of a collision between the earth and a comet.

Wherever they come from, they continue to inspire research and speculation. Recently, very tiny ones, *microtektites*, have been discovered in ocean-bottom sediments. Furthermore, these occur in the vicinity of the Australasian tektites and are of the same age, 700,000 years. What is even more intriguing is that their age coincides with a geomagnetic reversal and with a great change in fossil plankton forms. The Ivory Coast tektites which are about one million years old also fell at the time of a geomagnetic reversal. The thought that leaps to mind, of course, is that there is a weakening of the earth's magnetic field during a reversal, and that under those conditions, objects from space have a better chance of entering our atmosphere, objects such as comets, meteoritic "trash" from the moon, and cosmic rays which might cause widespread evolutionary changes. It will be interesting to see where further investigations into tektites and microtektites will lead.

The Earth from Space

As we plunge headlong into the fascinating exploration of space and its remote objects, we may forget that looking back can be helpful also. The United States Department of the Interior and NASA. are planning to do just that—take a good look at the earth from space using a variety of remote-sensing equipment. In addition to viewing the earth with regular cameras using the visible spectrum, studies will also be made using infra-red, radar, and gamma rays. The first Earth Resource Technology Satellite (ERTS) is expected to be launched in 1972, while the Interior Department's Earth Resource Observation Satellite (EROS) will come later.

The data from these satellites are expected to be helpful to geologists in several ways. In mapping the surface of the earth, photos from space cover a great deal more territory per photo than do airplane photographs. Thus, it required only eighteen photos from the Apollo 6 spacecraft to map a corridor from the Pacific Coast south of San Diego, along the southern boundaries of Arizona, and New Mexico, across Texas, including El Paso, Abilene, Ft. Worth, and Dallas, to Shreveport, Louisiana, more than 364 million square kilometers (140,000 square miles) of territory. Because each photo covers so much area, it enables geologists to study structural relations of rocks in a large setting. Aerial photos, of course, are still needed for more detailed work. Structures favorable for mineral prospecting, structures determining ground-water occurrence and flow, and those of engineering importance can all be spotted on space photographs.

ERTS and EROS will be very important for scientists in other fields, too, in their attempts to inventory all the world's natural resources. Some idea of the importance of these satellites can be gained from the knowledge that large-scale mapping from airplane photos has been going on for some fifty years, and only 5 per cent of the earth's land surface has been mapped in that time.

Lunar Geology

It may not seem too surprising, in view of the preceding paragraphs, that the moon has been mapped in considerable detail. We have had telescopic photographs for some time, and the early lunar probes have provided us with ever more detailed photographs of the moon's surface. Maps of the topography of both the visible and the normally invisible sides of the moon, then, do exist.

What may seem slightly incomprehensible is that there are *geologic* maps of the moon and even a time scale. This was initially worked out in 1962 by the United States Geological Survey and is shown below (Fig. A-3).

How can a geologic map be made entirely from photographs, without any ground checking? And why? To answer the second question first: the attempt was made in order to provide as much information as possible about the Apollo landing sites. In addition, it will be some time (and some expense) before detailed ground mapping will be feasible.

The "how" of the geologic mapping is based on the law of superposition; if one type of deposit overlies another, it must be younger, assuming the deposits are undeformed. The first task was to see if there are deposits on the moon that could be differentiated. The area chosen was that around the crater of Copernicus, and here definite deposits were identified. First were the ejecta from the Copernican crater itself: a hummocky facies, a radial facies, and ray streaks each of which had a secondary crater at the end nearest the primary crater. It could be seen from the photographs that these ejecta facies were spread out on top of parts of other craters, Eratosthenes and Reinhold. These craters then must be older than Copernicus. Another sign of their age is the fact that their rays have disappeared, through what agency is unknown. Finally, the Eratosthenes and Reinhold craters were formed in the surface of the Mare Imbrium which in turn must be still older. Figure A-5 shows the geologic map of the Copernican region and gives a description of the stratigraphic units. Figure A-5 shows the sequence of events as suggested by this mapping. As can be seen, the mapping is done entirely on the basis of what actually be observed; i.e. it is purely descriptive, and no attempt has been made to map according to the possible origin of the various features.

Similar maps have been made for the Humorum Basin and the Orientale

Fig. A-3 The first time-scale worked out for lunar geology.

Period	Epoch	Events
Copernican		Formation of ray craters.
Eratosthenian		Formation of craters of which rays are no longer visible.
Imbrian	Archimedean	Extensive deposition of mare material of the Procellarum Group. Formation of post-Apenninian craters older than at least part of the Procellarum Group
	Apenninian	Events related to the formation of the Mare Imbrium basin.
Pre-Imbrian		Not yet formally divided.

After Eugene M. Shoemaker, "The Geology of the Moon," *Scientific American*, December 1964.

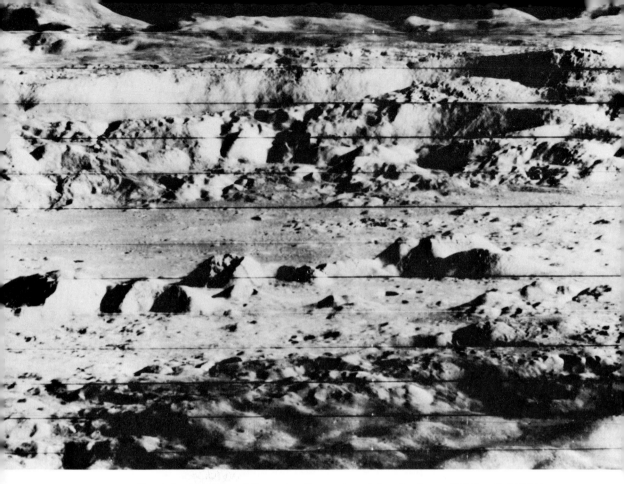

Fig. A-4   Close-up of the crater Copernicus taken by Lunar Orbiter II (NASA).

Basin, as well as the Imbrium Basin. Time relationships can be established within each basin's area, but the time relationships among the basins cannot be deciphered, unfortunately, for the contacts have been obscured by later events. The relative ages of individual craters can be estimated by the amount of erosion that has occurred. Erosion? But the moon has no atmosphere and no water. True enough, yet erosion takes place nevertheless. The main agent of lunar erosion is meteoritic bombardment. Repeated impacts fracture the consolidated rock and produce the lunar regolith which varies in thickness from place to place. The constant churning of the regolith by this means is known as "gardening" and is probably responsible for the disappearance of crater rays with time.

Lunar Surface Features

Craters are by far the most distinctive and most studied features on the face of the moon. A typical youthful crater of moderate size is roughly circular in shape with a simple bowl-shaped profile. The rim usually is raised sharply and has large blocky boulders on it. It has a blanket of ejected material surrounding it, the innermost part of which is rough and hummocky in texture. The outer part of the ejecta shows lines radiating from the crater.

Larger craters are more complex. The walls are not continuous slopes but are terraced and descend from the rim to the floor in steps. One of the most common features of these larger craters is a peak or mound in the center of the floor. Another typical characteristic is the presence of secondary crater fields whose craters presumably were produced by ejecta from the primary crater. Close to the primary crater the secondary ones are shallow and asymmetrical; the trajectories presumably were low and the velocity relatively slow. Ejecta thrown higher into the air would travel farther from the primary crater and strike the surface at a nearly perpendicular angle. These secondary craters are thus round and difficult to distinguish as being secondary.

Some craters are of an entirely different type. Their interiors appear to have been flooded with a viscous fluid material. In addition, they have very smooth rims and gently sloping, rather than steep, walls. The radial ejecta blanket and the secondary craters are absent in this type.

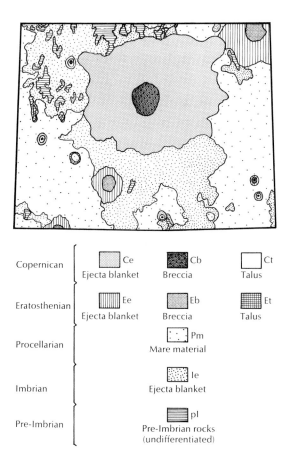

Copernican

Ce
Ejecta blanket

Cb
Breccia

Ct
Talus

Eratosthenian

Ee
Ejecta blanket

Eb
Breccia

Et
Talus

Procellarian

Pm
Mare material

Imbrian

Ie
Ejecta blanket

Pre-Imbrian

pI
Pre-Imbrian rocks
(undifferentiated)

Fig. A-5 Geologic map of the Copernicus area (E. M. Shoemaker and R. J. Hackman, "Stratigraphic Basis for a Lunar Time Scale" in Kopal and Mikhail, eds., *The Moon*, London, Academic, 1962.)

If we are to reason by analogy, we must now see what kinds of craters are formed on the earth. In spite of the more intensive erosion on our planet, we can distinguish several kinds of natural craters, in various stages of youthfulness and advancing age. Meteor Crater in Arizona immediately comes to mind as a prime example of a meteorite impact crater, but there are others, such as Flynn Creek crater in Tennessee and the Sierra Madera structure in Texas which show an even greater affinity to moon features in some respects.

Volcanic craters and caldera are, of course, another kind of naturally occurring holes-in-the-ground. Some are formed by subsidence and collapse and others by gas explosion with the formation of great quantities of ash. Manmade craters which should be considered include explosion craters produced by nuclear means or by TNT and those produced by missles and pellets.

A thorough comparison of moon and earth craters such as Mutch gives in his book, *Geology of the Moon*, shows that while some moon craters and some earth craters have some features in common, there is not enough evidence to prove the origin of lunar craters. Although this may be frustrating, we can probably say that the origin of lunar craters is not an either-or proposition, but that some were formed by impact and some by volcanism. Studies now in progress are revealing features of terrestrial craters that are not visible on photographs. These will be what lunar geologists will be looking for from future manned flights and include the sedimentary structure of the raised rims. Many impact craters on earth show overturned flaps of sediments forming their rims. Other things to look for are (1) distinctive minerals formed only by the specialized conditions of impact metamorphism, and (2) the arrangement of sediments in ejecta blankets.

Features other than craters appear to have a volcanic origin too, and these are generally associated with the maria. Among these are long narrow depressions called *rilles*. All straight and some arcuate rilles are most easily explained as the result of faulting. Branching and sinuous rilles are another matter entirely. One theory is that they were formed by streams at an early stage of the moon's history when there may have been water there. In their general pattern, however, they lack most of the characteristics of streams on earth. Another suggestion is that they were formed by lava streams, and a third that they are the result of a *nuée ardente* type of eruption where quantities of small particles are mobilized by the gases escaping from fractures or small craters.

Other possibly volcanic features include plateaus similar to the great basalt plateaus of the northwestern United States and the Deccan of India. Lobes of what appear to be lava flows indicate that the maria may be almost entirely floored with basalt. Also in this category are the Marius Hills which consist of low domes, steep domes, and clusters of domes on a smooth plateau. The low domes have been interpreted as shield volcanoes and the steep domes as plugs of more viscous material.

The highland areas of the lunar surface are much more complex than the maria regions, but they are important because highlands account for about

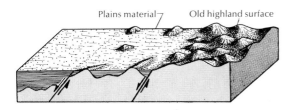

Plains material　Old highland surface

Fig. A-6 Schematic block diagram showing movement of structural blocks and superposition of materials. (Mutch, *Geology of the Moon,* Princeton, Princeton University Press, 1970.)

three-fourths of the surface; there are almost no maria on the dark side of the moon. Some relationships have been tentatively worked out, as shown in Figure A-6. Lineations have been interpreted as high-angle faults separating blocks which show variations in the amount of their relief. High structural blocks are rugged and may represent an old, intensely cratered surface. Blocks of intermediate height are not so rugged and may be the old surface partly flooded with volcanic material. The low smooth blocks may have been completely inundated by lava.

The Interior of the Moon

Just as seismology can provide much information about the interior of the earth, so scientists hope it will do the same for the moon. The Apollo 12 mission left a seismometer on the moon which recorded more than a dozen natural seismic events between the time of its emplacement and the man-made impact of Apollo 13's booster. That impact pointed up a basic difference between the earth and the moon, for it produced vibrations that lasted for over four hours. While this is a mystery still, it appears that several other bits of information can be interpreted. All the natural seismic events appear to be due to impact and not to moonquakes. This means that further seismic work on the moon will depend heavily on man-made impacts and explosions. These need not be done by manned missions, however, once the seismic recording devices are set up. It also appears that the moon materials at the surface continue without change to depths of 18 to 40 kilometers (12 to 25 miles).

Early results from magnetometer studies of the moon have been interpreted to indicate that the moon has a core. The moon has no measurable moonwide magnetic field, but lunar rocks show paleomagnetism, implying the presence of a magnetic field at some time in the past. The Apollo 12 experiment has revealed a localized magnetic field within 200 kilometers (124 miles) of the Apollo site, and such small areas may be responsible for the observed paleomagnetism.

It was an interesting coincidence that about the time the film "2001: A Space Odyssey," with its buried monolith, was released, *mascons* were discovered on the moon. Local speeding-up and slowing-down of the Lunar Orbiter spacecraft indicated areas where the moon's gravitational field was not what it should be. These strong positive anomalies indicate massive concentration

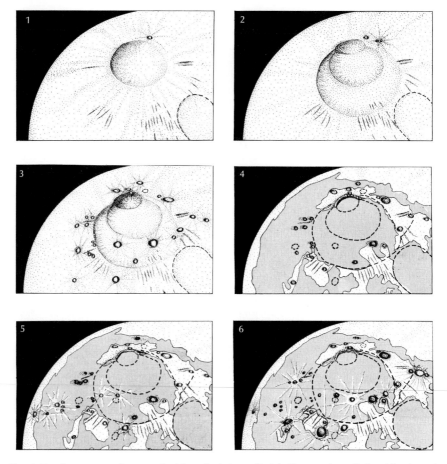

Fig. A-7   Sequence of events in Mare Imbrium (E. M. Shoemaker, *Scientific American*, 1964.)

(mascons) of denser material located under some of the circular maria. It has been suggested that mascons are embedded meteorites and also that they are thick lava beds, denser than the rocks that make up the highlands.

Apollo Data

The Apollo 11 landing site is on the floor of Mare Tranquillitatis between ejecta rays associated with either of two major craters, Alfraganus or Theophilus. At a distance of 400 meters (1300 feet) is a minor sharp-rimmed ray crater; another, a shallow, steep-walled crater, lies about 60 meters (200 feet) away. A maria site was also chosen for the Apollo 12 mission.

The Apollo 11 astronauts were scheduled to spend thirty-five minutes taking their documented rock samples. Geologists watching from earth experienced some anxiety as preceding tasks took much longer than planned until finally only three and a half minutes remained for this task. The rocks were eventually collected, however, and upon their return to earth were divided into (1) fine-grained crystalline igneous rocks showing vesicles, (2) medium-grained crystalline igneous rocks with larger cavities, and (3) breccias. The vesicles and cavities in the igneous rocks indicate a volcanic origin. The mineral composition of the volcanic rocks is generally the same as terrestrial basalt, although some have more ilmenite (an iron-titanium silicate). The absence of any liquid inclusions or hydrated minerals (those with water as part of their chemical structure) indicates that these rocks formed under anhydrous (without water) conditions and have remained dry since their crystallization. The breccias show much evidence of shock metamorphism—fracturing, shock-induced melting to glass, and distinctive shock structures within crystals. Rocks from the Apollo 12 site are more varied and include gabbro and diabase as well as basalt.

Fig. A-8   The far side of the moon photographed by Lunar Orbiter III. Maria, such as Tsiolkovsky, shown here, are much rarer on this side.

Fig. A-9 This photograph by Lunar Orbiter III, of a southern part of Oceanus Procellarum, along with other photographs, leads geologists to conclude that volcanic processes do operate on the moon.

The presence of glassy beads "naturally leads to a comparison with tektites, those glassy objects that occur in strewn fields throughout the South Pacific and are thought by some to have reached the earth following ejection from the Moon during one or more major impact events. The spheroidal, globular, and dumbbell shapes of lunar beads closely parallel these of tektites and microtektites" (Mutch, 1970). Their chemical conpositions, however, may be slightly different, although this does not necessarily prove that the tektites did not come from the moon.

The short core samples obtained from the Apollo 11 mission showed no stratification, probably because of the gradening effect at that shallow depth. The longer cores (70 centimeters) from the Apollo 12 mission do show layering, including what may be a volcanic ash bed.

One of the hoped-for results of lunar exploration is that new light will be shed on the origin and history of the solar system, especially that of the earth and the moon. To this end the radioactively determined ages are of special interest. The breccias composed of the oldest rocks near the Apollo 11 site were dated at 4.6 billion years which agrees nicely with the extrapolated age of the earth, suggesting that the two had a common, simultaneous origin. The igneous rocks, however, from the Apollo 11 mission were dated at about 3.7 billion years by both potassium/argon and rubidium/strontium methods. This difference implies that the fragmented rocks are not simply the result of the breaking up of the original crust, in which case they would all be the same age, but that new material has been added from meteoritic accretion or volcanism or both.

The igneous rocks at the Apollo 12 site were first dated at 1.7 to 2.7 billion years by the potassium/argon method, a billion years younger than the Apollo 11 rocks. This seemed appropriate since this site had fewer craters than the previous one and so appeared to be younger. The rubidium/strontium dates which followed, however, were on the order of 3.4 billion years. Some scientists have inferred that there was a 200- to 300-million-year episode of mare filling some 3.5 billion years ago, but the cause and mechanism of these events remain a mystery. Dates from other maria will add interesting information which may answer some questions and will undoubtedly raise many more.

Mars

Telescopic observations of the moon led to relatively accurate maps of its surface features, maps which have been refined but not substantially changed by satellite photographs. Mars, also, has been observed telescopically for several centuries, but each observer, it seems, draws a map differing, sometimes considerably, from all the others. Thus the pictures that have been taken of Mars by the Mariner space probes are especially interesting.

The following descriptions are based on photographs returned from Mariner 6 and Mariner 7 which were launched thirty-one days apart in early 1969. Cameras with wide-angle lenses and narrow-angle lenses were used in each

Fig. A-10   Meteor Crater near Winslow, Arizona. Does the entrenched river resemble a lunar rille? (Photograph by A. M. Bassett.)

spacecraft; the wide-angle pictures were 100 times better, as regards resolution, than ones taken from the earth, and the narrow-angle photographs were
1000 times better. Altogether 202 complete pictures were taken, some from
about 1,716,000 kilometers (a million miles) away, ranging down to 3500
kilometers (2180 miles) away.

Whereas Marine 4 in its 1965 flight had shown only a cratered surface much like the moon's, the more recent photos revealed two other types of terrain. One is a smoother plains type with almost no craters, and the other is an exceedingly jumbled and chaotic landscape.

Small Martian craters are bowl-shaped and fresh-looking, similar to many primary impact craters on the moon. Larger ones are flat-bottomed and appear to have been eroded. None have the lava flooded appearance that many lunar craters show, and there are not any other indications of volcanic activity. Limited ejecta blankets have been observed around Martian craters but no rays or secondary crater fields. These features may have been present but have subsequently been removed or obscured by erosion. In addition, there is no evidence of the kind of terrestrial tectonism which results in folds and mountains.

The Hellas desert is a circular area some 1600 kilometers (100 miles) across which is typical of the plains type of terrain, as it seems to have no craters at all. It has no counterpart on the moon. Scientists have assumed that the smooth area was once cratered and that some depositional agent caused the craters to be buried or that they have been worn down by erosion. One theory is called the pink popcorn hypothesis. It suggests that the Hellas desert was initially formed by an asteroidal impact. This event may have generated enough heat to melt the surrounding rocks and thus caused the release of gases. This sudden outgassing would leave the rocks in popcorn-sized bits which could be blown about by the wind but which would be too heavy for the wind to blow out of the impact area. The pink part of the hypothesis stems from the pink color occasionally shown by the Hellas desert.

The chaotic terrain consists of short ridges and valleys oriented every which way. It was very few craters and so is quite unlike the heavily cratered highland areas of the moon; nor does it appear to have any counterpart on the earth. It looks most like terrain that would result if the underlying material had been removed, causing a jumbled collapse of the surface material.

The arrangement and behavior of the polar ice caps suggest that they are composed of frozen carbon dioxide, a plausible interpretation since the atmosphere of Mars is composed primarily of carbon dioxide; water is quite rare.

"Canals" are those enigmatic features that so many observers on earth have mapped and which some have even seen forming an extensive network over the Martian surface. They have been cited as evidence of some form of intelligent life on the planet, so the search for them on Mariner photographs was of great interest. While some canal-like forms do appear on the far-encounter pictures, closer pictures do not show them, and it is thought that they may be simply "the chance alignment of randomly distributed dark patches." The existence of life on Mars becomes less and less probable the more we know of the planet. The fact that water is so rare and that it appears to have been rare throughout the planet's history makes it highly unlikely that life, at least as we know it, exists there.

Where Do We Go from Here?

Some have even questioned the existence of intelligent life on earth, when we spend billions on our probe of space, and so grievously neglect the welfare of our own planet and its inhabitants. Perhaps the conflict of interest can best be summarized by these two quotations.

Thomas Mutch (1970) depicts the fascination of lunar exploration thus: "It may be that the ultimate justification for man's interest in the Moon is also the simplest. The pages of history abound with accounts of exploration: new continents, new frontiers, mountains, ocean depths. The forces which drive man to an exploration of the world around him are so elemental that they defy rational analysis. In turning our attention to the Moon, is it not likely that we are really inexorably driven to take part, however vicariously, in yet another journey of exploration?"

On the other hand, the *New Scientist* points out in an editorial: "If it were not already glaringly obvious, the latest resignation from NASA—that of Dr. Eugene Shoemaker, the lunar mission's principal geologist—underlines the fact that scientific research is low on the list of priorities for the Apollo effort. The real priorities are international politics and show business, with science providing dubious scaffolding to justify the bill of 24 billion dollars. . . . During the week [of the Apollo 13 mission], 2000 children in Washington alone were unable to go to school because they did not have warm enough clothes. Eight million American children cannot read, and a third of the country's black teenagers are unemployed. . . . It is, of course, possible to dismiss such contrasts as naive, amusingly idealistic, or irrelevant, and to urge a more robust, pragmatic view. But to do so is to annihilate that subtle blend of intellect and moral sense which is Man's most precious contribution to life on earth."

Selected References

Colwell, R. N., 1968, Remote sensing of natural resources, Scientific American, vol. 218, no. 1, pp. 54–69.

Glass, B., and Heezen, B., 1967, Tektites and geomagnetic reversals, Scientific American, vol. 217, no. 1, pp. 32–38.

Leighton, R. B., 1970, The surface of Mars, Scientific American, vol. 222, no. 5, pp. 27–41.

Mutch, T. A., 1970, Geology of the moon, Princeton University Press, Princeton, N. J.

O'Keefe, J. A., 1964, Tektites and impact fragments from the moon, Scientific American, vol. 210, no. 2, pp. 50–57.

Appendix B

Geology and Evolution

Today we take the concept of evolution for granted. We speak of the evolution of the landscape, the evolution of an idea, such as democracy, the evolution of a civilization. When we use this method of thinking, we mean that each stage in the development of these things is derived from an earlier stage, and that these changes from stage to stage took place very slowly, a little at a time. What we do not often stop to realize is that this way of looking at things stems almost entirely from Charles Darwin's great synthesis about the organic world, a synthesis little more than a century old. Evolution is now so much a part of our thinking that we find it hard to imagine an intellectual climate where almost everyone believed that the earth and all its plants and animals were created at one time and in their present form, and all for the delectation of Man, essentially an unchanging world.

The Evolution of Evolution

The theory of evolution did not spring full-blown from one man's mind without any precursors at all. Theories evolve, too, and this one is no exception. We referred above to Darwin's "synthesis" and that is just what it was, a putting together of the results of other men's works, to which he added some ideas of his own. Furthermore, he took much of the theory out of the realm of pure speculation, and by extensive experimentation started the great work of confirmation of the theory that still goes on today.

Fig. B-1 Fremontia fremonti (Walcott), a trilobite from the Lower Cambrian Latham Shale, Marble Mountains, Cadiz, California. Length, 4.5 inches. (Photograph by Takeo Susuki,)

While most of the people in Darwin's time did believe in an unchanging world, this was not true of many of the scientists. Since the early 1800's the questioning spirit of the age had challenged the old beliefs, and by 1818, or fifty years before Darwin's first publication on the matter, all the elements necessary to the theory were in existence in the scientific body of knowledge. Hutton had published his uniformitarian views of the development of the earth's landscape with their emphasis on gradualism and the vastness of geologic time. Fossils were known and accepted as extinct life forms, and distinctive assemblages of fossils were used to tell different rock strata apart. What the scientists could not see was that there were transitional forms. It appeared to them that a group of organisms were suddenly created, existed for a while, and then mysteriously disappeared. The rapid extinctions could, of course, be accounted for by catastrophes.

The great problem was the origin of the new species which arose to replace the old ones. As the fossil record broadened, some transitional forms were discovered, and Lamarck, a predecessor of Darwin's, had even conceived the idea of an evolutionary development. The champions of catastrophism were very powerful and exceedingly persuasive, however, so much so that Lamarck's evolutionary hypothesis was not taken seriously for many years.

It remained for Charles Darwin to put the parts together into a broad theory of evolution and to add to it a method of creating new species. This mechanism for evolution Darwin called *natural selection.* Another Englishman, Alfred R. Wallace, working quite independently, had also discovered the principle of natural selection. When each man found out about the other's work, they published simultaneously in 1858 two short papers. The following year Darwin published his much longer and very important paper *The Origin of Species by Means of Natural Selection, or the Preservation of Favoured Races in the Struggle for Life.* In it he stated that each species of plant and animal produces random individual variations. Those variations which enable the organism to live better within its enviornment will be perpetuated because such organisms will live long enough to breed and produce many offspring which have inherited those variations. Individuals whose variations are ill-adapted to the environment will either die before breeding in extreme cases, or will produce fewer offspring. Thus are changes accomplished gradually by natural selection. Thus, also, new species gradually arise, as the adaptations continue in one direction, favored by the environment, until the differences between the evolved organisms and the original forms are such that interbreeding is impossible. A succinct way of stating it is, "The Darwin-Wallace principle of evolution through natural selection indicates that all organisms have evolved over a long period of time through gradual change from common ancestors" (Berry, 1968).

Darwin had seen the results of the selective breeding of domesticated plants and animals and had experimented at length with them himself. Only those organisms with a certain desirable characteristic (such as larger kernels in an ear of corn) are allowed to interbreed, and thus, through control by the husbandman, these traits are emphasized and strengthened. Darwin then observed that there were similar variations among the organisms in wild populations, and he realized that competition for survival was an even sterner breeder-controller than the husbandman. Thus he arrived at natural selection as the mechanism for the evolution of the organic world.

This theory has followed the path of all theories; it has been refined and modified, particularly with respect to selection. Work by geneticists, for example, showed that Darwin's concept of selection through individuals did, in fact, operate in a negative way to eliminate unsuitable characteristics. When genetic variation operated through a whole population, however, creative adaptations were possible.

Darwin and Wallace did not stir up very much controversy with their short papers nor did Darwin's longer work. But when he extended the theory to include man, he discovered that he had upset a great many people. Darwin himself could see no discrepancy between his ideas and religion, but theologians thought otherwise, and the issue continued to be controversial for many years. Now, however, as evidence from many areas, including genetics, has mounted, evolution is commonly accepted. Hutton's concept that the natural world can be explained by natural causes without divine intervention has traveled a rocky path from its application to the inorganic world, to the world of plants and animals, and finally to that unusual animal, man.

The Geologic Uses of Evolution

The Geologic Time Table

Aside from the reference to Hutton and uniformitarianism, all this seems not to have much to do with geology. In actual fact, however, evolution through natural selection is the basis of the geologic time table. At first, distinctive fossil assemblages were used to distinguish different layers of sedimentary rocks. As we saw in Chapter 1, this was the important discovery of the British canal-builder William Smith. He could determine from which layer of rock a representative collection of fossils had come, and he proceeded to use his discovery to make the first geologic map of Britain. Subsequently, distinctive fossil assemblages were discovered in many rocks in Europe.

When scientists tried to determine which rocks were older and which were younger, they ran into trouble. One set of fossils, to them, had no relation to any other set, but were apparently separated by some violent catastrophe. Attempts to set up time relationships were made, nevertheless, based on the law of superposition, which worked very well as long as the strata could be followed geographically without interruption, or with only minor gaps. But how was one to determine the age relations of rocks separated by great distances?

Fossil assemblages, then, were not enough, and the determination of true age relationships had to await the development of *fossil successions*. Armed with Darwin's theory, scientists could now follow the development of a shell, for example, from an early primitive form, through various transitional forms, to a later, usually more complex, form. A rock that had an early form of an organism was clearly older than rocks containing later forms. Furthermore, all rocks that had the early form, no matter how far apart those rocks were geographically, would have to be the same age.

Thus, the distinctive fossil assemblages could differentiate between, say, Cambrian and Ordovician rocks, but the fossil successions made it possible to say that the Cambrian rocks are older than the Ordovician rocks. In this way our geologic time table came into being. It is still being improved, and is the basis of almost all geologic work. Without the theory of evolution, and the interdisciplinary science of paleontology, it could not exist.

Ancient Environments

As an historical science, geology is concerned with paleoenvironments and paleoecology. One of the aims of geology, in unraveling the earth's history, is to reconstruct the conditions that existed at various times in the past. Without a knowledge of the way plants and animals evolve, such reconstructions would be far more difficult and much less complete, although still many details undoubtedly escape us. For example, in tracing the evolution of a particular shelled animal, we may find that it changes very rapidly when it first becomes a distinct species. Then, as it reaches a form that enables it to function well in its environment, when it reaches an equilibrium with that environment, it may change very little or not at all. Then another period of rapid change probably indicates that the environment has changed, and those natural variations which may have been suppressed in the old environment are now favored in the new one. This second spurt of rapid adaptation will continue until an equilibrium has been reached with the new environment. If the environmental changes are too rapid and the organism cannot adapt fast enough, it will become extinct.

We know, from observing present life forms and the conditions in which they live, that certain structural features are indicative of a type of environment. A horse, for example, has teeth and jaw structure which enable it to graze efficiently. It also has a body structure that enables it to run fast, which is its primary means of defense against predators. If we find fossil teeth similar to a horse's, we can conclude that the fossil animal also grazed, and if, in addition, we find a similar body structure, we can reasonably conclude that the fossil environment was open grassland. Function and adaptation both are aspects of evolution which aid geologists in their work.

Life of the Past

Now for a hurried glance at the diverse patterns life took as it evolved through ages past. This brief survey of the history of life is included to give an insight into this record, which should be part of any thoughtful person's background. This is the story of our shared heritage and of our kinship with all living things. If the lesson that all men are brothers is ever to be learned, it is to be learned from this story in the rocks of the origins of our common humanity.

The Origins of Life

No one knows where, or when, or under what conditions, exactly, life originated on earth. The search continues, however, and recent experiments have had intriguing results. We know, for example, that for all its diversity of form, living matter is principally made up of the elements hydrogen, oxygen, carbon, nitrogen, sulfur, and phosphorus, so we can safely assume that these elements were present in considerable quantities at the time of the origin of life. We also assume that our atmosphere was different from what it is now. The outer planets of the solar system have atmospheres made up of hydrogen, helium, methane gas (CH_4), and gaseous ammonia (NH_3). They have been able to retain these light gases because, being larger than the earth, they have stronger gravitational fields. It is not out of line to assume, then, that the earth may once have had a similar atmosphere which has since been lost. The development of living forms has probably changed the atmosphere, too.

When a mixture of ammonia, methane, and hydrogen has a steady flow of steam passed through it, and is at the same time subjected to a high-energy electrical spark, such as might be produced by lightning, it turns into a red, turbid liquid within a week. This liquid is found to contain a mixture of amino acids, some of the building blocks of organic tissues. Subsequent similar "primitive earth" experiments have produced almost all the principal building blocks of living systems.

If such an organic broth did exist in the primitive ocean, as seems likely, we still do not know the exact mechanism by which an organism capable of reproducing itself was developed. Certainly this mechanism seems nearer discovery now, with our knowledge of the composition, structure, and behavior of large organic molecules, than at any time in the past.

We do know that life is immensely old. We know, too, that for the greater part of the record the pages are blank. We are finding, however, that the pages are not so blank as we thought. For a long time, the evidence of fossil life in the more than 3900 million years represented by Precambrian rocks consisted of indirect clues, such as the traces of burrows, tracks, and trails. Recently, however, important discoveries have been made. One is in the Gunflint Chert in Minnesota and Ontario. Here fragments of organisms have been found

Fig. B-2   Reconstruction of a Middle Cambrian sea floor. At lower right and far left are colonies of tube-like sponges. A jellyfish floats just left of center. Swimming in the foreground and center are two kinds of trilobites, and several trilobite-like forms. (Courtesy of the Smithsonian Institution.)

which were probably bacteria and blue-green algae and which are about 2 billion years old. The presence of pigmented organisms indicates that the process of photosynthesis was already developed. Similar fragments have since been found in Precambrian rocks 3 billion years old.

Late Precambrian fossils have been found in Australia, fossils which include animal forms as well as plants. The animal impressions resemble jellyfish and worms which are alive today. There are also impressions which are unlike any known organism.

A remarkable thing is that with the start of the Cambrian (Fig. B-2), the great difference between rocks above this time boundary and those below is that the strata above are often richly fossiliferous while those below are not. All that we know is that at the beginning of the Cambrian representatives of all the principal *phyla* into which we divide the animal kingdom had appeared, with the exception of the vertebrates, and in the plant world none as yet grew on land.

A factor sometimes cited for the lack of fossils in Precambrian rocks is that the humble animals of that day, if they were akin to modern flagellates, protozoans, and jellyfish, were animals without shells. Only when external hard parts, such as the shells of mollusks and the chitinous protective carapace of arthropods, appeared could fossils be preserved in quantity. Very likely it is this essentially world-wide appearance of shelled organisms at the beginning of the Cambrian that is the fundamental difference between the Precambrian and the Paleozoic.

Why shelled animals should suddenly appear however, remains a riddle. Perhaps the concentration of lime in the sea became great enough to be available for use in making shells. Possibly competition became severe enough to place a premium on the development of protective devices.

Paleozoic Life

In the beginning of this long era, the only life we know of lived in the sea. The continents of that remote time stretched inland incredibly bleak and barren from the coast, with no vestige whatever of green plants. Assuming the earth then had a global climatic pattern, not too unlike today's, very likely there were places where the rainfall was 100 inches or more. How vastly different such a barren, deeply rilled landscape would be from the verdure-blanketed slopes of the recent tropical rain forest.

The life swarming in the early Paleozoic seas would have appeared modest to us, but nonetheless it represented a prodigious step forward from its humble beginnings. Typical creatures of that time were the *trilobites*. Most were less than 7.6 centimeters (3 inches) long, although some did attain a length of .6 meter (2 feet). These curious animals, distantly related to such things as hermit crabs, were very highly organized to have existed at the dawn of recorded life. They had complex, multilensed eyes (like those of flies and other insects), segmented bodies, elaborate sensory antennae, multiple legs, bodies differentiated into head, thorax, and tail, and a central nervous system. In their heyday they successfully exploited the available environments in the sea. Some swam freely on the surface, some dwelt in shallow waters near the shore, others scavenged and burrowed in the mud at the bottom.

More abundant than trilobite remains in early Paleozoic rocks, especially throughout the strata of the Atlantic and Middle Western states, are fossil *brachiopods* (Fig. B-3). A common name for some of these is lamp shells, a name based on their fancied resemblance to the ancient olive-oil lamps of

Fig. B-3 Zygospira modesta Say. Brachiopod shells in a rock of the Ordovician Maysville Formation, Cincinnati, Ohio. (Courtesy of the Smithsonian Institution.)

Fig. B-4   Pteraspis. One of the several ostracoderms, which were jawless verte-brates that lived during the Silurian and Devonian Periods. The posterior of the body was covered with a pattern of small scales while on the anterior part there was an unjointed armor shield. The ostracoderms were mostly less than one foot long. (From the film "The Dinosaur Age," courtesy of Film Associates of California; art work by A. D. Nellis.)

Mediterranean lands—the symbol we use for scholarship. Brachiopods apparently had roughly the same habitant that mollusks do today.

Gradually, through the Paleozoic, the seas began to be thronged with animals and plants, which, though extinct now, have living relatives. Among these were the forerunners of *sponges*, and they lived in great profusion at the start of the era. Abundant, too, were *corals, sea scorpions*, some of gigantic size, 2.7 meters (9 feet) or so long, *mollusks*, including varieties related to the squid and octopus, but encased in shells with a maximum length of 4.5 meters (15 feet), and *crinoids*. These last, the so-called sea lilies, are animals that grow upward from the sea floor on long, segmented, limy stalks. There they wave to and fro in submarine currents like forests of flowering plants.

A great step forward is recorded in the Ordovician with the appearance of fragmentary *fish* remains. Fish were well established by the Silurian (Fig. B-4), and their evolution and dispersal throughout the seas, rivers, and lakes was rapid. The first varieties appear strange to us. They were small sluggish creatures without true fins and the bodies of some were encased in bony plates, rather than being covered with scales as most modern fish are. A leading survivor of the ancient fish of the past are the sharks, and these arose in the Devonian. Look closely at a representative of this remorseless clan some day and you will be impressed with the fact that its tough, leathery skin is studded with thousands of teeth-like denticles, rather than scales, and these give the whole animal a rough, sandpapery surface.

Presumably at some time in the Devonian, a revolutionary step occurred, and this was the advance of the vertebrates from the sea onto the land. Credit for this achievement goes to the animals which appropriately enough we call the *amphibians*. A typical amphibian leads a life reminiscent of an amphibious operation—involving a move from water onto the land. Most amphibians are born in water from eggs laid and fertilized in water. Many go through a phase —tadpoles are an example—in which they fundamentally are fish, without legs and breathing by means of gills. Then, like toads and frogs, they develop legs and learn to breathe with lungs. Henceforth, in the case of toads and salamanders, they are land creatures.

Very possibly the amphibian's ancestry is from one of the lobe-finned Paleozoic fish, the *crossopterygians*, whose fins are muscular and have a central bony axis. Amphibian evolution may have gone through a phase like the living lungfish of the Southern Hemisphere. These curious, in-between creatures survive drought by gulping air into their swim bladders, rather than by circulating oxygenated water across their gills, and by going into a long hibernation sealed off in mud burrows along dried-up stream courses.

Vertebrates were not the only animals to make the transition from sea to land. Even such obscure beings as snails crossed the barrier, and the arthropods were notably successful. Not only are we afflicted by such things as scorpions and spiders in this group, but their close relatives, the insects, contest with us for supremacy. They outnumber all other land dwellers many times, including as they do perhaps as many as 800,000 species.

Plants, too, gained a foothold on the land and were established by the Devonian. These earliest varieties lacked true roots and leaves. Most reconstructions of the Carboniferous landscape show a swampy scene, much like the tropical rain forest of today, peopled by sluggish, squatty amphibians, ranging up to 3 or 4 meters (10 to 15 feet) long, being bothered by enormous insects, including "dragonflies" with wing spreads of 73 centimeters (29 inches) or so, droning through the trees. To many people the forest probably would look tropical, because it would present such a dark green wall, as the forests of New Guinea do today. There were no flowering plants in the Carboniferous, nor deciduous trees whose foliage changes with the march of the seasons, especially with the brilliant coloring of fall so cherished by dwellers in cool-temperate lands.

Some of the coal-making trees were large. *Lepidodendron*, with a narrow, tapered trunk and paired branches, grew to heights of perhaps 46 meters (150 feet). *Sigillaria* had no lower branches and sported feather-dusterlike ones on top. Both trees had curious eye-like patterns on their trunks which really were scars left by the branches which they shed. *Cordaites*, the forerunner of today's conifers, was a tree that would look moderately familiar to us. It had a narrow trunk surmounted by a crown of narrow, strap-like leaves up to several feet long.

This is the forest whose dead branches, trunks, leaves, and spores accumulated in the swamps of the Carboniferous, and in the course of centuries was converted into the coal of western Europe, the Atlantic states, and the Mississippi valley. Coal of the western states, such as Alaska, Utah, Colorado, and New Mexico, was deposited much later—a great deal of it in the Cretaceous.

Coal of Carboniferous age is found in latitudes far north and south of where forests even remotely resembling the ones of that period grow today. Examples are the productive coal mines of Svalbard (Spitzbergen), and the visible coal seams of Antarctica. A look at an economic geography map of commercial coal fields shows that most of them are in temperate rather than tropical lands. A further suggestive factor is that the present-day environment where peat accumulates is in boggy ground in such cool, rainy regions as Ireland, Scotland, and Scandinavia.

The coal forest was not necessarily tropical. However, it almost surely grew in an equable climate to judge from the lack of growth rings and of deciduous trees, and from the presence of large insects and amphibians, both of whose sensitivity to cold temperatures is pronounced.

Toward the close of the Paleozoic the climate became more rigorous. In the American Southwest where Texas, New Mexico, and Oklahoma are today, it was arid enough that thick beds of evaporites, such as salt, gypsum, and potash, accumulated. In the Southern Hemisphere, wide areas, such as a large part of the Republic of South Africa, are considered by many to have been glaciated.

From the evolutionary point of view the most noteworthy innovation of the Permian was the ascendancy of the *reptiles*, who appeared in the Pennsylvanian. In a time of aridity, the advantage is distinctly with the reptiles, in contrast to the amphibians. Reptilian eggs are laid directly on land; there is no necessity for an infantile stage spent in water; and finally, as adults, a dry scaly surface is an enormous advantage over a skin that has to be kept moist much of the time, as with amphibians.

The Permian reptiles were wide-ranging—great numbers are known from

Fig. B-5 A landscape of the Permian Period. Dimetrodon, the large fin-backed reptile with a blunt head, was carnivorous. Note the mammal-like differentiation of the teeth. A single Edaphosaurus, an herbivorous fin-back with a smaller, more pointed nose, stands in the center. The smaller lizard-like reptiles to the left are Casea; the two small specimens with boomerang-shaped heads at extreme lower right are Diplocaulus, an amphibian. A clump of horsetails is at the middle right; in the left background is a group of primitive conifers. (Courtesy of Chicago Natural History Museum; Charles R. Knight, artist.)

Fig. B-6 Tyrannosaurus, a giant Cretaceous carnivorous reptile, attacking the horned dinosaur, Triceratops. Tyrannosaurus, one of the last of the dinosaurs, was the greatest land-living flesh-eater known. Standing 20 feet high, it had jaws armed with saber-like teeth. The plant-eating Triceratops was protected by its three horns and by the bony frill covering its neck. Both of these dinosaurs lived in North America at the end of the Cretaceous Period, 60 to 70 million years ago. (Courtesy of Chicago Natural History Museum; Charles R. Knight, artist.)

the Southern Hemisphere, including some with skeletal traits that are prophetically mammalian. Chief among these attributes are differentiated teeth, such as we have in our molars, incisors, and bicuspids. Permian reptiles included the strange-looking *Dimetrodon* (Fig. B-5) from Texas, with a grotesque, sail-like spine running the length of its back. More familiar to us would have been turtles, which had appeared even at this distant date. How remarkable it is that such seemingly obtuse creatures as turtles and tortoises could have survived so many arrows of fate, while apparently far more gifted creatures vanished into oblivion.

Mesozoic Life

This was a truly medieval time in the long history of life on earth. It was a time when the emphasis for survival was placed upon brute force, limited intellect, and the development of armor to a degree never to be repeated in the vertebrate world.

This was the era of the *dinosaurs*, and nearly all of us have seen pictures, movies, or outdoor statuary of them—as at Rapid City, South Dakota. Their skeletons are impressive when we see them mounted in the United States National Museum in Washington, D.C., the American Museum of Natural History in New York, the Carnegie Museum of Pittsburgh, the Chicago Museum of Natural History (Fig. B-6), and the Denver Municipal Museum. Probably one of the more exciting opportunities to witness something of this life of the past is to visit the museum area in Dinosaur National Monument in northeastern Utah. Their bones can be seen in place in the rocks, and much of their story can be learned from displays in the Park Service museum as well as in the museum in the nearby town of Vernal.

Fig. B-7   Stegosaurus. This plated dinosaur is among the earliest of the known ornitischians (reptiles with a bird-like pelvic-girdle) and comes from the upper Jurassic. The edges of the peculiar vertical plates on the back were thin; the bases were thickened and embedded in the animal's back. What function these plates had is not known; perhaps they were protective, or they may have served some physiological function. (Courtesy of Chicago Natural History Museum; Charles R. Knight, artist.)

The dinosaurs (from the Greek words *deinos sauros*, meaning terrible lizard) were the dominant vertebrates of the Mesozoic. The name is more than a century old and no one would advocate changing it, but many of them bear no resemblance to lizards and many were far from terrible. In the beginning of their reign, in the Triassic, some were very small, scarcely larger than chickens. Typically, in this early part of the Mesozoic, some varieties scurried around on their hind legs; like kangaroos, they had long tails for balancing. Many of the early dinosaurs had three toes, which we know because scores of their tracks, looking for all the world like ones left by an enormous flock of running turkeys, are preserved in the red Triassic sandstones of the Connecticut Valley.

The golden age of the dinosaurs was the Jurassic and the early Cretaceous. In these periods they reached their greatest size and achieved an astounding diversity of forms and adaptations. Largest and most ponderous were the four-legged, swamp-dwelling, herbivorous dinosaurs, such as *Diplodocus*.

These weighed up to fifty tons, and with their long necks and tails were as much as 24 meters (80 feet) long. Their stumpy, pillar-like legs and general build suggest that they were swamp dwellers, possibly spending much of their time immersed in water as the hippopotamus does today.

In any natural population a balance is quickly struck between herbivorous animals and the carnivores preying upon them. Among meat-eating animals few could be guaranteed to strike more terror than *Tyrannosaurus rex*. This fearsome beast was perhaps 6 meters (20 feet) high and had an overall length of 15 meters (50 feet). Since it stood on enormously powerful clawed hind legs it needed a heavy tail for balancing. The front legs were dwarfed; in fact, they were far out of scale with the rest of the animal. The ferocious head was nearly 1.2 meters (4 feet) long and large and powerful jaws were sown with what to its prey must have appeared like a forest of dragon's teeth.

One of the more obvious means of survival for the less bloodthirsty land-dwelling dinosaurs was the typical medieval solution of retiring inside a defensive redoubt to resist a siege. This took the form among the dinosaurs of the addition of bony plates, spines, spiked tails, and so on, until some outdid the noble knight in the age of chivalry with their weight of armor.

Stegosaurus (Fig. B-7) was an imposing exemplar of this philosophy. This ponderous animal had curiously mismatched legs, with the rear pair much longer than the front; the result was that the beast must have presented a strangely humpbacked appearance with its head forced down to ground level. This probably did not matter too much since there was so little in it. Although stegosaurs weighed about ten tons, their brains weighed little more than 70 grams (2.5 ounces)—about the size of a walnut. To protect this dwarfed intellect, as well as the spinal cord, a double palisade of large triangular plates extended the length of the back, and the tail ended in a set of four fierce-looking spines.

Ankylosaurus, a squatty, four-legged dinosaur superficially resembling the modern armadillo, carried this defensive approach for survival to the ultimate degree. It clanked along furbished with studs, spines, and bony plates, culminating in a knobby lump of bone at the end of the tail, which could be wielded as a club, much like the medieval warrior's "morning star."

Triceratops, a Cretaceous dinosaur, combined both offensive and defensive elements in its anatomy. They were stocky animals, rather like a modern rhinoceros, but with three long and sharply pointed horns. Unlike rhinoceros horns, which are made of felted hair, these horns were solid bone. They were fused at the base into a solid shield of bone, much like an Elizabethan ruff, which protected the upper spine, heart, and lungs from a frontal assault. No triceratopsian outlived the end of the Cretaceous; and this might mean that hardware alone is not enough to guarantee survival—some brainpower is needed, too.

The Mesozoic reptiles are an excellent example of adaptive radiation, because they successfully invaded some part of every available environment;

Fig. B-8   Skeleton of Ichthyosaurus, a common marine reptile of the middle Mesozoic era. These marine reptiles were similar to modern porpoises in size and habits. They had no neck, nor any caudal fin. The four legs were modified to become like paddles. The numerous teeth had a labyrinthine structure characteristic of the labyrinthodont amphibians and the primitive colylosaurian reptiles—possible clue to the ancestry of these highly specialized marine reptiles. (Courtesy of the Smithsonian Institution.)

land, sea, and air. In this last regard they equalled, if not surpassed, the achievements of the mammals, because the only representative of our class in the aerial world is the bat. Flying reptiles collectively go under the name of *pterosaurs* (winged lizards), and their wings resemble those of bats more than they do those of birds. The leading edge of their wing for nearly half its length is made up of an enormously extended little finger. Just as birds do, pterosaurs differed greatly in size, ranging from small ones the size of sparrows up to veritable gliders, with wing spreads of 8 meters (27 feet), that ranged far out over the shallow, chalk-accumulating seas covering Kansas in the Cretaceous.

Seagoing reptiles achieved an extraordinary degree of success, considering the very real limitations of their physiology. Mesozoic turtles reached lengths of 3.6 meters (12 feet), which is quite a contrast to their diminished successors of today. The *ichthyosaurs* (Fig. B-8) (fish lizards), perhaps 7.6 meters (25 feet) or so long, were as fully adapted to life at sea as their mammalian counterparts, the modern porpoises. Ichthyosaurs had streamlined, fish-shaped bodies, although their legs were rounded out into flippers rather than fins. Unlike whales, but like fish, their tails were vertical. The backbone ran along the lower edge, and thus differed from fish whose backbone spreads out in a fan. Ichthyosaurs had narrow, teeth-studded jaws that from a distance made them look like swordfish. Their eyes were large and were encircled by a bony ring made up of wedge-shaped plates.

A remarkable adaptation was in the birth of their young. Since ichthyosaurs were fish-shaped they could not lay their eggs on land, as turtles do. Rather the eggs were hatched internally, as is demonstrated by embryonic ichthyosaurs found preserved as fossils inside their mother's rib cage. Possibly the baby ichthyosaur was born tail first, and slowly enough that it learned to swim before being released to make its way in the world. Unlike mammals there was no umbilical cord, nor were they nursed. A characteristic distinguishing the reptilian from the mammalian world is the total lack of any parental interest in the offspring.

Other maritime reptiles, such as the *plesiosaurs*, played a role approximately equivalent to that of seals and sea lions. They had barrel-shaped bodies, long necks and flippers, and only slightly modified front legs. Still others looked like mythological sea monsters—or perhaps even medieval dragons *sans* wings. *Elasmosaurus* was such a one, with a spiny back and long tail. Its remains are found in the former sea floor deposits of Kansas that are also the final resting place of the giant-winged pterosaurs.

Naturally there has been speculation as to why so diversified and seemingly successful animals as the Mesozoic reptiles perished. The odd thing is that not all did. Quite sizable reptiles survived to our day; among them are crocodiles and alligators, turtles and tortoises—including the immense ones of the Galapagos Islands—and pythons, lizards, and iguanas.

Possibly climatic changes at the end of the Cretaceous may have contributed to dinosaurian extinction. Certainly it appears more than coincidental that two great groups of organisms dominant in the world today—flowering plants and mammals—should have achieved hegemony during the time the dinosaurs vanished.

The earliest mammals, whose remains date back to the late Triassic, were an unprepossessing lot. The majority were rodent-like creatures, perhaps the size of small rats. They are difficult to tell from reptiles when only skulls are available. Teeth, as we have seen, are a diagnostic skeletal element; reptilian teeth are not differentiated according to function, while mammalian ones are. The number of mammalian teeth is fixed—the typical number is forty-four in placental mammals, and since we have only thirty-two, we have moved quite a distance along an evolutionary path in this regard.

Other differences between reptiles and mammals are well known. Mammals do not have scales, and having varying amounts of bodily hair. Their body cavity is divided by a diaphragm, and they have a four-chambered heart. Most significantly, they can maintain a constant body temperature—ours centers around 37° C. (98.6° F.). This is an enormous advantage in severe climates—reptiles are immobilized when temperatures drop too low and suffer from heat strokes when it is too hot since they do not perspire. This means that the geographic range of reptiles is sharply restricted when compared to the mammals. There are a number of large mammals that can tolerate cold temperature and whose range extends far into arctic regions, such as the blue whale, walrus, polar bear, musk ox, reindeer, and caribou.

Another significant difference is that reptiles are hatched from eggs, while mammals are born alive. Among the more primitive mammals, such as the *marsupials*, which include the opossum and the Australian kangaroo, the young are born prematurely and then placed in a pouch until they are fully formed. In *placental mammals* the embryo is carried full term in the womb.

Fig. B-9 A Jurassic landscape. The primitive toothed bird, Archaeopteryx, is in the middle foreground. Two individuals perch on a cycad frond, and two others swoop down toward two midget dinosaurs inspecting a primitive crustacean. Note the claws on the front of the wings of Archeopteryx. Above, and in the background, are several flying reptiles, Rhamphorhynchus. (Courtesy of Chicago Natural History Museum; Charles R. Knight, artist.)

Since new-born mammals are nursed and there is a degree of interest and solicitude shown in their welfare—an aspect of life totally lacking in the reptilian world—mammals have an immensely greater advantage for survival during their most vulnerable period. Incidentally, there is a correlation between the degree of intelligence and the length of parental care—the more intelligent mammals taking the longer time to mature.

The flowering plants, or *angiosperms*, appeared early in the Cretaceous. Their advent provided a wholly new source of food, a wide variety of nuts, fruits, seeds, and cones, which were possibly seized upon by the emerging mammals and neglected by the rather inflexible dinosaurs. With the ascendancy of the angiosperms the world took on an increasingly familiar aspect. The forerunners of the *sequoia*, the giant tree of the western forest, were widely distributed around the Northern Hemisphere. Another holdover from the Cretaceous is the *ginkgo*, a native of the Orient, but a tree whose rounded, pale greenish leaves enliven many a city park. *Cycads*, stumpy, palm-like, extraordinarily slow-growing trees, flourished widely. Other plants which diversified the formerly somber green forest were the forerunners of such familiar things as magnolia, willow, oak, laurel, and a wide variety of palms and conifers.

The angiosperms very possibly originated in the tropical uplands of the earth and spread outward into temperate lands. Their appearance had a profound effect upon the insects, whose role in plant fertilization—the bees are an example—is widely known.

Birds, too, gained their ascendancy during the Mesozoic. A remarkable fossil find was made at Solenhofen, Bavaria, in 1877, when the bones and feathers of a wonderfully preserved bird, *Archaeopteryx*, which was the size of a pigeon, were recovered (Fig. B-9). The fact that *Archaeopteryx* had feathers, and thus was not a pterosaur, was established beyond doubt; yet its anatomy still preserved many reptilian characteristics, including teeth along the beak, claws on the wings, and a long nonbird-like tail.

Evolutionary changes had not ceased in the sea while these developments were occuring on land. They were far too numerous to detail here, but many familiar forms of marine life gained a foothold in the Mesozoic. The bony, flexible-scale-covered fish, the *teleosts*, the dominant kind today in rivers, lakes, and shallow seas, achieved the ascendancy they now hold.

Distinctive among Mesozoic marine invertebrates were the *ammonites*. Ammonites are coil-shelled mollusks, and resemble the chambered nautilus of the tropical seas. The name comes from the ram-headed god Ammon, in the ancient Egyptian pantheon, and at first glance typical ammonite shells resemble the coiled horn of a mountain sheep. The animal lived in the outermost of the chambers, and when it outgrew this last one a new one was added. The distinctive difference between a nautilid shell and an ammonite's is in the intersection of the partition wall with the outer shell. In a nautilus the junction is smooth, in an ammonite it is extremely complex; much as though the partition were too large for the shell and had crumpled along the junction line. The result is the development of a pattern at the intersection, as involved as the intricately sutured way the bones in our skull are joined. The animal resembled a squid or octopus because it had imposingly large eyes, a mouth like a parrot-beak, and a collection of tentacles with suction discs—ten as contrasted to the customary eight for an octopus.

Near the end of their geologic life span some ammonite shells have the most aberrant form imaginable. Some are partially straight, then coiled; others were coiled so that finally the animal was turned back on itself and died. Still others, the *Baculites*, developed nearly straight, long, tapered shells, much like an old-time dunce cap (except for a tiny coil at the very end), and thus reverted to nearly the same pattern as the original nautiloids of the early Paleozoic. By the close of the Mesozoic all ammonites were gone.

Cenozoic Life

The climate in the beginning of the Cenozoic over much of the earth seems to have been warmer and more humid than today. Broad expanses of North

America were cloaked with a subtropical forest that included trees such as figs, breadfruit, magnolias, sassafras, palms, and palmettos.

Gradually through the Cenozoic this humid, subtropical aspect diminished, and the landscape acquired an increasingly modern appearance. In what now are temperate North America and Europe, oak, beech, maple, chestnut, spruce, fir, and pine succeeded the original forest. An important floral event in the mid-Cenozoic was the appearance and proliferation of the grasses. Although grass seeds have been recovered as fossils from Cretaceous rocks, grasses did not reach their peak until much later.

It would be nearly impossible to describe our debt to this humble plant. Grasses are the most widely distributed of the flowering plants, and include at least 5000 species. It would be hard to think of a region so desolate that no grass will grow. Not only is grass directly valuable as the food for nearly all grazing animals, but from wild grasses were derived a host of domesticated grains, such as wheat, rice, barley, and corn. Two other grasses play so vital a role in the tropics that it would be difficult to imagine this part of the world without them; these are bamboo and sugar cane.

Geologically, grass is enormously significant in controlling erosion. Grasslands are virtually impregnable to ordinary erosional processes, but when the sod is stripped away the resultant gullying may be spectacular. There was a time when the western prairies were truly a sea of grass, and our literature— as well as movies and television—has been telling for decades of conflicts between cattlemen and sodbusters.

In retrospect, it is interesting to speculate on what erosional processes were like, and what the appearance of the world may have been before the grasses evolved. This is much the same problem we were confronted with in reconstructing a mental image of the Precambrian terrain without land plants.

The yielding of forest to grassland in the mid-Cenozoic had a profound effect upon the evolution of the land mammals. This gave an advantage to the plains dwellers and to grazing animals as the horse, camel, and bison, as well as scores of cursorial animals, much like the ones we see in photographs of the African savannah.

In North America and in Europe the climate became cooler, with some oscillations back and forth, and many of the prominent deserts of the world began to develop in the later Cenozoic. This was especially true of the western United States, where the growth of intermontane desserts correlates closely with the rise of the Sierra Nevada and the blocking of rain-bearing winds from the west.

Near the end of the Cenozoic, in the Pleistocene, the climate of the world underwent one of its more stringent modifications, the ice age. As we learned in Chapter 12, there were at least four major advances of continental ice sheets, and the most drastically affected areas were in northwestern Eurasia and the northern part of North America. This was also a time of crustal disturbance—there very possibly is a causal relationship between the two events—

and the world's higher mountain ranges, such as those of Canada and the United States, the Cordillera de los Andes, and the Himalayas, were elevated to their present heights.

Sea level rose and fell in rhythm with the retreat and advance of glaciers on land. This rising and falling of the sea was vastly important in the distribution of land animals, including the wanderings of our prehistoric ancestors. When sea level stood low, Britain and Ireland were joined to Europe, Ceylon to India, and many of the islands of Indonesia were connected to one another as well as to the Asiatic mainland. Perhaps the most important of these land bridges was the one across Bering Strait which connected the New with the Old World. Across this isthmus between Siberia and Alaska trooped a parade of creatures. Migrating from North America westward marched the zebra, the camel, and the horse; eastbound came the elephant (at least four kinds), the bison, the mountain goat and sheep, the moose, the elk, and the musk ox, and lastly the ancestors of the diverse races of people we call American Indians.

Any attempt to describe the panorama of Cenozoic mammalian evolution in so brief a span as remains in this chapter would be foolhardy. It is nearly as hopeless a task as trying to describe the contents of the world's largest zoo in words, and at the same time make it interesting. Fortunately, because of the proximity to us in time of the evolution of the Cenozoic mammals, their fossil record is more complete than for animals of earlier eras.

In the beginning of the Cenozoic most of the mammals were small and the majority of them were forest dwellers. It almost appears that the death of the dinosaurs left a temporary vacuum the mammals at first were reluctant to fill. The earliest mammals included insect eaters, small hedgehog-like creatures, shrews, raccoon-like animals, and opossums.

Very quickly a number of distinctive lines appeared. The first unquestioned member of the horse family lived in the Eocene, *Eohippus* (meaning dawn horse), a graceful, slender-legged animal with short, even teeth, and a small head. A distinctive difference from the modern horse was the feet; the front had four toes, the back, three.

Rodents flourished in the Eocene forest, and were perhaps more common then than they are today. Carnivores prospered, too, and in this early age appeared the precusors of such later types as the cat and dog families.

In the Eocene the two landmasses of North and South America were separated, and in the Southern Hemisphere the evolution of distinctive animals such as the sloth, the armadillo, and the anteater went on quite independent of the rest of the world. Another curious world apart was Australia, where, through early isolation, the vastly more primitive marsupials did not have to contest with more advanced placental mammals for survival. Marsupials filled most of the available ecological niches, even producing a dog-like carnivore, the so-called Tasmanian wolf.

Two bizarre forms in the early Cenozoic fauna were (1) the archaic mam-

Fig. B-10   Brontotherium, a large Oligocene titanothere, in the middle and back-ground; at right foreground, Hyaenodon, a primitive carnivore; and in the left foreground, giant tortoises. Brontotherium, distantly related to the horse, was the largest (8 feet at the shoulder) of the titanotheres, a group whose evolutionary history was marked by an increase in size and the development of horns; the titanotheres nevertheless remained primitive in other respects, particularly in the dentition and feet. (Courtesy of Chicago Natural History Museum; Charles R. Knight, artist.)

mals and (2) the giant birds. The first were hoofed animals, with stump-like legs, large ungainly bodies much resembling those of the rhinoceros, and heavy, low-browed skulls supporting a variety of knobby horns (Fig. B-10). A typical archaic mammal was *Uintatherium* (beast of the Uintas—which are mountains in northeastern Utah), and, judging from the brain case, his I.Q. did not exceed that of the recently extinguished dinosaurs by very much. Like the dinosaurs, the early Cenozoic uintatheres were an unsuccessful experiment in which brute strength and feeble intellect lost out to more agile, aggressive, and intelligent competition. The birds were an uncanny element in the landscape of the early Cenozoic. With no effective carnivores preying on them, some reached enormous size and at the same time were wingless. *Diatryma* was one of these monster birds in the Eocene, and reconstructions of it are like something remembered dimly from a childhood nightmare. This appalling bird stood about 2.1 meters (7 feet high), its large skull and beak measured approximately .45 meters (1.5 feet) long, and it had termendously powerful scaly lower legs and claws. It must have been a frightful thing for an early mammal to encounter. *Phororhacus*, with a great hooked beak, was another utterly savage looking individual from the middle Cenozoic of South America. In the later Cenozoic and on into modern days, flightless birds characterize the Southern Hemisphere, where they are relatively free from carnivores. Among them are ostriches in South Africa, rheas in Argentina, emus and cassowaries in Australia, and penguins in Antarctica and the South American coastal islands. Most remarkable of all were the moas of New Zealand—4-meter (13-foot) giants which were exterminated by the Maoris as recently as the sixteenth century.

Beginning in the Oligocene and continuing into the Miocene, the animals of North America and Europe acquired an increasingly modern aspect. Some of the more bizarre and primitively organized earlier mammals died out, and with the thinning out of the primeval forest and the spread of grassland, grazing herbivores came into their own.

Merychippus, a typical horse of this time, was the size of a small pony. It ran on graceful, nimble legs which terminated in a three-toed foot. Only the middle toe was functional and the nail had already evolved into a hoof. Incidentally, the atrophied trace of the useless second and third toes can still be identified in the modern horse as the splint-like cannon bones on either side of the hoof. In response to the plains environment, *Merychippus'* teeth grew high-crowned and developed elaborately enameled surfaces to cope with the harsh prairie grasses—grass is a plant that has silica in its composition. In harmony with the lengthening of legs, there was a corresponding lengthening of *Merychippus'* neck and muzzle. In fact, the jaw had lengthened and the gap that is such a characteristic feature of the modern horse appeared between the incisors and the rest of the teeth.

The camels, also a distinctively North American animal, were undergoing a parallel evolution. Two varieties were *Procamelus*, without a hump and about the size of a sheep, and *Alticamelus*, a strange-looking version with a long, giraffe-like neck which enabled it to browse on the branches of trees.

Other mid-Cenozoic mammals displayed the variety we associate with these, our distant relatives, today. The ancestors of the whales started on a way of life that would take them to the most distant seas of the earth. Forerunners of the deer, antelope, and bison emerged and were preyed on, then as now, by carnivores—clearly differentiated in this distant day into members of the cat (lions and tigers) and the dog (wolves, coyotes, bears) clans.

One of the most appealing of animals is the elephant—perhaps because of our nostalgic recollections of circus parades—although mid-Cenozoic elephants would look unfamiliar to our eyes. Some had tusks only on their lower jaws, and these curved sharply downward at right angles to the jaw. No one knows their function, but possibly they were used for gouging out roots. Others had four tusks; two above and two below.

No one knows where the lineage of elephants had its start, but there is some evidence it may have been in the vicinity of the valley of the Nile. Should this be true, they started from here on an immense journey that was ultimately to take them under such alien skies as those of Alaska, Siberia, Europe, and North America. Two great families emerged: the *mastodons* and the *mammoths*. The mastodons are the smaller of the two, and became extinct during the Pleistocene. The living elephants are related to the mammoths.

The Pliocene Epoch, which merged with the Pleistocene, saw the rise of the living, or only very recently exterminated, animals of the earth. The nature of the boundary between the Pliocene and Pleistocene is still being argued, but the beginning of the general advance of the continental glaciers southward across the prairies of the Middle West and radially outward from the Scandinavian highlands somewhere between 1 and 3 million years ago is a logical event to separate the two epochs.

To return to the history of the horse, the Pliocene form, *Pliohippus*, was larger than his Miocene progenitor, and only slightly smaller than his Pleistocene descendant, *Equus*. Both varieties had single hoofs, and the vestigial remnants of the second and third toes essentially had vanished. The main difference is in the teeth, with those of *Pliohippus* being smaller and with less intricately involuted enamel partitions within the tooth than those of *Equus*. Horse teeth grow upward during the modern horse's life until it is about thirty-four years old. Then, with no further growth, the teeth rapidly disappear and death is likely to result from starvation. The ability to estimate a horse's age from his teeth is a knack that stood the old-time horse trader in good stead.

From the Pleistocene on, the story of the horse takes on a truly epic character. The evolutionary development from a small, multi-toed, browsing animal the size of a sheep, up to a single-toed grazing creature the size of an Indian pony took place largely within the Western Hemisphere. Then for no reason now known, the horse vanished from its former home, crossed the Bering land bridge, and appeared in Eurasia.

Where the horse was first domesticated is uncertain, but its impact on the slowly emerging Eurasian world was profound. For one thing, it gave the Eurasian man a beast of burden—something the American Indian sorely lacked. For another, the horse added a completely new dimension to warfare, with perhaps the greatest mobility being realized by the Tartars when they swept everything before them from Korea to the gates of Warsaw.

Although the Golden Horde achieved amazing mobility, the Tartars lacked the ability to make a stand and fight from horseback. This "achievement" awaited the invention of such a simple device as the stirrup. With the stirrup it was possible to strike a sweeping blow with a sword, and with the sword came the knight in armor, and with him came not only the Age of Chivalry (from Old French, *Chevalier*, one who rides a horse) but the social pattern of the medieval world.

The horse's wanderings were not ended, for, brought to the New World by Hernando Cortes, some escaped and re-established themselves in what had been their ancestral home. The impact of the feral horse on the life of the Indian—especially the plains tribes—can only be described as explosive. Some tribes accepted the horse; others did not—and quickly came to rue the day. Among the more successful horsemen were the Comanches, who were the scourge of the frontier in the Civil War era. They ranged in pillaging bands all the way from Durango in Mexico north into central Kansas. The Sioux and other northern tribes were equally redoubtable foes on horseback, as General Custer, for one, learned on a bitter day in June 1876.

Other animals evolved rapidly in the Pliocene and into the Pleistocene—many of them were vastly more numerous than now, as well as being far larger and more formidable. An impressive example is the elephant. In addition to the mastodon, at least four major varieties of mammoth roamed the northern continents of the world. In the far north, or in lands marginal to the Pleistocene ice sheets, both the mammoth and the mastodon were covered with thick, coarse hair. They lived and died in Siberia in such vast herds that for centuries their tusks, dug out of the tundra, constituted one of the world's sources of ivory.

We know very clearly how the woolly mammoth looked (Fig. B-11). For one thing, his curious hump-backed profile was drawn repeatedly on the walls of caves by his contemporary, stone-age man. For another, his remains, frozen in permafrost, complete even to his last meal, undigested in his stomach, have been recovered in Siberia and Alaska. The strange, dome-like top of his head and the hump on his back were reservoirs of fat, which, together with his covering of hair, helped him survive the arctic chill.

The southwestern United States was the domain of the imperial elephant —a creature that would have gladdened the heart of the departed Barnum— standing 3 to 4 meters (12 to 14 feet) high, with enormous recurving tusks. He, too, was a contemporary of early man in the New World.

Other animals of the latest Cenozoic were the woolly rhinoceros, the bison —larger than the ones which darkened the western plains in pioneer days— the ubiquitous musk ox, reindeer and deer in great profusion, and their inevitable predators, wolves, lions, tigers, and bears. The cave bear, an exceptionally large and powerful animal, contended with our ancestors for occupancy of caves and shelters in Europe during the ice age.

A paradox of our generation is that one of the more revealing insights we have into this savage world of the ice age is in the heart of a major American metropolis. To someone driving down Wilshire Boulevard in Los Angeles, attempting to survive the hurtling traffic dimly seen through smog, the glittering facade of nearly identical, "international" style buildings is an insipid vista when compared with the once more colorful scene on the same site.

Tarry for a moment a few blocks west of La Brea Avenue; there, in a small park surrounded by the ziggurats of today, some memories of this bolder past are recalled for us. Dotted through the park are concrete statues of some of the vanished animals of the Pleistocene (Fig. B-12). These are reconstructions of animals whose skeletons were recovered from tar seeps once interrupting the surface of the then barren plain where Hollywood now sprawls.

The La Brea tar pits were active in 1769, when the first white men, members of the Portola expedition on their way northward to find a harbor of refuge for the Manila Galleon, passed by them in search for a way out of the Los Angeles basin. Later, when tar was first spread on the dusty streets of the "City" of Our Lady, Queen of the Angels, enormous numbers of bones were sifted out of the pits excavated on Rancho la Brea (Brea means tar in Spanish), and they still remain as dank, scum-filled depressions—perhaps not notably different from their appearance in the late Pleistocene. Then, scores of animals were lured to their death when they sought to drink the water that formed a thin film over the deadly tar beneath.

Most intriguing of the animals recovered were sabre-tooth tigers. These were stocky, powerfully built cats with broad and heavy shoulders. Their

Fig. B-11 Woolly mammoth and woolly rhinoceros in a Pleistocene glacial land-scape. Both of these animals have been recorded in the cave paintings of southern Europe and were hunted by early man. The woolly mammoth, a member of the same family as the modern-day elephant and more distantly related to the mast-odon, is known to have been covered by a dense coat of hair because frozen car-casses of this beast have been found in the tundra of Siberia and Alaska. (Courtesy of Chicago Natural History Museum; Charles R. Knight, artist.)

lower jaws were hinged to drop far back, which enabled them to strike a deadly stabbing blow with their long, sharp-edged incisors.

Other exotic denizens of southern California a few tens of thousands of years ago were the sloths (Fig. B-13). These outlandish-looking, witless creatures evolved quite independently in South America, and when the two continents were rejoined in the late Cenozoic, they wandered northward to North America.

The skeleton of one of them that was found in a cave in Virginia was de-scribed by no less a personage than President Jefferson, who had the bones spread around the White House when he was studying them. Few Presidents since then have had the inclination, or found the time, to carry on an inde-pendent paleontological investigation. Jefferson believed the sloth remains were the skeleton of an immense lion, probably living somewhere in the wilds of the Lousiana Purchase, because, in his own words, "Such is the economy of nature, that no instance can be produced of her permitting any one race of animals to become extinct. . . ."

The ground sloths resembled great clumsy bears with long recurving claws. Their coarse, shaggy hair covered a tough hide studded with little bony pellets, or scutes. Sloths lived in the southwestern United States contemporaneously with the early Indians, and may have survived almost into our day, since a settler in the Cape Horn region in the early 1890's found a sloth hide together with signs of human occupancy in a cave bordering the appropriately named Inlet of the Ultima Esperanza, or Last Hope Inlet.

Another distinctive member of the Pleistocene of Rancho La Brea was *Teretornis*, an enormous, carrion-eating bird with a wing spread of 3.6 meters (12 feet), the largest flying bird ever found. The California condor, its diminished successor, even now is an unforgettable sight soaring on motionless wings thousands of feet aloft in the summer sky.

Thus, evidence from the La Brea pits, as well as other Pleistocene graveyards around the world, poses an unanswered problem similar to the great dying out of the dinosaurs at the end of the Cretaceous. In the case of the Cenozoic mammals it must not have been the rigors of glaciation; most of them survived these, only to perish with the advent of the modern climate.

The wanton destructiveness of mankind is sometimes named as a cause. Although men are remorseless destroyers of wild life, and even within recent generations a sobering list of man-induced exterminations can be compiled, it is extremely doubtful that stone-age man was that effective. The Indian, for example, had little impact on the buffalo population compared with the deadly slaughter accomplished by white hunters in the decades following the Civil War.

Origin of Man

When the men of western Europe, having gained momentary technical ascendancy with gunpowder and the sailing ship, set out to explore the rest of the world they brought back animals, plants, and people as samples of the wonders they beheld on distant shores. With increasing knowledge of the diversity of the plant and animal kingdoms, attempts were made to create an orderly biological classification. Then the problem arose: where do we fit in? There was not only the question posed by the great diversity of the living races of men, but that of our relationship to the anthropoids, such as chimpanzees, gorillas, orangutans, and the like. Awareness of this resemblance goes back

Fig. B-12 Smilodon, the great saber-tooth cat, attacking a bison mired in the tar seeps of La Brea tar pits, Los Angeles, California.

The acme of the saber-tooth evolution, that began in the Oligocene, was attained in Smilodon during the Pleistocene. This saber-tooth cat was as large as a modern lion and preyed upon slow-moving animals. His powerful neck and long sabers were completely specialized for destroying his prey, but when his supply of slow animals became scarce in the late Pleistocene he was unable to compete with his swifter cousins in the chase of the remaining more speedy game animals, and he became extinct. (Courtesy of Film Associates of California.)

Fig. B-13 Statues of the giant ground sloth in Hancock Park, Los Angeles. These animals, which are represented by many remains in the La Brea tar pits, were common in the southwestern United States in the late Quaternary time, and were contemporaneous with early man.

Their closest relatives are the armadillos and the giant anteaters, and they are characterized by peg-like teeth with which they chewed leaves of trees and bushes. They grew to the size of a small elephant and lumbered along on the sides of the hind feet and upon the knuckles of the front feet. (Courtesy of Film Associates of California.)

to the beginnings of recorded history. According to Green (1959), Ennius, who lived from 240 to 169 B.C., wrote, "How much doth the hideous monkey resemble us!"

Linnaeus (1707–78), founder of the binomial classification of organisms that biologists employ today, recognized this kinship and placed man in the same order with the apes and, of all things, sloths.

Throughout this same period in the formative years in the growth of natural science, increasing numbers of finds were made of human remains and of stone-age implements throughout western Europe. Oddly enough, there was no general appreciation of the significance of these artifacts up to the beginnings of the nineteenth century. This seems strange, because as far back as the Renaissance the knowledge was widely shared that the Romans, for example, had subjugated people far more primitive than they. Furthermore, in the Age of Discovery, explorers in distant parts of the earth encountered people using stone weapons and tools. This was sometimes explained on the basis that they were degenerate tribes who had wandered far from the center of civilization, perhaps even before the flood.

The beginnings of a stratigraphic sequence of the relics of ancient man were worked out first in Denmark. This classification, familiar to educated people —the Old Stone Age, the New Stone Age, the Bronze Age, and the Iron Age— was established by Christian Jürgensen Thomsen in the 1830's when he was director of the Royal Museum in Copenhagen. No one visiting Copenhagen today, after taking in the sights of Tivoli, should fail to see the amazing collection of implements, clothes, people, and so on in the museum. Denmark in prehistoric times was unusually favored because of the immense supply of amber from the Baltic—which was traded the length of Europe, even to the Mediterranean, across well-established Alpine routes—and also for its supply of flint which was used for spear and arrow heads.

However, cultural relics and chronologies erected on them pose the same problems fossils do when we attempt to use them in correlating sequences of events in distant areas. Similarities between the methods and philosophy of paleontology and archeology are close, and each of these sciences can learn much from the other.

While such a relative succession as stone, bronze, and iron has local validity in Europe, it is not of universal application. The American Indians, with few exceptions, never left the stone age until they were suddenly brought in contact with relatively advanced forms of the Iron Age in A.D. 1492. Even in Europe, one age did not end abruptly and another begin; stone implements were still handy to have around the house, and were far less expensive than bronze when it was first introduced. The same holds true for iron; it was used as long ago as 3000 B.C., but did not really reach central Europe until about 800 B.C. Such, however, was the conservative nature of the military mind that long afterward the legions were still fighting with the traditional Roman short bronze sword while the barbarians were equipped with weapons of iron. An enormous advantage of an iron sword is that a man can strike with it, whereas with bronze he can only stab.

Puzzling as his cultural remains were to our ancestors, the actual skeletons of early man were even more so. In the first place, very few of them were found, or are very likely ever to be discovered for that matter. Above all else the founders of our race must have had a certain amount of sly cunning. They were not likely to be trapped in great numbers on the sandy flood plains of large rivers, as happened repeatedly to grazing animals. Also, the number of prehistoric peoples, to judge from the estimated pre-Columbian population of North America, was insignificant compared with the dense agglomerations we find in parts of the world today. Then, too, it is a rare environment in temperate climates where bones do not quickly decay. Fortunately for us, early man was a contemporary of the ice age in Europe. Storm, cold, and driving rain were no strangers to him, with the result that caves were a highly prized refuge—even if title to them had to be disputed with the cave bear.

Since limestone is a relatively common rock in many parts of Europe, caves are fairly abundant there, the most renowned being in the Spanish Pyrenees and the Dordogne Valley of France. Fortunately, the cave people were not compulsive tidiers. They would not have been the best of all possible room-mates, and any notion on their part of sanitation was nonexistent. In some caves which were more or less continuously inhabited, debris on the floor may be 21 meters (70 feet) or so deep. Such piles of litter yield records of people and their possessions extending back tens of thousands of years.

Apparently among the first unquestioned representatives of the now-vanished Neanderthal race to be described was a skeleton recovered from a quarry in the Rock of Gibraltar in 1848. Like the first of almost everything it was ignored and forgotten until more publicized remains of the same race were discovered in the Neander Valley near Düsseldorf in West Germany. Since the description of this find was published in 1858, only one year before Darwin's *Origin of Species* appeared, the possibility of this being some kind of extinct prehuman was recognized almost immediately.

A debate soon broke out between those who were persuaded that here indeed was a subhuman type, and those who believed this was no more than the skeleton of some unfortunate, imbecilic victim of an advanced case of rickets.

Today we know this unkind interpretation was not true. More than thirty sites, some including a fair number of individuals, have been found from Iran to the coast of Britain. We also have an idea of how Neanderthal man looked. He looked tough. Perhaps many of the reconstructions made have not dealt charitably with him, but he did lack a certain amount of grace. He was short, with long arms, and he had a barrel-like chest with powerful shoulders. He was bent forward more than we are because his backbone lacked the four curvatures that give spring to our backs and allow us to hold our heads upright. Neanderthal man had a low forehead, a broad and rather flattish face, a powerful jaw and yet a receding chin. Whether or not he was as furry as reconstructions often show him to be is unknown. In spite of the fact that this description may remind you of some of your acquaintances, he was not of our race, but vanished somewhere between 30,000 and 50,000 years ago. Despite his rugged appearance, he was far from a cretin—his brain size was about the same as ours, 1200 to 1600 cubic centimeters. He had learned to make fire, to make stone axes and spears, and to bury his dead.

For a brief while Neanderthal man lived contemporaneously with his successor, Cro-Magnon man, and then vanished as completely as the great creatures—the cave bear, the aurochs, the great elk, the mammoth, and the mastodon—who had been his companions and his adversaries in this shadowy world before the dawn of ours.

The first recorded discovery of the Cro-Magnon people was at Aurignac, France, in 1852. There seventeen skeletons were found, carefully collected, and buried in the village cemetery, thus becoming lost to science. A later find, in 1868, of five skeletons in a rock shelter known as Cro-Magnon in France was recognized for what it was, and this discovery gives its name to this remarkable race.

They would have been regarded as a handsome people by our standards. They were tall and powerfully built—taller, for example, than the Celts or other modern European races that followed them—and their cranial capacity averaged larger than ours. As an identifiable race they lived from 35,000 to around 5000 B.C., and are deserving of our lasting admiration for the extraordinary level of their artistic achievement.

As we learned earlier, the cultures associated with the vibrantly lifelike paintings in the caves at Altamira, Spain, and Lascaux, France, have a carbon-14 age of 15,000 years, and this makes them very likely the handiwork of the Cro-Magnon people who also developed the technique of fashioning stone tools and weapons to a degree of perfection never equaled since.

What happened to the Cro-Magnon race remains a mystery. They apparently went into a decline in the closing days of their history, and may well have been overrun by the Mediterranean and round-headed Alpine races that invaded Europe at the beginning of historic time. Some traces of the Cro-Magnon physique are said to survive in the people of southwestern France, in Brittany, and perhaps among the Basques.

However, both the Neanderthal and the Cro-Magnon people were enough like ourselves to leave little doubt of our kinship. What of their predecessors? What connection is there between our first human ancestors, and the higher primates, such as gorillas, chimpanzees, and baboons?

Here we may become lost in a maze of fact, of fancy, of speculation, and of special pleading. Several things, however, do appear to be reasonably certain. First, we are physically related to the animals of the world. Zoologically, we are classified as primates; we are mammals; and we belong to the great group of vertebrates, or animals with backbones.

Second, we have unique specializations that distinguish us physically from our closest relatives, the anthropoid apes. Among these distinctive features are our upright posture, which is related to the four curvatures of the spine, our basin-shaped pelvis, and our relatively short arms—quite different from the gorilla, for example, whose arms dangle as far as his knees.

Loss of bodily hair is a distinctive feature and a puzzle since it has disappeared chiefly from the back, whereas in animals that have lost hair as an adaptation to a hot climate, it disappears on the underside first. In northern races the loss of skin pigmentation is a specialization, too.

Our facial angle is much greater than that of the anthropoids. This means that a line drawn touching most of our face is nearly vertical, around 85°, while for a chimpanzee, whose jaw is longer and forehead lower than ours, it is only about one half as great, or 43°. The anthropoid jaw moves more nearly transversely, while ours has a much greater rotary capability. Watch a dedicated gum chewer some time to see this attribute displayed. Our reduction in the number of teeth, from the normal mammalian forty-four down to as low as thirty-two, is a primate rather than a human characteristic. However, the reduction in use, or in size, of premolars, and occasionally incisors, is an indication of a continuing trend.

Our opposable thumb is not a uniquely human attribute, but it is an extraordinarily important trait. It gives us an ability to pick up and handle objects, ultimately to shape them to our needs, and thus with weapons and tools in our hands to achieve supremacy over our fellow creatures, as well as over much of our physical environment.

The physical characteristic most significantly separating us from our distant cousins is articulate speech. This ability to communicate with one another, to convey ideas, to express emotion, to achieve the co-operation that only human beings can, is the tangible attribute that most clearly sets us apart. No other single attainment of the organic world compares with it, because through this medium our common humanity has been achieved and the world of the spirit is made possible. Our nearest relatives, the apes, may seem remarkably human at times, but the difference between anthropoid behavior and that of the most primitive men is wide and fundamental.

Should one look for a driving force responsible for the acquisition of these social traits differentiating the anthropoid from the human way of life, it probably came from the development of hunting as a mechanism for survival. As soon as men learned to hunt, a social organization was essential, and a means of communication became necessary. That these lessons were learned long ago is demonstrated by the association of the skeletons of early man with the fire-charred remains of Pleistocene mammals. Fully 50,000 years ago, men were skilled and deadly hunters.

With the background of the discoveries of cave people of Europe and with the growing interest of nineteenth-century naturalists in Darwin's theory, it is not surprising that a search began for the so-called missing link. One of the searchers was a young doctor, Eugene DuBois, whose conviction was that the remains of the progenitor of the human race would be found in a warm climate, much like the habitat of the larger anthropoids today. His urge to visit the tropics was strong enough for him to join the Netherlands Army as a surgeon and be shipped out to Sumatra. He had no success in his search there, but when word reached him of fossil discoveries on Java he journeyed there, and after the customary disappointments he made a series of remarkable discoveries near Trinil in the years 1890–93. He found the top of a primitive man-like skull, a jaw bone, a number of teeth, and a femur. To these he gave the name of *Pithecanthropus erectus* (erect ape-man), and although he was correct in reasoning that he had found the remains of a primitive man-like creature, he was subjected to so much ridicule that he locked his finds away until as recently as 1923. Fortunately, DuBois's work was continued by a German paleontologist, G. H. R. von Koenigswald, from 1930 to the beginning of World War II, when he was interned by the Japanese. Von Koenigswald found enough more partial remains of *Pithecanthropus* to enable us to know what this primitive man-like being looked like. He walked upright; he was shorter than most of us; his brain case was low and flat, with a capacity of only 700 or 800 cubic centimeters; and he had very large and beetling eyebrows. His teeth were large, his lower jaw was powerful and extended forward; yet, like Neanderthal man, he lacked a chin. The animals and plants associated with his remains indicate a cool and relatively rainy climate, and this is generally correlated with the second interglacial stage. How long ago this actually was depends on what the length of the Pleistocene may have been.

In 1941, very shortly before his internment, von Koenigswald made an amazing discovery—the true significance of which has never been evaluated. This was the finding of fragments of an immense lower jaw—at least one inch thick near the base—and several enormous teeth. The same sort of human teeth, many times larger than ours, also showed up in apothecaries' shops in Hong Kong and elsewhere as "dragon's bones." Ground up, they were much prized as a cure-all for a broad spectrum of ailments.

Some day we may know the significance of these massive molars. To some archeologists they were puzzling traces of a vanished race of giants. Were the same proportions to hold with these teeth as do with ours, these people would have been 3.3 meters (11 feet) tall. To counter this argument it has been pointed out that the teeth of orangutans are larger than ours, yet their body size is much less.

The bones of a much better known race of ancient people were recovered from cave deposits at Chou-k'ou-tien about thirty-five miles southwest of Peking. There, in a low range of limestone hills honeycombed with caverns, an immense quantity of fossil bones has accumulated. In 1923 some human teeth were recovered by two paleontologists, J. G. Anderson of Sweden and Otto Zdansky of Austria. Through the years, with the support of the Rockefeller Foundation and others, and under the devoted leadership of Davidson Black, a young professor of anatomy at Peking Union Medical College, an immense amount of material was recovered. When Black died in 1934 the work went on under the direction of Franz Weidenreich, from the University of Chicago, and Weng Chung Pei, a Chinese paleontologist.

The bones of the primitive people found there are now assigned to the same general race as Java man, but put in a slightly different species, *Pith-*

ecanthropus pekinensis (Black). They had the same massive jaws, strong teeth, low-crowned brain cases, and exceptionally prominent brow ridges. All told, parts of more than forty people were recovered, so that a convincing record was available of this early Pleistocene race. Primitive as they were, they possessed weapons and tools and used fire. They may have been either head hunters or cannibals since much of the material consists of fire-burned skulls with the basal part broken open.

This remarkable collection appears to have vanished forever. On December 5, 1941, when the threat of a Japanese advance on Peking appeared imminent, the fossils were turned over to the legation guard of U.S. Marines for safe-keeping. On December 8 the men of the detachment on their way to the coast at Ch'ingtao were captured by the Japanese. To this day no record has come to light of the unhappy fate of Peking man.

In recent years the most challenging finds of early man have come from South Africa. The first discovery, that of a child's skull, was made in 1924 by Professor R. A. Dart of the University of Witwatersrand. This ape-like fossil was recovered from a filled-in limestone cavern. Its discovery was greeted with disbelief tinged with derision, but when the authenticity of the so-called Taungs child, or "Dart's Baby," was certified by the leading South African anthropologist, Robert Broom, it was recognized as a distinctly different prehuman, *Australopithecus africanus* (Dart), which means African southern ape. Since then, through the devoted labors of Dr. Broom, a number of discoveries have been made in caves and quarries in the Transvaal; at Swartkrans in 1949, Sterkfontein in 1947, Makapansgat in 1947 and 1949, and Komdraii in 1958.

Even more ancient hominid remains were discovered in 1960 at Olduvai Gorge in Tanganyika by the archeological team of Dr. and Mrs. L. S. B. Leakey, and to these earliest tool-makers they gave the name of *Zinjanthropus*. Fortunately, the fossil remains are buried in volcanic ash layers whose age is determinable by the potassium-argon method. A provisional age of 1,750,-000 years was assigned to them by G. H. Curtis and J. F. Evernden of the University of California.

Although a variety of names have been given the various hominids in this South African assemblage, they do show enough resemblance to one another that collectively they are called the *Australopithecines*, or southern apes. What they really were has not been determined satisfactorily. They were short, perhaps around five feet tall; they walked upright and their skulls were well balanced. They had large and powerful jaws and teeth and low and receding foreheads. Their teeth were man-like, but their brain capacity was more akin to the apes, averaging only around 600 cubic centimeters. Whether they knew anything of fire, or how to shape implements or weapons has not been established completely. Although these ancient beings of South Africa were far removed from us, nonetheless the resemblances to us turn out to be greater than the differences.

There are many other kinds of primitive men in addition to these, but the study of their peculiarities and the relationships of one to the other, and possibly to us, properly belongs in the domain of archeology. The essential purpose of this brief dicussion has been to demonstrate that physically we appear to be related to ancestors more primitive than ourselves. That the evolution of the modern races of mankind required a long time also seems certain. We do know that men lived during the ice age, and were contemporaries of many of the Pleistocene mammals, now vanished from the earth.

This, then, brings us up to the present. The earth, the sea, the air, and all living things, including man, are intimately interrelated. This abbreviated account of our knowledge shows some of these relationships. Many questions remain still unanswered; hopefully enough will be answered in time to ensure the conservation of the earth and the preservation of man.

Many of these questions have been with us a long time—in essence they are the same as those asked millennia ago by God speaking out of the whirlwind to Job:

38:4 "Where wast thou when I laid the foundations of the earth?
38:16 "Hast thou entered into the springs of the sea?
Or hast thou walked in the recesses of the deep?
38:18 "Hast thou comprehended the earth in its breadth?
Declare, if thou knowest it all.
38:25 "Who hath cleft a channel for the waterflood,
Or a way for the lightning of the thunder;
38:28 "Hath the rain a father?
Or who hath begotten the drops of dew?
38:29 "Out of whose womb came the ice?
And the hoary frost of heaven, who hath gendered it?"

And perhaps the answer given then is equally valid today:
"Speak to the earth, and it shall teach thee."

(Job 12:8)

Selected References

Berry, W. B. N., 1968, Growth of a prehistoric time scale, W. H. Freeman and Co., San Francisco.

Greene, J. C., 1959, The death of Adam, The Iowa State University Press, Ames, Iowa.

Laporte, L. F., 1968, Ancient environments, Prentice-Hall, Inc., Englewood Cliffs, N.J.

Simpson, G. G., 1967, The meaning of evolution, Yale University Press, New Haven, Conn.

Wald, George, 1954, The origin of life, Scientific American, vol. 191, no. 2, pp. 44–53.

Glossary–Index

Aa: rough blocky basalt, 84, 100

Abrasion, glacial, 309

Absolute age: age measured in years, 5, 6

Abyss: the deep floor of the sea, 124

Abyssal hill: a small hill, probably volcanic, on the deep-ocean floor, 290

Abyssal plain: a very flat area of the deep-ocean floor, 291

Afar Triangle, 487, 488

Agassiz, Louis, 299

Aggrading stream, 206

A-horizon, 173, 174

Alabaster: a variety of gypsum, $CaSO_4 \cdot 2H_2O$, 61

Alluvial fan: a fan-shaped deposit of rock debris at the basin edge of desert mountains, 125, 350

Alpine glaciation: glaciation having its origins in mountains and generally following pre-existing stream valleys, 308

Ammonite, 557

Amphibian, 549

Amphibole, 57, 59

Andesite: a fine-grained volcanic rock containing feldspar and ferromagnesian minerals but little or no quartz, 95, 99

Andesite line: the boundary, generally surrounding the Pacific Ocean, between andesitic volcanic eruptions and basaltic eruptions, 484

Angiosperm, 556

Angle of repose: the maximum slope at which sand grains will stand without sliding by gravity, 134

Angular unconformity: a buried erosion surface developed on a series of tilted sedimentary beds, 414, 415, 417, 419

Anhydrite: an anhydrous form of gypsum, $CaSO_4$, 61, 144

Ankylosaurus, 553

Anorthosite: a variety of gabbro composed almost entirely of coarse plagioclase crystals, 99

Antarctic ice cap, 326

Anticline: a fold having its limbs dipping away from each other and its oldest beds, after erosion, nearest the central core, 390, 391, 392, 393

Apollo data, 532

Appalachian Mountains, 395, 465, 466, 468

Aquifer: a rock layer which can easily hold a large quantity of ground water and through which ground water may easily move, 234

Aquifuge, 234

Archaeopteryx, 556, 557

Archipelagic apron: a very flat area of the ocean floor around volcanic islands, 291

Arete: a razor-sharp rock wall separating two glacial cirques, 315, 319

Arkose: a sandstone containing angular fragments of feldspar, 142

Artesian well: a well in which the water is under pressure because a confined aquifer has been penetrated, 234, 235

Ash: volcanic debris the size of dust or sand, 66, 104

Asthenosphere: a low-density layer in the upper mantle, 481, 489

Atmosphere, 34

Atoll: a ring-shaped island made up of coral skeletons, 288, 289, 290, 291

Atomic crystals, 54

Atomic number: the number of protons in the nucleus of an atom, 46

Atomic weight: the number of protons and neutrons in the nucleus of an atom, 46

Augen: eye-like clots of resistant material in metamorphic rocks, 375

Augite: a ferromagnesian mineral, Ca(Mg, Fe,Al)(Si,Al)$_2$O$_6$, 57, 59

Aureole, 107, 373

Australopithecus africanus, 572

Axial plane, 389, 391

Axis, 389, 391

Backswamp, 212

Backwasting: the retreat of a slope adjacent to a stream valley by maintaining an angle parallel to its original one of equilibrium, 210

Bajada: a surface built up of coalescing alluvial fans, 352

Barchan: a crescent-shaped sand dune, 365

Barrier beach: a long narrow sand bar paralleling the coast but separated from it by a narrow body of water, 262

Basalt: a dark, fine-grained volcanic rock, 37, 83, 84, 95, 99

Base level of erosion: the level at which stream erosion ceases, usually sea level, 209

Batholith: a body of plutonic rock with a surface area of at least 64 square kilometers, 106

Bauxite: an ore of aluminum, Al$_2$O$_3$ · 2 H$_2$O, 177

Bayou: an abandoned section of river channel which forms a crescent-shaped lake, 216

Beach, 262

Bearing, 392

Bed, 126

Bed load, 204

Bend, 213

Benioff Zone: a plane of deep earthquakes which dips below a continent, 481

Bergschrund: a crescent-shaped crevasse at the head of an alpine glacier, 309, 310, 322

Berm, 166

B-horizon, 173, 174

Biotite: a dark mica, K(Mg,Fe)$_3$AlSi$_3$O$_{10}$ (OH)$_2$, 57, 58

Blackhawk, California, slide, 189, 190

Block mountains: mountains which owe their elevation to differential movement along faults, 460, 461

Body waves: earthquake waves that follow paths within the earth rather than on its surface, 443

Bolson: a desert basin bordered by mountains whose lower slopes are partially buried in their own debris, 353

Bomb, 105

Bonding: the method by which ions are held together to form molecules, 47

Borax, 145

Bottomset beds: horizontal strata at the bottom of a deltaic deposit, 135

Bouguer, Pierre, 28, 32

Bowen, N. L., 95

Bowen reaction series, 95

Box corer, 277, 279

Brachiopod, 547

Braided stream pattern, 217, 218

Breaker, 253, 254, 255

Breccia: a conglomerate made up of angular fragments, 105, 140

Breeder reactor, 513, 514

Bryce Canyon National Park, Utah, 132

Building materials, 499

Burner reactor, 513, 514

Calcareous tufa: a limy deposit formed by evaporation around a spring or lake or by ground water, 145

Calcite: CaCO$_3$, 40, 43, 54, 57, 60

Calc-silicate rocks: those rocks resulting from the metamorphism of limestones containing clay and sand, 381

Caldera: a volcanic depression much larger than the volcanic vents within it, 67, 68, 69, 79

Caliche: a crust-like cap of limy rock deposited by ground water, 145, 174

Calving: the breaking off of icebergs from the sea end of a glacier or ice sheet, 323

Cambrian, 8, 18, 546

Canal, 537

Capacity: the potential load that a stream can carry, 205

Capillarity, 163

Capillary fringe: a band of thread-like extensions of water extending a short distance upward from the water table, 231

Cap rock: a rock layer of such low permeability that it prevents the upward movement of oil, 503

Carbon-14, 9, 11

Carbonate rocks: rocks that are chiefly compounds of calcium or magnesium carbonate (calcite or dolomite), 145

Carbonation: the chemical combination of minerals with carbon dioxide, 170

Carboniferous, 18, 549

Carlsbad Caverns, New Mexico, 237

Catastrophism: the belief that geologic history consists of major catastrophic events, usually of a supernatural origin, 22

Cavendish, Henry, 28

Cementation: the deposition of a soluble substance in the open spaces of a sediment until a sedimentary rock is formed, 137

Cenote, 240

Cenozoic, 8, 17, 18, 557

Chandler Wobble, 439

Chemical precipitates, 137

Chemical weathering: the decomposition of rock by solution, oxidation, hydration, and/or carbonation, 169, 171

Chert: a very dense, hard, non-clastic rock composed of microcrystalline silica, 146, 147

Chilean earthquake, 1960, 427

C-horizon, 174

Chute: an abandoned and partially filled river channel, 213

Cienaga, 351

Cinder cone: a volcanic cone built up of loose volcanic fragments, 80

Cinders: volcanic debris the size of small pebbles, 104

Cirque: a horseshoe-shaped, steep-walled, glaciated valley head, 312, 314, 319, 322

Clastic: fragmental, 137

Clastic sedimentary rocks, 138

Cleavage: the tendency of a mineral or rock to break in one or more preferred directions, 39, 48, 54

Cleaver: an ice-sharpened, over-steepened ridge between two glaciated valleys, 314, 319

Climatic changes, 333

Coal, 147, 499, 552

Coasts, 260

Col: a glacially-produced gap in a dividing ridge of a mountain range, 315

Color, 56, 128, 132

Columbia basalt plateau, 88

Columnar jointing, 101, 103, 398

Compaction: the squeezing together of particles in a sediment during lithification, 137

Competence: the size of particles a stream can transport, 205

Complex mountains: a composite class of mountains consisting generally of igneous and strongly deformed sedimentary and metamorphic rocks, 460, 464

Composite cone: a volcanic cone composed of interbedded lava flows and pyroclastic material, 78

Compressional force, 414

Conchoidal fracture: a type of fracture, particularly characteristic of glassy materials, which leaves a shell-like mark, 92

Concordant intrusion: an igneous intrusion whose form parallels the pre-existing stratification, 109

Concretions: round solid bodies formed during lithification of the surrounding sediments, generally by accretion around a nucleus, 135, 138

Cone of depression: an area of the water table where it has been locally lowered by heavy pumping, 236

Conglomerate: a sedimentary rock which consists of rounded fragments the size of pebbles, cobbles, or boulders, 37, 127, 140, 141

Contact metamorphism: a variety of metamorphism in which heat plays the principal role, 373

Continental glaciation: the covering of vast areas of the earth's surface by an ice sheet, 305, 322

Continental rise: the sediment-laden boundary between the continental slope and the deep-ocean floor, 284, 469

Continental shelf: the part of the sea floor which extends from the shore zone to a depth of 100 fathoms, 124, 280

Continental slope: the part of the sea floor that extends from 100 fathoms to 2000 fathoms, generally with a steeper gradient than the continental shelf, 124, 282

Convection: the mass movement of parts of a fluid because of differential heating, 478, 488, 490

Copernicus, 527, 529

Coquina: a sedimentary rock composed of a felted mass of sea shells, 133

Coral, 548

Cordaites, 549

Core: the inner 3380 to 3540 kilometers of the earth, 450

Correlation: determining whether events in two separate areas occurred at the same time, 12

Country rock: the rock surrounding an igneous intrusion, 106

Covalent bonding: the method of molecule formation in which atoms share pairs of electrons in order to complete their outer electron shells, 47, 49

Crater, 528

Crater Lake, Oregon, 69, 74

Craton: a wide expanse of crystalline rocks generally forming the stable interior parts of continents, 467

Creep: the slow movement of the soil mantle downslope, 183, 184, 185, 186

Creep: small movements on a fault, 438

Cretaceous, 8, 18, 552

Crevasse: a crack or fissure in a glacier, 303, 304, 322

Crinoid, 548

Cro-Magnon man, 569

Cross-bedding: an original stratification not parallel to the prevailing stratification, 134, 136

Cross-cutting relationships: a geologic assumption which states that invading rocks are younger than those invaded, 12

Crossing, 214

Crossopterygian, 549

Crude oil, 499

Crust: the outer shell of the earth, from 30 to 50 kilometers thick, 450

Crystal chemistry, 44

Crystal form, 50

Cubic close packing, 51

Curie point: the temperature at which a crystallizing substance first becomes magnetic, 32

Current ripples, 134

Cut bank, 215

Cut-off: the new channel of a river which has cut through the neck of a meander, 216

Cycad, 556

Cyclopean steps: great cliffs interrupting the longitudinal profile of a glacial valley, 318

Dana, James Dwight, 465

Darwin, Charles, 16, 288, 541

Decomposition: weathering in which chemical processes are dominant, 158

Deep-focus earthquake, 446

Deep Quest, 274

Deep-seated intrusive bodies, 105

Deepstar, 274, 275

Deflation: excavation of large basins by the wind, 360

Degradation: the erosion of a stream channel by an underloaded stream, 206

Delta: a triangular deposit of sediment formed at the mouth of a river or stream, 125, 134, 217

Delta-flank depression, 218

Density, 27, 453

Desert pavement: a veneer of pebbles left in place when the wind removes the finer material, 357

Devonian, 8, 18, 548

Diabase: a rock of basaltic composition in which feldspar laths are embedded in a groundmass of irregularly shaped augite crystals, 100

Diatom, 147, 148, 149

Diatomite, 149

Differential weathering: the production of an uneven surface by the faster erosion of less resistant rocks or parts of rocks, 154, 157, 158

Dike: a hypabyssal, tabular intrusive body which is discordant, 79, 109, 110, 111, 114

Dilated mud: a mud with 80 to 90 per cent porosity, 233

Dinosaur, 551

Diorite: a medium-dark, coarse-grained plutonic rock containing feldspar and ferromagnesian minerals but little or no quartz, 95, 98

Dip: the angle that a bed is inclined below an imaginary horizontal plane, 386, 387, 389

Diplodocus, 552

Dip-slip fault: a fault in which most of the movement has occurred in the direction of the dip, 401, 404

Discharge, 202

Disconformity: a buried erosion surface above and below which the rock layers are essentially parallel, 414, 417

Discordant intrusion: an intrusion that cuts across the stratification of the rocks it intrudes, 109

Disintegration: weathering in which mechanical processes are dominant, 158

Distributaries: the smaller branches into which a river divides when it reaches its delta, 218

Dolomite: a mineral composed of CaMg $(CO_3)_2$; also a rock composed almost entirely of this mineral, 60, 145, 148, 374

Doubly-plunging folds, 392, 393

Downwasting: a process through which the steepness of slopes adjacent to a stream valley gradually diminishes through soil creep, gravitative transfer, and the decomposition of rocks, 210

Drift: unstratified, widely scattered glacial detritus, 299

Drumlin: elliptical, low, rounded hills deposited by continental glaciation, 327, 329

Dune, 125, 362

Dunite: an igneous rock composed almost wholly of olivine, 60

Dynamic metamorphism: the type of metamorphism in which directed pressure is the dominant process, 373, 375

Dynamo-thermal metamorphism: the type of metamorphism in which both heat and directed pressure are operative, 376

Earthflow: a form of gravity mass movement faster than creep yet slower than a mudflow, 185, 187

Earthquake Information Bulletin, 441

Earthquake prediction, 437

Earthquake-resistant buildings, 436

Echo-sounder: an acoustic device for measuring the depth of the ocean floor, 277

Effluent stream: a stream to which water is added from the water table, 231

Elastic deformation: temporary deformation of a rock after which the rock returns to its original size and shape, 388

Elements, 42

Eluvial zone: the surface layer of a soil from which the soluble constituents have been leached, 173

Embayed coast: a coast in which the sea extends inland, sometimes for long distances, 260

End moraine: a hummocky, crescent-shaped, rocky ridge looped around the snout of a glacier, 319, 322, 327

Environments, 544

Environments or deposition, 124

Eocene, 8, 18, 19, 559

Epeirogeny: vertical movements of large continental masses or ocean basins, 416

Epicenter: a point on the earth's surface directly above the focus of an earthquake, 433, 445, 447

Eratosthenes, 26

Erg: large desert areas covered with sand dunes, 362, 364

EROS, 526

Erosion: the processes that disintegrate or dissolve and remove material from the earth's surface, 309, 348, 357

Erosional cycle, 355, 356

Erratic: a large boulder which has been transported by ice, sometimes to areas of completely different bedrock, 299

ERTS, 526

Esker: an elongate, narrow, sinuous ridge of glacial till, sometimes with rude stratification, 327, 329

Eugeosyncline: the more active, generally seaward part of a geosyncline, usually containing some volcanic rocks, 468

Eustatic sea-level changes: changes of sea level caused by a change in the quantity of sea water rather than the elevation or depression of land masses, 257

Evaporites: rocks that result from the evaporation of water containing dissolved solids, 144

Evolution: the principle that each stage in the development of life is derived from an earlier stage and that the changes from stage to stage take place gradually, 541

Exfoliation: the peeling off of concentric shells from a rock surface, 168

Extrusive rocks: the fine-grained volcanic rocks, 96

Faceted spur, 319

Facies: the over-all characteristics of a sedimentary unit, including mineral composition, grain size, fossils, etc., 150

Facies change, 151

Fault: a break in rock along which movement has taken place, 399

Fault breccia: a jumbled mass of broken rock in the vicinity of a fault, 400

Fault scarp: a cliff which is the surface expression of a fault at depth; rarely, the fault itself, 398, 399

Fault types, 402

Feldspar, 43, 57

Fenster: a hole, caused by erosion, in an overthrust sheet, exposing younger rocks below, 404

Ferromagnesian minerals: dark, rock-forming minerals which contain Fe and Mg, 57, 58

Fetch: the distance over which wind friction can operate on ocean waves, 251

Fire fountain, 82, 83

Firn: granular, recrystallized snow, 302

First-motion studies: the determination of the direction of the earliest movement along a fault during an earthquake, 481

Fish, 548

Fissility: the ability of a rock to split along well-developed and closely spaced planes, 143

Fission, 513

Fissure, 82

Fissure eruption: a volcanic eruption in which large quantities of lava flow from long cracks rather than from single vents, 88

Fjord: a stream valley glacially excavated below sea level and subsequently partially inundated by sea water, 297, 318

Flint, 147

Flood plain: a flat surface adjacent to a stream over which the stream spreads in time of flood, 125, 212

Flow banding: a streaked pattern in volcanic rocks, which is the result of differential concentration of material in the viscous lava just before solidification, 97, 98

Focus: the actual spot at which an earthquake occurs, 433

Fold, 389

Folded mountains, 460, 464

Foliated rocks, 373, 376

Foliation: banding in metamorphic rocks, related to the parallelism of the minerals, 39

Foraminifera, 149

Foreset beds: non-horizontal beds deposited at the forward slope of a delta, 135

Fossil: the traces or remains of a once-living form, exclusive of historic time, 12, 16, 131, 132

Fossil assemblage: a collection of fossils normally found in a particular bed, 543

Fossil succession: the gradual replacement of one fossil assemblage by another, 544

Frequency of waves, 250

Frost heave, 162

Frost wedging, 161

Fusion, 513, 515

Gabbro: a dark, coarsely-crystalline plutonic rock, containing a lot of ferromagnesian minerals, some feldspar, and no quartz, 95, 99

Galileo, 30

Gardening: the constant churning of the lunar regolith by meteoritic bombardment, 528

Genetic classification of rocks, 36, 37

Geochemical prospecting, 178

Geode, 135

Geodesy: a branch of surveying concerned with the size and shape of the earth, 27

Geoid: the shape of the earth including its gross irregularities but not its topographical ones, 27

Geologic map, 21

Geologic time table, 8, 16, 543

Geomagnetism: ancient magnetic conditions of the earth, 32

Geosyncline: a large depression which subsided over a long period of time and which gradually filled with a great thickness of sedimentary and sometimes volcanic rocks, 465

Geothermal development, 245, 513

Geyser, 242, 243

Gilbert, G. K., 461

Glacial deposits, 125

Glacial erosion, 309

Glacial stages, 331

Glacial surges, 307

Global expansion, 491

Global tectonics: the study of the broad structural features of the earth and their causes, 470, 473

Glomar Challenger, 275, 482

Gneiss: a moderately foliated, crystalline metamorphic rock, 39, 373, 377, 379, 381

Gohna, India, rockslide, 189

Gondwana, 478

Gouge: finely ground rock along a fault zone, 400, 404

Graben: the dropped blocks between normal faults, 401, 406

Grade of a stream, 205

Graded beds: sedimentary layers which have coarse sediments on the bottom changing to finer and finer sediments at the top, 126

Graded stream: a stream which has attained an equilibrium among its discharge, channel shape and size, velocity, and load, 206

Gradient: the slope of a stream, 207

Grain sizes, 139

Grand Canyon, 416, 417

Granite: a light-colored, coarsely crystalline rock containing feldspars, quartz, and sometimes a trace of ferromagnesian minerals, 37, 41, 95, 96, 97, 106, 172

Granitization: the formation of granite in place from other rocks through metaphoric processes which may or may not produce actual melting, 107, 382

Gravitation, 28

Gravitational gliding: the formation of overthrust structures by slipping along a tilted bedding plane, 414

Gravity: the property of acceleration which the earth produces in a freely falling body, 30

Gravity anomaly: an area of the earth's crust where the force of gravity is different from what would be expected, 32

Gravity coring, 276

Graywacke: a variety of conglomerate which contains fragments of dark igneous and metamorphic rocks set in a silty matrix, 142

Great Salt Lake, Utah, 367

Ground moraine: deposits blanketing a formerly ice-occupied area, 329

Groundmass: the finely crystalline background material of porphyritic rocks, 93

Gulf of California, 487

Gumbotil: a gray, clayey soil, sticky when wet, hard when dry, of glacial origin, 331

Guyot: a submarine volcano with a truncated summit, 288

Gypsum: $CaSO_4 \cdot H_2O$; also a rock composed almost entirely of the mineral, 38, 57, 61

Halite: $NaCl$, 47, 48, 50, 57, 62

Hall, James, 465

Hanging valley: a valley whose lower course has been removed by glacial erosion so that it ends in a waterfall, 317, 319

Hardness, 55

Hardpan: a cemented layer of iron oxide in the B-horizon of a soil, 174

Hawaii, 79, 80, 82

Hawaiian Islands, 78

Heat flow: the amount of heat escaping from the earth, 480

Hebgen Lake earthquake, 430, 431, 432

Hekla, Iceland, 64

Hellas desert, 537

Hells Canyon, Idaho, 209, 223

Himalaya Mountains, 28, 29

Horn: a jagged, sawtoothed pinnacle which is the remnant of a mountain that has undergone extreme glacial erosion, 315, 316

Hornblende: $Ca_2Na(Mg,Fe)_4(Al,Fe,Ti)_3Si_6O_{22}(O,OH)_2$, 40, 57, 59

Hornfels: a dense, hard, non-layered rock which is generally the product of contact metamorphism, 373, 374

Horst: the higher block between normal faults, 401, 406

Hot springs, 242

Hutton, James, 22

Hydration, 170

Hydraulic mining, 512, 513

Hydroelectric power, 512

Hydrosphere, 34

Hydrothermal metamorphism: metamorphism

in which chemical fluids are the dominant factor, 373, 374

Hypabyssal rocks: rocks that crystallized at intermediate depth so that their textures are intermediate in coarseness between those of plutonic and volcanic rocks, 108

Hypotheses, multiple working, 21

Ice age, 323, 334, 558

Ice cap, 322, 325

Ice sheet, 322

Iceberg, 323

Iceland, 88

Ichthyosaurus, 554

Igneous rock: a rock which has solidified from a silicate melt, 37, 39, 90, 96

Illuvial zone: the layer of a soil into which substances leached from overlying soil accumulate, 173

Index fossil: a fossil with a short geological history yet which achieved a wide geographic distribution and underwent rapid evolutionary changes, 16

Indicator rock: a glacially transported rock which, because of its distinctive characteristics, may be traced back to its area of origin, thus indicating the direction of movement of the glacier, 328

Influent stream: a stream which is above the water table and thus contributes to the ground-water supply, 231

Inselberg: the remnant of a mountain, most of which has been leveled to a pediment by erosion in an arid region, 356

Interglacial interval, 332

Interior drainage: a stream pattern in which the streams arising in an area also end there and do not travel to the sea, 346

Intermediate-depth intrusion, 108

Intrazonal soil: a soil whose nature is determined by the local environment, 175

Intrusive rock: a plutonic or hypabyssal rock, 96

Ion: an electrically unbalanced atom, 47

Ionic bonding: the formation of a chemical compound by the transfer of electrons between atoms, 47

Island arc: an island chain commonly found around the Pacific Ocean and usually associated with a deep trench, 287

Isomorphous series: a series of minerals which have the same appearance and nearly the same crystal form but whose chemical composition varies systematically, 44

Isostasy: the theory that all large portions of the earth's crust are in balance as they float on a denser layer underneath so that less dense areas are higher than more dense ones, 449, 474

Isotopes: elements having the same atomic number but different atomic weights, 46

Jasper: a red variety of chert, 147

Joint: a fracture in rock along which no movement has occurred, 393

Jurassic, 8, 552, 556

Kanat, 229

Karst: a rugged landscape developed by solution of the underlying limestone, 239

Kettle hole: a depression in glacial debris caused by the melting of a buried block of ice, 321, 327

Kilauea, 78, 84, 87

Kimberlite: a hypabyssal rock composed dominantly of ferromagnesian minerals and which is the source rock for South African diamonds, 112

Klippe: the remnant of an overthrust most of which has been removed by erosion, 404

Krakatoa, 65

La Brea tar pits, California, 563

Lagoon: a broad stretch of shallow water separating a barrier island from the mainland, 263, 264

La Jolla submarine canyon, California, 283

Lake Bonneville, 367

Lake Lahontan, 367

Laminae: thin layers of sedimentary rock, 126

Laminar flow, 199

Land bridge, 559

Landslide, 191

Lapilli: volcanic ejecta about an inch in diameter (4 to 32 millimeters), 104

Lateral moraine: a debris ridge on the flank of an alpine glacier, 319, 320, 321, 322

Laterite: red-brown tropical clay, 176

Laurasia: a pre-drift landmass comprising what is now Europe, Asia, North America, and Greenland, 478

Lava, 37

Left-lateral fault, 406, 412

Lehman Caves, Nevada, 238

Lepidodendron, 549

Levees, natural: low embankments built up along many rivers during floods, 212

Life, origin of, 545

Limb of a fold, 389

Limestone: sedimentary rock, usually of organic origin, having the general composition $CaCO_3$, 38, 148

Lithification: the process of changing a sediment into a rock, 137

Lithosphere: the solid portion of the earth, 34, 35

Lit-par-lit: bed-by-bed structure in which a plutonic rock alternates with metamorphic rocks, 107

Load of streams, 203

Loam: rust-stained, sandy clay, 173

Loess: fine, wind-transported glacial debris, 328, 362

Longitudinal dune, 365

Long-period seismic waves, 443
Lunar geology, 526
Lunar time-scale, 527
Luster, 55
Lyell, Sir Charles, 18

Magma: a silicate melt, 37, 91
Magnetic field, 33
Magnetic reversal, 33, 34
Magnetism, 32
Mammal, 555
Mammoth, 561, 563, 564
Man, 565
Mantle: the layer of the earth between the crust and the core, 448, 449
Maps, geologic, 21
Marble: the metamorphic equivalent of limestone, 373, 380
Mare Imbrium, 532
Marginal sea, 285
Marine-built terrace, 258, 262
Mark Twain, 214, 224
Mars, 535
Marsupial, 555
Mascon: a mass concentration of very dense material located under some of the moon's circular maria, 531
Mass, 27
Mastodon, 561, 562
Mauna Loa, Hawaii, 78
Meander: a broad curving bend of a river, 213, 214, 215
Mechanical weathering, 159
Medial moraine: a dark band of rocky debris down the middle of an alpine glacier, caused by the combining of two lateral moraines when two glaciers join, 319, 321, 322
Mercalli Intensity Scale, 433
Mercury, 516
Mesozoic, 8, 17, 551
Metallic bonding: a type of chemical bonding in which the electron clouds around the atomic nuclei merge so that the closely packed nuclei share equally a great crowd of electrons, 48
Metallic luster, 55
Metamorphic rock: a rock which is the product of heat, pressure, and chemical activity so that some or all of its minerals are recrystallized and may show preferred orientation, 38, 370
Metasomatism: the transformation of one rock into another with a different chemical composition, 382
Meteor, 524
Meteor Crater, Arizona, 524, 530, 536
Meteorite, 523
Meteoroid, 524
Mica, 57
Microtektite: a very tiny glassy object occurring in sea-floor sediments and believed by some to have come from outer space, 525
Mid-ocean ridges: a world-wide system of ridges on the ocean floors which sometimes are located approximately equal distances from the coasts, 285, 287, 491
Migmatite: mixed igneous and metamorphic rock, 106, 373, 382
Mineral: a naturally occurring substance having a fairly definite chemical composition and characteristic physical properties, 43
Mineral fuels, 510
Minerals, rock-forming, 57
Miocene, 8, 18, 19, 560
Miogeosyncline: the less active, generally landward part of a geosyncline, usually containing no volcanic rocks, 468
Mississippi River, 199, 217, 219, 220, 221
Mississippian, 8, 17
Mohorovičić Discontinuity: the boundary between the crust and the mantle, 448, 456
Mohs scale of hardness, 55
Mollusk, 548
Monadnock: a residual hill or mountain on a peneplain, 210
Monument Valley, Arizona, 394
Moraine: hummocky debris left by a melting glacier, 319, 320, 321, 322, 327, 329
Mount Mazama, 69
Mount Pelée, 70
Mount Ranier, Washington, 78
Mount St. Helens, Washington, 78
Mud cracks, 131
Mudflow, 186, 188
Mudstone, 143
Multiple working hypotheses: a method of approaching geological problems by considering simultaneously many possible explanations, 21
Muscovite: a mica, $KAl_3Si_3O_{10}(OH)_2$, 54, 57, 58
Mylonite: a metamorphic rock whose original minerals have been finely crushed and then recrystallized into a very hard rock, 357, 375

Natural gas, 499
Natural levees: low embankments built up along many rivers during floods, 212
Natural selection, 542
Neanderthal man, 568
Negative anomaly: an area of the earth's crust where the force of gravity is less than expected, 32
Neogene, 8, 19
Névé: granular recrystallized snow, 302
Newton, Issac, 28
Nonconformity: an erosion surface developed on crystalline, non-stratified rocks and overlain by younger beds, 414, 417
Nonfoliated rocks, 373, 380
Nonmetallic luster, 55

Nonrenewable resources, 508

Normal fault, 401, 404, 406

Nuclear energy, 513

Nuée ardente: a rapidly moving, very hot cloud of volcanic fragments and gasses, 72, 530

Oahu Island, Hawaii, 81

Oblique fault: a fault in which the total movement has both vertical and horizontal components, 400

Obsidian: volcanic glass, 92, 93, 98, 102

Oligocene, 8, 19, 560

Olivine: $(Fe,Mg)_2SiO_4$, 57, 60

Oölite: a limestone made up of minute spherical grains of $CaCO_3$, probably the result of direct precipitation, 146

Open-pit mining, 509

Ordovician, 8, 18, 548

Organic deposits, 138

Organic reef, 506

Organic sedimentary rocks, 147

Orogeny: the processes of mountain-building, 416

Orthoclase: feldspar, $KAlSi_3O_8$, 43, 54, 57

Oscillation ripples, 134

Outwash plain: a broad sheet of water-laid but glacially derived sediment, 327

Overthrust: a thrust fault with a low dip whose movement may be measured in kilometers, 404, 414

Oxidation, 169

Pahoehoe: smooth, ropy lava, 84, 86, 100

Paleocene, 8, 19

Paleoclimate, 477

Paleogene, 8, 19

Paleomagnetism: ancient magnetic characteristics of the earth, 33, 480, 484

Paleontology: the study of ancient life, 15, 474, 482

Paleosol: a fossil soil, 178

Paleozoic, 8, 17, 547

Panama Canal landslide, 192

Pangea: Wegener's pre-drift continent, 473

Pedalfer: a soil in which soluble material does not accumulate; may be enriched in Al_2O_3 and Fe_2O_3, 175

Pediment: bedrock surface gently sloping away from low desert mountains, 355

Pediplane: the arid equivalent of a peneplain; an extensive pediment, 357

Pedocal: a soil which contains such soluble substances as calcium and magnesium, usually with carbonates and sulfates, 175

Pedology: the study of soils, 175

Pegmatite: coarse-grained plutonic rocks usually occurring as dikes, 498

Pelagic sediments: ocean-bottom sediments consisting of very fine particles, usually tiny marine animals, continental and meteoritic dust, and volcanic ash, 292

Peneplain: a nearly level widespread plain, near sea level in elevation, and created by stream erosion, 210

Pennsylvanian, 8, 17

Peridot: the gem variety of olivine, 60

Period of a wave: the length of time required for two crests of two troughs to pass a fixed point, 250

Permafrost: permanently frozen ground, 163, 166

Permeability: a rock's capability of having a liquid transmitted through it, 233

Permian, 8, 18, 550

Permo-carboniferous glaciation, 334

Petroleum, 499

Phenocryst: one of the larger crystals in a prophyritic rock, 93

Phreatic explosion, 87

Pillar, 239

Pink popcorn hypothesis, 537

Pipe dredge, 273

Piston corer, 278

Pithecanthropus, 571

Plagioclase: feldspar, $NaAlSi_3O_8 \cdot Ca\ Al_2Si_2O_8$, 44, 57

Plains coasts, 262, 264

Plastic deformation: the permanent folding and bending of rocks, 388

Plastic flow, 306

Platform, wave-cut, 262, 265, 266, 268

Playa: a desiccated desert lake bottom, 346

Playa lake: an ephemeral desert lake, 346

Pleistocene, 8, 18, 19, 558, 561

Plesiosaur, 555

Pliny the Younger, 73

Pliocene, 8, 19, 561

Plunge, 392

Plutonic metamorphism: metamorphism believed to occur deep within the crust under conditions of very high pressure and elevated temperature, 373, 381

Plutonic rocks: rocks that form at great depths in the earth, 37

Pluvial: a period of increased rainfall, 332

Point of a stream, 213

Polar wandering, 481

Polarity: the northness (positiveness) and southness (negativeness) of the earth's magnetic poles, 33

Polish, glacial, 306, 307, 311

Pompeii, 73, 77

Pool of a stream, 214

Porosity: the portion of the total volume of a rock that is pore space, 137, 232, 233

Porphyritic texture: texture showing crystals of two markedly different sizes, 93, 94

Postglacial climatic changes, 333

Potash, 145

Precambrian, 8, 485, 546

Precession of the equinox, 30

Primary seismic waves, 444

Profile, soil, 173

Profile of equilibrium, 207
Pterosaur, 554
Pumice: obsidian which has been so dilated by volcanic gasses as to become a petrified froth, 102, 104
Push waves, 444
Pyroclastic rocks: rocks composed of fragmental volcanic debris, 102
Pyroxene, 57

Quarrying, glacial, 309
Quartz, SiO_2, 41, 42, 43, 57
Quartzite: the metamorphic equivalent of sandstone, 373
Quaternary, 8, 19
Quick clay: a glacial clay that changes suddenly from a solid to a fluid state when it is jarred, 193

Radiolaria, 147
Reach of a stream, 213
Reactors, 513
Recent, 8, 19
Recycling, 517
Red Sea, 472, 486, 487
Refraction, 252
Regional metamorphism: metamorphism in which recrystallization is brought about by heat and directed pressure, 373, 376
Regolith: the mantle of rock particles on the surface of the earth, 159
Relative age: age as compared to the age of some other object or event, 5, 10
Renewable resources, 508
Replacement, 147
Reptile, 552
Reusable resources, 510
Reversal, magnetic, 33, 34
Reverse fault, 402, 405, 408
Rhyolite: a generally light-colored, fine-grained volcanic rock containing much feldspar, quartz, and very little ferro-magnesian material, 95, 97
Ria coast: an embayed coast whose indentations were shaped by stream action before invasion by sea water, 261
Richter Scale of Magnitude, 433, 435
Right-lateral fault, 406
Rille: a long narrow depression on the moon's surface, 530, 536
Ripple marks, 122, 133, 363
Rock: a heterogeneous (usually) aggregate of minerals, 41
Rock chart, 36
Rock glacier, 185
Rockfall, 188
Rockslide, 187

Saber-tooth tiger, 563, 565
St. Pierre, Martinique, 70, 71
Salt, 38, 144
Salt dome, 507

San Andreas fault, 400, 406, 413, 425, 426, 438, 439, 488, 489
San Francisco earthquake, 421
Sand dunes, 125, 362
Sand transport: the shifting of sand by longshore currents parallel to the coast and predominantly in a constant direction, 264
Sandstone: a sedimentary rock composed of sand grains which have been cemented together, 37, 128, 141, 142
Satellites, 526
Satin spar: a type of gypsum, $CaSO_4 \cdot 2H_2O$, 61
Saturation, 230
Scale models, 415
Schist, a metamorphic rock formed from a variety of other rocks by recrystallization under directed pressure and moderately high temperature, 373, 377, 378
Scree: the accumulation of blocks of rock in a long apron at the base of a steep slope, 162, 164, 165
Sea cave, 267, 268
Sea floor, 285, 291
Sea-floor spreading: the theory that portions of the ocean floor move away from the mid-ocean ridges as new sea floor is created there, 480, 483
Sea level, 257, 298, 559
Seamount: ocean-floor volcanoes which do not rise above sea level, 280, 287
Secondary seismic waves, 444
Sedimentary rocks: rocks made up of fragments of pre-existing rocks or chemically precipitated in water, 37, 39, 123
Seiche: an earthquake-caused wave which sloshes back and forth in a closed basin such as a lake or reservoir, 432
Seismogram: the record of earthquake waves produced by a seismograph, 433, 441, 443
Seismograph: an instrument which records the waves set up in the earth by an earthquake, 433, 441
Selenite: a variety of gypsum, $CaSO_4 \cdot 2H_2O$, 61
Sequoia, 556
Serpentine: $Mg_3Si_2O_5(OH)_4$, 373, 374
Shadow zone, 449, 452
Shake waves, 444
Shale: a sedimentary rock composed of very fine particles, 38, 131, 142, 143
Shear, 444
Sheet flood, 350
Sheeting, 398
Shelf break: the outer edge of the continental shelf, 280
Shells, energy-level, 46
Shield: a wide expanse of crystalline rocks generally forming the stable interior parts of continents, 120, 467
Shield volcano: a volcano with a gently

sloping profile composed entirely of lava flows, 79

Shore zone, 124, 256, 259

Shoreline classification, 257

SIAL: the light, primarily granitic upper layer of the earth's crust, 448

Sigillaria, 549

Siliceous rocks, 147

Sill: a tabular body of hypabyssal rock essentially parallel to the stratification of the enclosing rocks, 79, 109, 114

Silurian, 8, 18, 548

SIMA: the heavier, denser, mostly basaltic lower layer of the earth's crust, 449

Sink-hole: solution pit in a limestone terrain, 240

Sinter, 147

Sizes of sedimentary particles, 139

Slate: the metamorphic equivalent of shale, 39, 373, 376

Slickensides: the polished grooves on an actual fault surface, 400, 401

Slip: the actual, as distinguished from the apparent, direction of movement on a fault, 401

Slip face, 134, 364

Slip-off slope, 215

Sloth, 564, 566

Smith, William, 14, 543

Snake River, Idaho, 209, 223

Soil, 173

Soil profile, 173

Soilfall, 188

Solar energy, 511

Solar system, 25

Solar-topographic concept, 338

Solifluction, 183

Solution, 169, 203

Specific gravity: the weight of a substance compared to the weight of an equal volume of water, 56

Spit, 262, 263

Stack, 248, 262

Stalactite: an icicle-like pendant of travertine hanging down from a cave roof, 239

Stalagmite: a deposit built upward from a cave floor, 239

Stegosaurus, 552, 553

Steno, Nicolaus, 51

Steppe: a semi-arid region with a nearly continuous cover of grass and sometimes brush, 345

Stock: a body of coarse-grained igneous rock having a surface extent of less than 64 square kilometers, 106

Strata: layers, 38, 126

Stratification: layering, 126

Stratigraphic trap: an arrangement of sedimentary strata such that the oil in a reservoir rock accumulates beneath or against an impervious rock layer, 506

Strato-volcano: a volcano made up partly of lava flows and partly of pyroclastic material, 78, 79

Stream flow, 198

Stream transportation, 202

Strewn fields, 525

Striations, 306, 307

Strike: the strike of a bed is the compass direction of any line made by the intersection of the inclined bed with an imaginary horizontal plane, 386, 387, 389

Strike-slip fault, 405, 411

Sub-bottom profiling: an acoustic method for determining the topography and some of the geology of the ocean floor, 277, 281

Subdelta, 219

Subduction: the dragging down or sinking into the mantle of the leading edge of a crustal plate, 490

Submarine canyon, 283

Submarine valley, 283

Submersible, 274

Superposition: a geologic principle which states that younger rocks normally rest on top of older rocks, 12

Surf, 252

Surface waves, 443

Surge, glacial, 307

Surtsey, 290

Suspension in streams, 203

Swell, 252

Syncline: a fold whose limbs dip toward each other and whose youngest beds, after erosion, are nearest the central core, 390, 391

Talus: an accumulation of blocks of rock found in a long apron at the base of a steep slope, 162, 164

Tar pits, La Brea, California, 563

Tarn: a lake formed by scour in the rock beneath an alpine glacier, 313, 314

Tectonic map, 462

Tektite: a small glassy object that may have come from space, 525, 535

Tensional forces, 414

Terminal moraine: a hummocky deposit of unsorted material dumped at the end of a glacier as it melts, 319, 320, 322, 327

Terrace, marine-built, 258, 262

Tertiary, 8, 19

Texture, 91

Thrust fault, 403, 404, 408, 410

Thulean Plateau, Ireland, 89

Tillite: an unsorted mixture of sediment, glacial in origin, which has been lithified, 335

Time, geologic, 3

Top and bottom of beds, 150

Topset beds: horizontal strata at the top of a deltaic deposit, 134

Towhead: a low, sandy island that appears above the surface of a river at low water; an emerged sand bar, 214

Transform fault, 481, 489

Transportation, stream, 202

Transvection: a theoretical mechanism to account for continental drift by which the crustal plate is pushed from behind or dragged down in front, in contrast to the mechanism of convection in the mantle, 490

Transverse dune: a sand dune aligned at right angles to the prevailing wind direction, 365

Trap: a geological structure that retards the migration of oil and concentrates it in a limited space, 503

Travertine: a limy rock deposited from spring waters saturated with $CaCO_3$, 145

Triassic, 8, 18, 552

Triceratops, 551, 553

Trilobite, 540, 546, 547

Tsunami: an earthquake-caused ocean wave, 66, 254, 432

Tufa: a limy rock formed in springs and lime-saturated lakes with the aid of lime-secreting algae, 145, 367

Tuff: a pyroclastic rock composed of compacted volcanic ash, 74, 105

Turbidity current: a moving stream of water that, because of its high content of suspended material, sinks and flows as an undercurrent, 292

Turbulent flow, 199, 200

Turtle Mountain landslide, Alberta, 189

Tyrannosaurus rex, 551, 553

Uintatherium, 560

Unconformity: a buried erosion surface indicating a gap in the depositional record, 414, 416

Uniformitarianism: the principle that the earth today is the result of natural forces which can be seen acting now and is the product of a long, gradual development rather than a series of universal catastrophes, 22

Uranium, 6, 7, 510, 513, 514

U-shaped valley, 318, 319

Vaiont Dam, Italy, 181, 182, 195

Valley glaciers, 308, 325

Varves: the annual layering of very fine-grained laminae deposited on the bottom of cold-climate lakes, 130, 330

Veins: cracks and cavities filled by material solidified from hot, acqueous solutions, 498

Velocity, 200, 250, 444

Verde antique, 337

Vesicles: small holes in volcanic rocks where gas bubbles became trapped while the rock was molten, 101

Vesuvius, 73, 75

Vitreous luster: a glassy appearance, 56

Volcanic breccia: a volcanic rock composed of angular fragments that have been cemented together, 83, 105

Volcanic mountains, 460

Volcanic neck: the solidified magma of a volcanic conduit which has been exposed by erosion so that it forms a tower, 112

Volcanic rocks: rocks which crystallize at or near the earth's surface from lava, 37

Volcanism, 113

von Jolly, Phillip, 29

von Laue, Max, 53

Water table: the ground-water surface below which all openings are filled with water, 230

Water witching, 241

Wave base: the lower effective limit of wave transportation and erosion, 256

Wave-cut platform: a planed-off rock bench cut by wave action at the base of a receding sea cliff, 262, 265, 266, 268

Wave erosion, 254, 256

Wave length, 250, 251

Wave-reflection, 448

Wave-refraction, 252, 448, 452

Waves, 249

Weathering: the disintegration and decomposition of rock, 155

Wegener, Alfred, 473

Wegener's reconstructions, 475

Welded tuff: a volcanic ash which retained its heat long enough to fuse its particles together, 105

Wells, 234

Wentworth scale, 139

Widmanstätten figures: a lattice-like pattern of metallic meteorites caused by the intergrowth of two types of iron, 524

Williams, Howell 68, 70

Wind deposition, 362

Wind erosion, 357

Window: a hole, caused by erosion, in an overthrust sheet, exposing younger rocks below; same as fenster, 404

Wollastonite: $CaSiO_3$, 381

Xenoliths: angular inclusions of unmodified country rock in granite, 382

Xerophytes: desert plants with extensive roots, leathery leaves, and a large waterholding capacity, 344

X-ray crystallography, 53

Yellowstone National Park, 243, 244, 245

Yosemite National Park, 297

Zinjanthropus, 574

Zion National Park, 136

Zonal soils: mature soils with characteristics determined by the prevailing climate over a wide region, 175

Zone of aeration: the area above the water table in which the pore spaces are entirely or partly dry, 230